T0345203

DESIGN
for
MANUFACTURABILITY

How to Use Concurrent Engineering to
Rapidly Develop Low-Cost, High-Quality
Products for Lean Production

DAVID M. ANDERSON

DESIGN *for*
MANUFACTURABILITY

How to Use Concurrent Engineering to Rapidly Develop Low-Cost, High-Quality Products for Lean Production

CRC Press
Taylor & Francis Group
Boca Raton   London   New York

CRC Press is an imprint of the
Taylor & Francis Group, an **informa** business

A PRODUCTIVITY PRESS BOOK

CRC Press
Taylor & Francis Group
6000 Broken Sound Parkway NW, Suite 300
Boca Raton, FL 33487-2742

Library of Congress Cataloging-in-Publication Data

Anderson, David M. (Engineer)
    Design for manufacturability : how to use concurrent engineering to rapidly develop low-cost, high-quality products for lean production / author, David M. Anderson.
        pages cm
    Includes bibliographical references and index.
    ISBN 978-1-4822-0492-6 (hardback)
    1. Lean manufacturing. 2. Concurrent engineering. I. Title.

TS183.A57 2014
670--dc23   2013048176

Visit the Taylor & Francis Web site at
http://www.taylorandfrancis.com

and the CRC Press Web site at
http://www.crcpress.com

Dedicated to my loving and supportive wife, Lin.

Contents

SECTION I   Design Methodology

SECTION II   Flexibility

SECTION III Cost Reduction

SECTION V   Customer Satisfaction

SECTION VI   Implementation

SECTION VII   Appendices

List of Figures

Preface

This book shows companies how to design products that are manufacturable *the first time* and enables companies to quickly develop low-cost, high-quality products that satisfy customer needs *by design.*

It might seem obvious enough to ask: why would anyone do otherwise? Many companies think that because elements of the opening sentence are in the corporate goals and mission statements, this will automatically happen by decree. Therefore, why would any company need a book on design for manufacturability? Unfortunately, there are many reasons why products are not automatically designed for manufacturability.

Engineers are generally not taught DFM (design for manufacturability) or concurrent engineering in college. The focus is usually on designing for functionality. Further, they are typically trained to design *parts,* not *products* or *systems.* Many design courses don't even talk about how the parts are to be manufactured. And engineering students rarely follow their designs to completion to obtain feedback on their manufacturability.

Similarly, powerful computer-aided design (CAD) tools help engineers design parts, not products. Sure, CAD tools can assemble parts into products for analysis, but that does not generate the most creative product design, the simplest concepts, or the most optimized product architecture. Because engineering training and tools are more adept at part design, engineers and managers tend to skip the critical concept/architecture phase and "get right to work" designing parts. This behavior is reinforced by far too many managers, who want to see "visible progress," which may mean a quickly constructed breadboard which, after it "works," is drawn up and sent into production.

Product development management usually stresses schedule and cost, which, if not measured right, may further reinforce all the above suboptimal behavior. Pressuring engineers to complete tasks on schedule is really telling them to just throw it over the wall on time. In reality, the most important measure of schedule is the time at which the product has ramped up to stable production and is satisfying all the customers who want to buy it.

Similarly, cost metrics usually emphasize just part cost, assembly cost, and development budget, which are usually a small percentage of the only cost metric that matters—the selling price. Overemphasizing only these costs, just because they are the only ones measured, encourages engineers to specify cheap parts, cut corners, omit features, move assembly to low-labor-rate countries, and perform other shortsighted actions that make the product less desirable and ultimately more expensive on a total cost basis.

In addition, too often engineering education and computer tools emphasize individual efforts instead of teamwork. Further, college deadlines may be loose and, if not, the traditional college all-nighter might just compensate for procrastination. Traditional homework assignments issue all the data needed—not too much, not too little—and there is a single answer. Often, students don't even have to get the answer right, as long as they have the right *approach*. However, real life adds many constraints beyond functionality, such as cost, quality, and time to market. And the designers have to do all of this quickly and efficiently. Further, the designs have to be manufacturable. Very few individuals, especially right out of college, have enough experience to pull this off alone.

Fortunately, companies can compensate with multifunctional teams that have enough specialties to successfully address all the goals and constraints. Teamwork may never have been taught to or practiced by many engineers or managers, but their companies need multifunctional teams that can work together to design products for manufacturability.

One goal of this book is to present many improvements to current engineering practices, education, tools, and management. It shows the importance of thoroughly optimizing the concept/architecture phase, designing products as systems—not just collections of parts—and how multifunctional teams can accomplish this quickly. This book contains more than a hundred design guidelines to help development teams design manufacturable products. It shows how to design for Lean Production and build-to-order and to *design in* quality and reliability. The book has a big picture perspective that emphasizes designing for the lowest total cost and time to production when volume, quality, and productivity targets have been reached.

If engineers practice the principles of this book, they will be able to spend a higher proportion of their time doing fun, productive design work and less on change orders and firefighting.

READING SCENARIOS

Engineers: Read the whole book; be familiar with Chapter 7 and Appendix A

Team leaders and engineering instructors: Read the whole book; be familiar with Chapter 7 and Appendix A

Purchasing: Chapters 1, 2, 5, and 6; Sections 7.6 and 7.8; and Appendix A

Program and R&D managers: Sections I, III, and VI and Appendix A

Finance: Sections 1.3, 1.4, and 2.6; Chapters 6 and 7; and Appendix A

Marketing and portfolio planners: Sections 1.4 and 1.5; Chapters 2 and 3; and Appendix A

Managers, investors, and boards: Chapters 1, 2, and 3; Sections 6.1–6.3 and 11.5; and Appendix A

BOOK OUTLINE

Section I: Design Methodology

Chapter 1 introduces the concept of *design for manufacturability* and describes the problems that can be avoided when products are designed for manufacturability. It also discusses roles, focus, and how to overcome resistance, understand the myths and realities of product development, and motivate engineers to design for manufacturability, avoid arbitrary decisions, and *do it right the first time*. The chapter concludes with benefits of DFM.

Chapter 2 shows how to use *concurrent engineering* to develop products in multifunctional design teams. Such teams are most effective when they have early and active participation of *all* specialties. This chapter describes the problems when this does not happen and how to ensure availability of resources. Just as Chapter 1 showed that the majority of the cost is committed by the concept/architecture, the key to getting products quickly to market is thorough up-front work. Product development phases are presented with the tasks that enable good DFM, including: defining products to satisfy the *voice of the customer* with QFD (Quality Function Deployment);

optimizing the product architecture and strategies for operations and supply chains; raising and resolving the issues early; concurrently designing the product and processes; and launching quickly into production.

Chapter 3, "Designing the Product," focuses on thorough up-front work, optimizing the concept/architecture phase, and a wide scope of design considerations. The chapter also shows how to use creativity and brainstorming to develop better products and how to develop half-cost products.

Section II: Flexibility

Chapter 4 shows how to design products for Lean Production, build-to-order, and mass customization.

Chapter 5 offers effective procedures to standardize parts and materials, save time and money with off-the-shelf parts, search for them early before arbitrary decisions preclude their use, and implement a standardization program.

Section III: Cost Reduction

Chapter 6 emphasizes the importance of minimizing the *total cost* and then shows many ways to minimize total cost *by design*. It also shows why cost is hard to remove after products are designed.

Chapter 7 emphasizes the importance of quantifying all product *and* overhead costs and then shows easy ways to quantify total cost.

Section IV: Design Guidelines

Chapter 8 presents 27 design guidelines for product design, including assembly, fastening, test, repair, and maintenance.

Chapter 9 presents 51 design guidelines for designing parts for manufacturability. The chapter also has a section on tolerance step functions and how to specify optimal tolerances.

Section V: Customer Satisfaction

Chapter 10 shows how to *design in* quality and reliability with 34 quality guidelines and sections on mistake-proofing (poka-yoke) and designing to minimize errors. The chapter also explains that product quality is a function of the cumulative exponential effect of part *quality* and part *quantity*.

Section VI: Implementation

Chapter 11 shows how to implement DFM, including: determining the current state of how well products are designed for manufacturability; estimating how much could be improved by implementing DFM; getting management support and buy-in; arranging DFM training; forming a task force to implement DFM; stopping counterproductive policies; implementing DFM at the team and individual levels; and implementing standardization and total cost measurements.

Section VII: Appendices

Appendix A presents effective methodologies for product line rationalization to maximize resource availability for product development and increase profits immediately.

Appendix B lists the design guidelines without explanation to help DFM task forces create customized design guidelines and checklists.

Appendix C contains several useful forms for obtaining feedback from customers, factories, vendors, and field service.

Appendix D provides resource listings for the references that were cited the most in this book and information about the author's websites, customized in-house training, workshops, consulting, commercialization, and half-cost design studies.

PREFACE FOR INSTRUCTORS

This book can be especially effective for use as a textbook for a senior or graduate-level course on design for manufacturability and for company in-house training. It contains the latest material from the author's 27 years of in-house DFM seminars at manufacturing companies.

The book evolved from his experience initiating and implementing the DFM program for electronic products at Intel's Systems Group and teaching internal courses. That evolved into college courses on DFM at the University of Portland and later in the management of technology program at the University of California at Berkeley.

Various editions of this book have been used for courses at UC Berkeley Extension, Bemidji State University, Cleveland State University,

University of Colorado, University of Dayton, Eastern Michigan State University, Morehead State University, New Mexico State University, North Carolina State University, North Central Michigan State, Northern Illinois University, Oregon Institute of Technology (two campuses), University of Portland, San Jose State University, Sinclair College (part of joint program with University of Dayton), South Alabama University, Southern Methodist University, the St. Thomas University, West Carolina University, Washington State University (four campuses), the University of Wisconsin at Platteville, and Worchester Polytechnic Institute.

The industrial orientation of this book should give practical direction to college students to help them adapt quickly to the real world and design manufacturable products. Additional reading assignments can be selected from the references listed at the end of each chapter and in Appendix D, Section D.1.

This book can also be used to supplement courses on machine design, project design, system engineering, engineering management, engineering economy, value analysis, or management courses in business administration or mechanical, industrial, or manufacturing engineering.

A complimentary instructor package is available from the author that includes a college course outline, term project suggestions, homework, and exam questions with answers.

COMPANIES THAT USE THESE PRINCIPLES

DFM Books

The following companies have bought five or more copies of previous editions of this book (* means more than 50 copies; ** means more than 100 copies):

Allergan-Humphreys *	Bristol-Meyers Squibb
Applied Materials	Brooks/PRI Automation
Asyst Technology	Daikin McQuay
Bayer Corporation	EG&G Instruments
Beckman-Coulter *	Fisher Controls
Bio-Rad	FMC
Boeing	Freightliner
Boston Scientific *	Hewlett-Packard **

Ice-O-Matic
Itron
Hoeffer Scientific
Hollister
KLA/Tencor
Lam Research*
Loral**
Measurex
Moog Aircraft
Parker Hannifin
Physics International

Plantronics
PRI Automation*
Rainbird
Rantron
Smiths Aerospace
Spraying Systems*
Stanford Telecom
United Technologies Corp**
Watlow
W.L. Gore

In-House Seminars

Dr. Anderson has conducted seminars (see description in Appendix D, Section D.4) or provided consulting services for the following companies (number of seminars in parentheses):

Advanced Bionics
Ansitsu
BAE Systems (4)
Ball Aerospace
Bausch & Lomb
Beckman Coulter (3)
Becton-Dickinson
BOC (formerly Airco)
Boeing (4)
Bucyrus, a division of Caterpillar
Crane Merchandising Systems
Emergency One (3)
Emerson Electric (2)
FMC/JBT FoodTech (2)
Freightliner (2)
GE Energy & Transportation (3)
Hewlett-Packard (7)
Honeybee Robotics
Intel Systems Group (10)

Invivo, now Philips (3)
John Deere
L-3 Communications (3)
LG Group, Korea (4)
Loral (2)
Medrad
Moog Aircraft
NCR (2)
Northern Telecom
Plantronics (5)
PRI Automation-Robot Division
Qualcomm
Sloan Valve
Smiths Aerospace, now GE (4)
St. Jude Medical (2)
United Technologies (2)
Varian Medical Systems
Winegard (2)

About the Author

David M. Anderson, PhD, is the world's leading expert on using concurrent engineering to design products for manufacturability. Over the past 27 years presenting customized in-house DFM seminars, he has honed these methodologies into an effective way to accelerate the real time-to-stable production and significantly reduce total cost.

His book-length website, www.HalfCostProducts.com, presents a comprehensive cost reduction strategy (summarized in Section 6.3) consisting of eight strategies, all of which can offer significant returns as stand-alone programs and even greater results when combined into a synergistic business model. DFM is a key strategy because it supports most of the others. Dr. Anderson shows clients how to apply these strategies for cost reduction, ranging from half cost to an order of magnitude, which he teaches in customized in-house seminars, workshops, and design studies to generate innovative breakthrough concepts (see Appendix D).

In the management of technology program at the University of California at Berkeley, he wrote and taught the product development course twice. He wrote the opening chapter in the sixth volume of the *Tool and Manufacturing Engineers Handbook*. His second book on mass customization, *Build-to-Order & Mass Customization: The Ultimate Supply Chain Management and Lean Manufacturing Strategy for Low-Cost On-Demand Production Without Forecasts or Inventory*, is described in Appendix D.

Dr. Anderson has more than 35 years of industrial experience in design and manufacturing. For seven years, his company, Anderson Automation, Inc., built special production equipment and tooling for IBM and OCLI and did design studies for FMC, Clorox Manufacturing, and SRI International. As the ultimate concurrent engineering experience, he personally built the equipment he designed in his own machine shop. He has been issued four patents and is working on more.

Dr. Anderson is a fellow of ASME (American Society of Mechanical Engineers) and a life member in SME (Society of Manufacturing Engineers). He is a certified management consultant (CMC) through the Institute of Management Consultants. His credentials include professional registrations in mechanical, industrial, and manufacturing engineering and a doctorate in mechanical engineering from the University of California,

Berkeley, with a major in design for production and minors in industrial engineering, metalworking, and business administration.

Dr. Anderson can be reached via email: anderson@build-to-order-consulting.com. His websites are www.design4manufacturability.com and www.HalfCostProducts.com.

Section I

Design Methodology

1

Design for Manufacturability

Design for manufacturability (DFM) is the process of proactively designing products to (1) optimize all the manufacturing functions: fabrication, assembly, test, procurement, shipping, service, and repair; (2) ensure the best cost, quality, reliability, regulatory compliance, safety, time-to-market, and customer satisfaction; and (3) ensure that *lack of manufacturability* does not compromise functionality, styling, new product introductions, product delivery, improvement programs, or strategic initiatives and make it difficult to respond to unexpected surges in product demand or limit growth.

Concurrent engineering is the proactive practice of designing products in multifunctional teams, with all specialties working together from the earliest stages. Concurrent engineering with multifunctional teams is discussed in Chapter 2.

DFM and concurrent engineering are proven design methodologies that work for any size company. Early consideration of manufacturing issues shortens product development time, minimizes development cost, and ensures a smooth transition into production for the quickest real time-to-market.

Quality is *designed in* (Chapter 10) with concept and process simplicity, optimal tolerances, quality parts, mistake-proofing, concurrent design of robust processes, and specification of quality parts to minimize the cumulative effect of *part* quality on *product* quality.

Many costs are reduced with products that can be quickly assembled from fewer parts. Products are easier to build and assemble, in less time, with better quality. Parts are designed for ease of fabrication and commonality with other designs.

Products are designed for *Lean Production* and *build-to-order* with aggressive standardization (Chapter 5), elimination of setup *by design,* and the concurrent engineering of versatile products and flexible processes (Chapter 4).

Companies that have applied DFM have realized substantial benefits. Total cost and time-to-market can be cut in half with significant improvements in quality, reliability, serviceability, product line breadth, delivery, customer satisfaction, growth, and profits.

1.1 MANUFACTURING BEFORE DFM

Before DFM, the motto was "I designed it; you build it!" Design engineers worked alone or only in the company of other design engineers in "the engineering department." Designs were thrown over the wall to manufacturing, which then had the dilemma of either objecting ("But it's too late to change the design!") or struggling to launch a product that was not designed well for manufacturability. Often this delayed both the product launch and the time to ramp up to full production, which is the only meaningful measure of time-to-market.

Poor manufacturability raises many categories of cost to pay for launch difficulties, special equipment or modifications, difficult part fabrication, inefficient assembly, excessive part proliferation, laborious procurement, numerous changes, and many other overhead costs. These issues not only raise cost but also delay shipments. Problem product introductions may absorb so much effort that production of other products may suffer.

Lack of manufacturability also degrades quality, which, in turn, raises costs further and delays the real time-to-market. This is because products not designed for quality are unnecessarily complex, have too many parts from too many suppliers, require more difficult manual assembly, and may not be robust enough for consistent processing. Further, counterproductive "cost reduction" may compromise quality while, ironically, not lowering total cost.

Probably the most subtle effect (but most damaging in the long run) is that a series of problem product introductions drains resources (both people and money) away from new product development (NPD) and continuous improvement efforts that should be making product lines and factories more competitive.

Excessive proliferation of parts and products can make it harder to implement just-in-time, Lean Production, build-to-order, and mass customization.[1]

DFM may make the difference between a competitive product line and, in the extreme, products that are not manufacturable at all. The main causes of product failures are that costs are too high, quality is too low, introductions are too late, and stable production is even later, or, if the product is a big hit, production is unable to keep up with demand. *These are all manufacturability issues and therefore can be much improved by DFM.*

Before DFM, companies felt limited by trade-offs, such as "cost, quality, time-to-market; take two!" Some companies may not have been able to achieve two or one or even *any* of these goals if they rushed up-front work, repeated past mistakes because lessons were not learned, did not resolve issues early, did not design in manufacturability and quality, did not work from a clear and stable product definition, and did not design in low cost.

1.1.1 What DFM Is Not[2]

- DFM is not a late step that, once checked off, gets you through a design review or gate.
- DFM is not done only at the parts level; most opportunities are at the system architecture level.
- DFM is not done by the "DFM engineer."
- DFM is not to be "caught" later in design reviews.
- DFM is not an afterthought.
- DFM is not to be accomplished by changes.
- DFM is not done alone by engineers in their cubicles.
- DFM is not done by a "tool."

1.1.2 Comments from Company DFM Surveys

The following are verbatim comments from company surveys before DFM training. (The use of these surveys is discussed in Section 11.2.) When asked about the consequences of inadequate DFM, engineers and managers usually cite problems with quality, cost, delivery, profits, and competitiveness, which are tabulated in Section 11.2. The colorful comments convey what it is like to work in a company that does not design products well for manufacturability.

The consequences of inadequate DFM for delivery are
    "Line stoppers"
    "Parts do not assemble correctly"

"Endless engineering change orders"

"Much pruning, grooming, and tuning to get products out the door"

"Poor yield invariably results in late delivery or 11th hour miracles"

"When a problem is encountered the production line comes to a stop"

"Emergency change orders and redlines to keep manufacturing operating"

The behavioral hurdles to good DFM are

"Lack of DFM training," "Lack of DFM knowledge"

"Parts designed with no consideration of how it is to be built"

"'Over the wall' syndrome: after release, no longer engineering's problem"

"Never enough time to design parts right the first time; always enough time to do it over"

The attitude hurdles to good DFM are

"Tradition"

"Designer's limited knowledge of manufacturing processes"

"Reluctance to accept suggestions from suppliers regarding design issues"

"We don't seem to allot time to design systems properly up front, but we are willing to do it over later after a product is released"

The bottom line consequences, besides profitability, include:

"Unhappy customers"

"Sometimes problems get shipped to the field"

"Customers losing confidence in our products"

"Problems increase overall costs, resulting in loss of the ability to compete"

"Low product quality leads to poor customer satisfaction, poor performance, and eventually to high costs"

"Post-launch redesigns"

1.2 MYTHS AND REALITIES OF PRODUCT DEVELOPMENT

Resistance to DFM may stem from myths about product development. Here are the most common myths and the corresponding realities:

Myth #1: To develop products *quicker,* immediately move forward with detail design and software coding, and then enforce deadlines to keep design release and first-customer-ship on schedule.

Fact: The most important measure of time-to-market is the time to stable, trouble-free production, which depends on getting the design right the first time.

Myth #2: To achieve *quality,* find out what's wrong and fix it.

Fact: The most effective way to achieve quality is to *design* it in and then *build* it in.

Myth #3: To *customize* products, take all orders and use an ad hoc "fire drill" approach.

Fact: The most effective way to customize products is with the concurrent design of versatile product families and flexible processes. This is known as *mass customization.*[3]

Myth #4: *Cost* can be reduced by cost reduction efforts.

Fact: Cost is designed into the product, especially by early concept decisions, and is difficult to remove later.

1.3 ACHIEVING THE LOWEST COST

Figure 1.1 shows that by the time a product is designed, 80% of the cost has been determined.[4] And by the time a product goes into production, 95% of its cost is determined, so it will be very difficult to remove cost at that late a date. The most profound implication for product development is that *60% of a product's cumulative lifetime cost is committed by the concept/ architecture phase!* This is why it is important to fully optimize this phase, as will be shown in Section 3.3.

1.3.1 Toyota on When Cost Is Determined

The Toyota philosophy confirms this. "The cost of a [product] is largely determined at the planning and design stage. Not much in the way of cost improvement can be expected once full-scale production begins." "Skillful improvements at the planning and design stage are ten times more effective than at the manufacturing stage."[5]

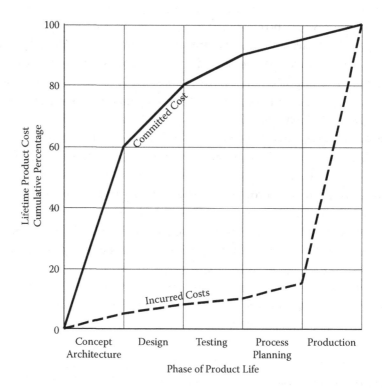

FIGURE 1.1
When costs are determined.

1.3.2 Ultra-Low-Cost Product Development

Ultra-low-cost product development techniques have been used to develop the $2,200 Tata Nano, $100 computers (for the One Laptop per Child Foundation), $35 cell phones, and low-cost medical products such as the Siemens Essenze, which "provide small clinics and rural hospitals access to quality healthcare services at a fraction of the cost of standard MRI equipment."[6]

In the *Industry Week* article by A.T. Kearney about ultra-low-cost product development, the author notes: "It is well known that over 70% of product cost is determined during the development phase and a substantial amount of additional costs are caused by engineering changes that happen late in the product development process."[7]

1.4 DESIGNING FOR LOW COST

The DFM techniques presented herein provide *proactive* ways to achieve much greater cost reduction than can be achieved by *reactive* approaches, such as "cost reduction," the problems of which are discussed in Section 6.1. This reality counters the myth that cost can be reduced by cost reduction efforts.

> *Cost is designed into the product, especially by early concept decisions, and is difficult to remove later.*

The book, *When Lean Enterprises Collide,* studied the competition and cost practices of several Lean Production companies in Japan. The author, Robin Cooper, has also written extensively on total cost accounting. His view on product development's role in determining cost: "Effective cost management must start at the design stage of a product's life because once a product is designed the majority of its costs are fixed."[8]

The main cost minimization opportunities are to simplify concepts and product architecture (Section 3.3), which determines 60% of the lifetime cumulative cost, as shown in Figure 1.1. A key part of this is optimal selection of versatile modules (Section 4.7), previous engineering (Section 5.18), and off-the-shelf parts (Section 5.19), which usually result in total cost savings while ensuring quality and reliability because they are proven parts that can be verified by their track records. Chapter 6 shows several ways to minimize cost through design. Applying these DFM techniques should enable companies to develop products at half the total cost, with special emphasis on the key points discussed in Section 3.8.

1.4.1 Design for Cost Approaches

Various approaches exist to determine the goals for costs and the pricing of products.

1.4.1.1 Cost-Based Pricing

The way this approach typically works is engineers design the product and then add up the parts and labor costs, which are usually their only cost focus. To that the company adds average overhead costs, selling costs, and profit to arrive at the selling price.

1.4.1.2 Price-Based Costing (Target Costing)

The target costing approach starts with a selling price that is estimated to be competitive. From that, profits, selling costs, and overhead are subtracted to determine the *target* cost for parts and labor.[9,10] A more advanced version would be to subtract profits and selling commissions from the selling price to determine the *total cost,* which would include parts, labor, and *all* overhead costs.

However, it should be kept in mind that costing/pricing policies are really *targets,* not *strategies* to design low-cost products. If product development teams do not know how to really design low-cost products, these costing/pricing approaches will have different consequences.

In the case of cost-based pricing, not knowing how to design low-cost products will result in higher-than-necessary costs, which will result in pricing that may be too high to be competitive. In the case of price-based costing, not knowing how to design low-cost products can result in the following dangerous scenario:

- Engineers design the product as they usually do—for functionality.
- Accountants add up the part and labor costs and apply the usual overhead costs.
- Management then realizes how much the part and labor costs are "over the target" and pressures the engineering, purchasing, and manufacturing departments to "lower the cost" (of the parts and labor).
- Design engineers and "value engineers" find it difficult to reduce cost after design (for reasons presented in Section 6.1), so they may be tempted to do some desperate things to achieve the part and labor targets, such as cutting corners, omitting features, or specifying cheaper parts. All of these changes incur the costs of making and documenting the changes. Moreover, cutting corners and cheaper parts will add to quality costs.
- The purchasing department might try low bidding of parts, pressuring suppliers, or changing suppliers for a slightly better purchase cost. However, while this might *appear* to lower part cost, it will most likely raise other costs and compromise quality (Chapter 6), delivery (Chapter 4), and collaboration with vendors to develop more manufacturable products (Chapter 2).

- Manufacturing, under enough pressure, might do some desperate things, such as outsourcing and moving manufacturing to "low-labor-cost" countries, which decreases responsiveness while not really reducing total cost.[11]

1.4.1.3 Cost Targets Should Determine Strategy

Cost goals should determine the approach, not exert pressures to "do the same thing, but better."

Key cost goals should be expressed to management and the design teams in degrees or even the desired percentage improvement. For instance, a 5% goal might be achieved with better diligence. A 20% to 30% cost improvement goal would need some serious application of all the DFM principles presented herein. Above 50% would require breakthrough concept innovation, because that is where most cost is determined.

So the degree of the needed cost reduction would determine how ambitious the product development approach should be and how well staffed and how well structured the timelines should be.

1.4.2 Cost Metrics and Their Effect on Results

This book shows how to design products with the lowest *total cost*, which is discussed in Chapter 7. If all costs are quantified, and DFM principles are followed, design teams will design to minimize total cost. On the other hand, if the only costs measured are parts and labor, the results will be counterproductive.

If the driving cost metric is *labor* cost, then decision makers may conclude that if labor rates offshore are one-fifth, then they will save four-fifths of the labor budget. However, this fails to consider that the *total* will go up much more, for reasons presented in Section 4.8 (on offshoring effects on product development, Lean Production, and quality) and Section 6.9 (on the cost of quality).

If the primary cost metric is *parts* cost, then engineers will be encouraged, even pressured, to specify cheap parts. Those "savings" will be more than cancelled out by the costs of quality (diagnoses, repair, scrap, retesting, etc.) and "firefighting," as shown graphically in Figure 1.2. This will compromise product development by:

Basing "cost" primarily on BOM cost *and* pressuring to lower "cost" or achieve "cost" targets result in:

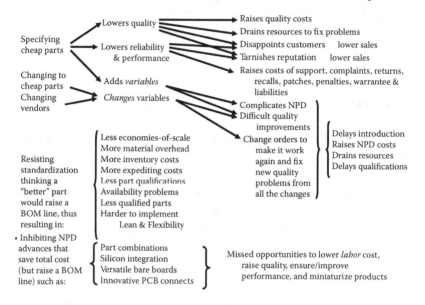

FIGURE 1.2
Hidden costs and consequences of cheap parts.

1. Draining engineering and manufacturing resources away from designing low-cost products to deal with problems caused by the cheap parts and write engineering change orders, some of which will, ironically, be to change to better parts.

2. Adding variables to the product and to the product development process. These extra variables will then require even more resources to work longer to get working prototypes into production, thus raising product development cost and delaying the product's introduction.

In addition to lowering quality, raising quality costs, and complicating product development, pressures to minimize part cost will discourage standardization, the value of which is discussed in the opening two pages of Chapter 5. Standard parts are better than what is needed for most applications and, thus, may appear to raise cost when only the raw part cost is considered. However, standardization results in an overall net cost savings from economies of scale, less inventory, and material overhead that, compared to oddball parts, can be 10 times less! Similarly, pressure to lower part cost may discourage innovative strategies to lower total cost, such as combining printed circuit boards, higher levels of silicon integration

(VLSI, ASICs, FPGAs, etc.), and flex layers (Section 3.1), which may appear on a bill of materials (BOM) line as an expensive part.

1.4.3 How to Design Very Low Cost Products

Quantify *all* costs. Without quantifying overhead costs, cost reduction strategies will focus on just parts and labor, because (1) parts and labor are only a portion of the cost and, worse, (2) shortsighted efforts to reduce parts and labor costs usually raise overhead costs much more.

Avoid policies that inhibit *real* cost reduction opportunities or drain resources, such as rushing up-front work, selling difficult high-overhead orders, not prioritizing engineering resources, and not correcting critical staffing gaps that inhibit concurrent engineering.

Understand that 80% of cost is committed by the design phase and 60% is committed by the concept and architecture phase, as shown in Figure 1.1. Given this, don't assume that a poorly designed product can be cost-reduced by high-volume mass production or automation.

Don't just look at lists of parts, because (1) that will cause you to miss opportunities at the architecture level, which determines 60% of the cost, and (2) substituting cheaper parts requires new product development resources and introduces many new variables that will lower quality, raise other costs, and delay product development itself, as shown in Figure 1.2.

Investigate what worked well and what caused extra expenditures on related programs—the "lessons learned." Also, raise and resolve all cost issues early, including cost of quality, which should be quantified.

Identify and prioritize cost challenges and cost reduction opportunities.

Preselect vendors and partners who will help design their parts. This will save much more money than designing parts in isolation and going for the low bidder.

Implement concurrent engineering in which complete multifunctional teams do all of the above.

Thoroughly search for standard off-the-shelf parts, *before arbitrary decisions preclude their use.* Then the product will be *literally* designed around them.

Select off-the-shelf parts by thoroughly searching acceptable *ranges* of candidate parts, instead of telling purchasing: "this is the spec I need; find a part that matches that spec" (see Chapter 5).

These substantial opportunities will be missed if an engineer designs for function, specifies the part desired, throws that over the wall to purchasing to look for it, and then concludes that off-the-shelf parts just won't work for their products so they will just have to design custom versions.

1.4.4 Cost Reduction by Change Order

Except for isolated low-hanging fruit, cost reduction after the product is designed is an ineffective way to lower cost because:

Cost is designed into the product: 80% of cost is committed by design, and by the time it gets to manufacturing, only 5% is left.

Cost is hard to remove later because so much is cast in concrete and so much is boxed into multiple corners. Thus, cost reduction usually focuses on parts, making systematic cost reduction almost impossible.

Cost reduction efforts on one product will not have the time or bandwidth to reduce any overhead costs, which may be more than half the cost but are rarely quantified. So the focus usually shifts to specifying cheaper parts, omitting features, beating up suppliers, switching to a new low bidder, letting labor costs dominate sourcing and plant location decisions, or cutting corners.

Cost reduction by cutting corners may cheapen the product and compromise the image or integrity of the product.

The changes will cost money, which exponentially increases with development time, so it may not be paid back within the life of the product.

The cost of changes rises drastically as the product progresses toward production. Figure 1.3 shows how the cost *for each change* escalates during the development of a major electronics product.[12] Thus, it

Time of Design Change	Cost
During design:	$1,000
During design testing:	$10,000
During process planning:	$100,000
During test production:	$1,000,000
During final production:	$10,000,000

FIGURE 1.3
Cost of engineering changes over time.

can be concluded that a very expensive and time-consuming way to implement DFM is through engineering change orders. And yet, that is what happens when DFM and all the other considerations are ignored in the early design steps.

The changes will cost time, especially if requalifications are required, which may delay the time-to-market. Changes may induce more problems, thus requiring further changes, which will involve expenditure of more hours, calendar time, and money, and possibly compromise functionality, quality, and reliability.

Another reason cost reduction cannot be counted on is that it may just not happen due to competing priorities, such as mandatory changes and designing new products.

Studies show that *cost reduction does not work.* Mercer Management Consulting analyzed 800 companies over five years. They identified 120 of these companies as "cost cutters." Of those cost-cutting companies, "68% did not go on to achieve profitable revenue during the next five years."[13]

Committing valuable resources to implement cost reduction strategies after design takes resources away from other more effective efforts in product development, quality, Lean Production, and elsewhere. If too many resources are committed to cost reduction, then:

1. There will not be enough resources available for real cost reduction through new product development. If this continues over time, the result will be little, if any, real reduction in cost, while such a drain of resources will impede new product development innovation.

2. It will prevent the transition from back-loaded efforts to the more effective front-loaded methodology, which uses complete multifunctional teams to design low-cost products right the first time.

3. The company will be lured into thinking it is doing all it can to lower cost, when, in fact, costs are not really being reduced and opportunities for real cost reduction are not being pursued.

In conclusion, do not attempt cost reduction on existing designs after receiving an order. Furthermore, cost reduction attempts, coupled with incomplete cost data, may discourage innovative ways to lower cost, maybe even thwarting promising attempts. If certain product lines have chronic cost challenges and individual cost reductions efforts do not work on a total cost basis, consider redesigning them for manufacturability.

1.5 CUTTING TIME-TO-MARKET IN HALF

Time-to-market should be measured using big picture measure, such as the time to target volume, quality target, productivity target, qualification, change orders completed, or customer acceptance, as shown in the middle graph in Figure 2.1. To compare to previous projects, metrics may have to be recomputed. Defining time-to-market as design release really means "throw it over the wall on time." This results in abandonment by design engineers and unfinished designs being released, both of which delay the real time-to-market: the time to stable production or customer acceptance.

Define the product methodically to avoid "changes" to satisfy customers. Make sure the product requirements are complete before the engineering begins or else the requirements may be poorly formed or the project will be delayed. In one survey, 71% of managers said that poor product definition caused product development delays, making it the top reason for delays.[14] Another survey of 153 companies[15] concluded that the biggest cause for product failure[16] was "unclear or continuously changing product definitions," and the next most common cause of failure was "product does not meet customer or market requirements."

Only take custom orders that can quickly and efficiently be designed and built, instead of taking all orders. Remember that all sales do not take the same amount of time and resources, have the same cost, or generate the same profit.

Find out how much actual time and total cost it will take for each contemplated customization and configuration. Ask for time and cost data before accepting the order. Get approvals from engineering, manufacturing, purchasing, and so forth for each custom order. Create databases with enough time and total cost data to generate accurate estimates, which can then be built into software configurators.[17]

Ensure all necessary specializations are available and active early to optimize all aspects of manufacturability from the very beginning. Ensure all resources are available for immediate deployment when they are needed throughout the product development.

Create realistic schedules based on a combination of market needs and new product development capabilities. If the market is moving fast, develop fast, efficient NPD (new product development) methodologies that will quickly get to stable production or customer acceptance. Make sure NPD resources are not spread too thin on too many projects and fire drills so that complete multifunctional teams can be fast and efficient. Encourage customers to order early, and streamline the sales and contract process. Plan ahead to start NPD as early as possible.

Begin decisively. Do long-term market research, industry analysis, and product portfolio planning to avoid getting behind and rushing to catch up. Teams should not procrastinate or wait for deadline pressures to get motivated. Encourage customers to place orders early through incentives for early orders or penalties for late or rush orders. Minimize the time to negotiate orders.

Learn lessons from past projects to avoid wasting time repeating the same mistakes (see Section 3.3). Raise and resolve issues early and avoid the delays of much more difficult change orders. Thoroughly resolve technical or functional challenges and issues early (see Section 3.3).

Work together well throughout the project with on-demand meetings, instead of saving issues for periodic meetings.

Work efficiently with the most efficient design and simulation tools and use product data management to document progress to ensure that everyone's work is based on the most current drawings and documents.

Purchase and outsource wisely, since low bidders selected for cost often cause delays.[18]

Avoid "creeping elegance" to endlessly pursue unnecessary refinements and enhancements.

Avoid administrative delays for design reviews, budget hold-ups, and loss of key people, temporary or permanent, to firefights on other projects.

Design thoroughly and complete all documentation because time-to-market can be significantly delayed by design gaps, glitches, and incomplete documentation. Significant, expensive launch or build delays can be prevented by up-front thoroughness, which is much more efficient in the relatively orderly design stage than in the panic mode that results when problems delay production. Design and documentation shortcomings not only delay that product but also take resources away from other products.

Avoid premature release. Don't allow bad metrics, such as development budget pressures or defining time-to-market as "design release," to allow or encourage premature designs to be released to manufacturing, because finishing the design under change control constraints will cost much more and take more calendar time, in addition to starving new product developments of operation resources.

Design for existing processes to eliminate the need to design, develop, and debug new production machinery. Obey all the design rules for all processing to avoid delays to either correct the designs or correct the problems every time products are built. Avoid redesigns, which will take time to design, debug, and build, which will delay the time-to-market.

Select proven materials, parts, suppliers, and vendors. Any glitches in these can delay the market launch. Thoroughly optimize material and part availability, including best-case sales scenarios, to avoid delays to find and incorporate alternatives.

Avoid excessively long supply chains, which can increase the calendar build time and are vulnerable to cumulative delays and shipping interruptions.

Proactively avoid compromising functionality, quality, cost, or manufacturability to get products out the door.

Thorough up-front work greatly shortens the real time-to-market and avoids wasting time and resources on revisions, iterations, and ramp problems. This is thoroughly discussed in Section 3.2 on the importance of optimizing the concept and architecture stages (the up-front work), citing the Lexmark model (Figure 3.1). This model graphically shows how the "concurrent" model (basically, doing everything recommended in this book) results in completing the production ramp 40% sooner than the "linear" model, which rushes right through the concept and architecture stages. The first half of Chapter 3 explains how to optimize the critical up-front work.

Finally, *mitigate the risk of changing market conditions* with fast product development.

1.6 ROLES AND FOCUS

Optimal company performance comes from whole-company synergy, where the whole company works together to develop products for manufacturability:

Engineering and manufacturing concurrently engineer products and processes, as discussed in Chapters 2 and 3.

Marketing works with the team from the earliest stages to define whole product families that satisfy the "voice of the customer."

The purchasing and materials groups support product development by nurturing vendor and partner relationships instead of looking for the lowest bidders; taking pressure off product development teams by shortening procurement times; encouraging part standardization; and prequalifying parts and vendors to optimize quality and delivery.

Finance quantifies total cost to support relevant decision making and arranges for appropriate overhead charges for new-generation products.[19]

A DFM task force incorporates DFM steps into the product development process and creates, issues, and updates a consistent set of design rules and guidelines (see more in Section 11.4).

1.6.1 Human Resources Support for Product Development

The human resources (HR) department should hire or develop good project leaders with team-leading abilities, as well as good team players. According to Sony:[19]

> "When hiring, you have to also be mindful of how well the new managers work with the rest of your existing team. Hiring a good team player is as important as hiring someone with the right expertise."[20]

HR should provide team-leader training for managers and team-building[21] workshops for engineers. At Nokia, "the focus was not just on recruiting but on 'marinating' that begins with orientation and ends with highly refined team training."[22]

Growing departments should not be allowed to raid other departments, which may weaken critical internal functions. HR should hire design engineers with experience in manufacturing, test, field service, sales, and so forth. It should give extra consideration to potential employees who have been users or worked for customers, suppliers, or regulators. Finally, HR should arrange for training for all aspects of product development (see Chapter 11 on implementation).

Senior management should work with HR to ensure that performance measurements encourage teamwork and support overall goals. The company must also ensure retention of talent, information, and complete teams during downturns, restructuring, and internal transfers, and strive to maximize internal continuity and minimize turnover.

1.6.2 Job Rotation

Company policy should also arrange job rotation to encourage cross-learning and informal communications, starting with placing new design engineers in manufacturing first. It can be argued that it takes new employees time just to learn their way around the company, and engineers on their first job need to learn even more. So while they are getting up to speed on how the company operates, they can be learning about the company's manufacturing practices.

> "At Honda, all entry-level engineers spend their first three months in the company working on the assembly line. They're then rotated to the marketing department for the next three months. They spend the next year rotating through the engineering departments—drive train, body, chassis, and process machinery. Finally after they have been exposed to the entire range of activities involved in designing and making a car, they are ready for an assignment to an engineering specialty, perhaps in the engine department."[23]

> At Samsung, "team-building practices include frequent meetings for all individual team members along with something as simple as conversations over a drink after working hours—for the purpose of exchanging internal communications or resolving conflicts."[24]

Nokia owes much of its success to encouraging its brightest managers to work in manufacturing. CEO Jorma Ollila says, "If you do well in manufacturing, you get a good career in Nokia."

1.6.3 Management Role to Support DFM

Senior managers and executives should understand these principles enough to execute the following advice.

Encourage innovation. Bill George, CEO of Medtronic, encourages "walking through the labs and learning about creative ideas before they get killed off" by the system, because "a growing bureaucracy is a huge barrier to innovative ideas and dampens creativity, no matter how much it spends on research and development. Leaders committed to innovation have to work hard to offset these tendencies, giving preference to the mavericks and the innovators and protecting new business ventures while they are in the fragile, formative stage."[25]

Plan the product portfolio and its evolution over time objectively to provide the greatest net profit over time, defined as all the financial gains minus all the costs.

Don't spread resources too thin by "taking all orders." Focus on selling the most profitable products and rationalizing away the "losers."

Ensure resource availability so that complete teams can form early. Don't waste product development resources trying to reduce cost after the product is designed. Preselect vendors so they can help the team design the parts they will build, which saves much more money than bidding.

Have realistic expectations compatible with product development methodology.

Encourage a high proportion of thorough up-front work through good product definition, early issue resolution, concept simplification, and architecture optimization. Avoid early deadline pressure that thwarts thorough up-front work.

Implement total cost measurements (Chapter 7) to enable prioritizing all activities, planning product portfolios, rationalizing products by real profitability, and relevant decision making.

Follow through so the team is responsible for transition into production and the team stays with the project until production has stabilized in volume, productivity, and quality. Paul Horn, who oversees research at IBM, says:

> "Everything we do is aimed at avoiding a 'handoff'—there is no 'technology transfer.' It is a bad phrase at IBM. Research teams stay with their ideas all the way through to manufacturing."[26]

Encourage job rotation. Nokia "encourages job rotation within the company—from Ollila's executive board to all levels of the workforce."[27]

Empower an effective project leader to make decisions as they need to be made, thus minimizing dependence on design reviews.

Encourage feedback and be receptive to all news about product developments. Create an open culture where issues can be raised and discussed early with the focus on issue resolution. When the current Ford Americas President Mark Fields came to Ford from IBM, *he was discouraged from airing problems at meetings unless his boss approved first!*[28]

Ensure ownership so that product development teams own all aspects of manufacturability and are accountable for the total cost and the real time-to-market.

Implement compensation and reward systems that encourage teamwork and big picture goals; change sales incentives from revenue to profit. Motorola's policy is that, "Corporate leaders must emphasize the need to partner and, as at Motorola, even overhaul how people are paid, rewarding those who promote partnering."[29]

1.6.4 Management Focus

Senior managers and executives should focus on the following:

- The activities and methodologies that lower cost and speed development, instead of relying just on goals, targets, metrics, reviews, gates, and deadlines. It is important for product development teams to have the right focus when developing products.
- Customers' needs, not on the company programs, competition, or technology.
- Proactive resolution of issues early rather than reactive resolution later.
- Problem avoidance rather than problem solving.
- Eliminating engineering change orders rather than streamlining the change control process.
- Core competencies and new and pivotal aspects of the design instead of reinventing the entire wheel and diluting resources with low-leverage activities.
- Product and software architecture, not just drawings and code.
- The design process itself, not on project control and management.
- How to optimize activities in the phases, not the gates or design reviews.
- On-demand discussions and decision making, not periodic meetings and reviews.
- Product design, not proofs-of-principle, breadboards, and prototypes.
- Optimizing product architecture, not just designing a collection of parts and subassemblies.
- Rapid production ramps in real production environments, not pilot production by prototype technicians or engineers.

- Time-to-stable-production, not time-to-design-release or first-customer-ship.
- Minimizing total cost, not just reported costs (labor and materials).
- Designing and building in quality and reliability, not by testing, inspections, or reacting to field problems.
- Compensation systems that encourage behavior that benefits the company, not departments or individuals.
- Activities that achieve major and lasting cost reduction (superior product development, Lean Production, quality programs, etc.), instead of cost reduction attempts that may compromise real cost reduction and delay the time to stable production.
- Activities that achieve goals, not the goals themselves. The movie *Jerry Maguire* made famous the phrase, "Show me the money," which was chanted in the part of the movie when neither the football player nor his agent were being "shown" any money because they were concentrating solely on the *goal* of making money rather than *what achieved the goal*, which was playing good football. The analogies for business:

The CEO says,	"Show me the profits."
The CFO says,	"Show me the market value."
The V.P. of sales says,	"Show me the sales."
The V.P. of marketing says,	"Show me more market share."
The V.P. of research says,	"Show me more patents."
The V.P. of purchasing says,	"Show me less part cost."
The V.P. of engineering says,	"Show me faster developments."
The V.P. of manufacturing says,	"Show me less assembly cost."

However, none of these goals will be met unless the company has effective ways to achieve them. Dr. W. Edwards Deming said, "A goal without a method is cruel."

Ironically, too much pressure to meet departmental targets without a real way to achieve them usually leads to counterproductive actions, like buying cheap parts (Section 6.11), which degrades quality. Similarly, moving manufacturing to low-labor-cost areas can actually increase other costs and compromise responsiveness and product development, as discussed in Section 6.1.

1.6.5 Successful or Counterproductive Metrics for NPD

Cost Measurement (Chapter 1)

Successful Metric	Counterproductive Metric
Total cost	Quantify only parts and labor, which may distort: • Make/buy decisions • Part/material selection • Costing, if overhead is averaged • Pricing, when good products subsidize bad ones and well-designed products subsidize hard-to-build products

New Product Development Completion (Chapter 2)

Successful Metric	Counterproductive Metric
Measure to stable production, customer acceptance, or time-to-revenue	Measuring to release or to ramp results in "throw over the wall on time" incomplete designs without thorough up-front work, which leads to: • Difficult resolution of design and quality issues • Lengthy ramps and delays to reaching volume production • Missed sales and/or disappointed customers

Sales or Development Incentives (Section 2.2)

Successful Metric	Counterproductive Metric

Profitability	Based on revenue or quantity of projects, which results in:
	• "Taking all orders," or
	• "Develop all products for all markets," which:
	  • Lowers profits by taking on low-profit/money-losing products or projects
	  • Drains NPD resources away from developing truly low-cost products that will actually sell better and make more money

Intermediate Deadlines (Chapter 3)

Successful Metric	Counterproductive Metric
Allot time for thorough up-front work for the best: • Time • Cost	Early deadlines to "show early progress" result in lack of thorough up-front work and suboptimal architecture, which misses opportunities for: • Substantial product cost reduction • Shorter product development, lower NPD budget, and quicker time to stable production • High-quality products by design

1.7 RESISTANCE TO DFM

Despite these problems, some companies and individual designers still resist DFM. Here are the most common reasons that people often cite:

1. Linear thinking

    Just let me get something working now; sometime *later* we (actually someone else) will take care of manufacturability, cost, quality, reliability, serviceability, variety, etc.
2. Misconceptions about time

    I don't have time to worry about manufacturability now; I've got deadlines to meet.
3. Misconceptions about constraints

    I'm a "blue sky" thinker; don't bother me with unnecessary constraints. Don't limit my design freedom.
4. Misconceptions about innovation

    I'm very creative, and don't want to be stifled even thinking about design for...whatever.
5. "I don't have time for DFM." However, if you don't have time to do DFM now, how will you ever find the time to do it later when it is much more difficult, maybe impossible?

However, manufacturability problems can delay the launch, cause availability problems and shortages throughout the product life cycle, delay rapid growth if the product is a big hit, degrade the quality of the look and feel of the product, and raise the cost so much that the price has to be raised or profit lowered, all of which will result in an unsuccessful product. Then there is the common resistance to different ways of doing things.

This section concludes with Murphy's law of product development: *If you don't consider manufacturability early in the design, it is very unlikely that it can be quickly and easily incorporated later.*

1.8 ARBITRARY DECISIONS

Designers may be tempted to think that fewer constraints result in more design freedom and many may resist DFM on those grounds. But, in reality, too few constraints may lead to the design equivalent of "writers' block." If every design decision has many open choices, the whole design will represent an overwhelming array of choices that can lead to design paralysis.

So, the designer breaks the impasse by making arbitrary decisions. Every arbitrary decision will probably make it difficult to incorporate other considerations later. And the further the design progresses (the more arbitrary decisions), the harder it will be to satisfy additional considerations.

Not considering all the goals and constraints at the beginning results in arbitrary decisions that eliminate solutions downstream.

Thus, another motto for product development teams should be:

No arbitrary decisions!

Figure 1.4 graphically shows a decision tree. The concept of decision trees applies to everything from product development to life in general. The tagline in TV ads for the show *Touched by an Angel* comes to mind:

In each moment lies a choice that can change the story of your life.

The process of developing products is a series of decisions, from deciding which products to develop to optimizing product architecture to designing parts. Every decision sends you down a certain branch of a decision tree. Every arbitrary decision will probably send you down the wrong path, which will limit subsequent choices. And more arbitrary decisions are even more likely to send you further down the wrong path. So instead of making methodical decisions to arrive at the desired point A, arbitrary decisions lead you to point B. And the farther you get into a design (and the more work is based on that path), the harder it is to make the changes to backtrack from point B to point A.

The following are common examples of arbitrary decisions that should be avoided in product development:

- *Product definition* is often arbitrary instead of systematically understanding the "voice of the customer," as shown in Section 2.11.
- *Markets in which to compete* may be chosen arbitrarily if overhead costs are averaged, thus obscuring the profitability of individual products. This may lead companies to compete and develop products in markets that, historically, may have not been the most profitable.

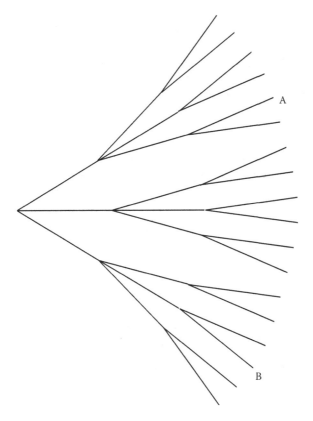

FIGURE 1.4
The decision tree.

- *Project milestone deadlines* are often set early in an arbitrary way (for instance, based on quarterly milestones) and underemphasize crucial architecture optimization.
- *Concept and architecture* are often based on previous or conventional concepts or the first idea to come to mind.
- *Technologies* may be chosen by techies who are impressed with the advertised functionality. However, technology decisions also need to include how proven the technology is, its track record, the total cost, the real availability, the production volume capabilities of the supplier, the financial strength of the supplier, contingency (Plan B) planning, and the risks, especially if there are multiple sources of risk.
- *Manufacturing* decisions, if arbitrary, may steer the design in the wrong direction and preclude the best processes and sources.

- *Outsourcing* decisions may arbitrarily draw a line between "us" and "them," which can result in outsourcing all parts, even when selective internal integration would save time and money and improve the manufacturability of the design (Section 4.8).[30]
- *Supplier* decisions may be made on part cost only and not include quality, quick turn prototype capabilities, production deliveries, and the value of assistance designing products.
- *Part selection* decisions may be based only on functional specifications and the cost of the part itself, and not include the impact of sourcing, quality, standardization, and the part's impact on total cost.
- *Order of design* may be chosen arbitrarily from the ground up or starting with the most obvious or most understood aspects instead of the most constrained aspects first, which may be the least obvious.
- *Detail design* decisions are often arbitrary if designers ignore any design considerations.
- *Tolerances* are often specified arbitrarily if designers do not understand processes, dimensional referencing, or relationships between multiple tolerances. When in doubt, many designers specify unnecessarily tight tolerances, instead of methodically specifying and dimensioning tolerances, as discussed in Guideline Q12 in Section 10.1.
- *Overhead allocation algorithms* usually allocate overhead arbitrarily based on labor, processing, or material costs, which does not provide relevant computations of the total cost (Chapter 7).
- *Styling* decisions which are made arbitrarily can compromise manufacturability by unnecessarily complicating designs, tooling, and manufacturing operations. Don't allow styling to be thrown over the wall. Instead, the whole team should work together to create styles that both look good and are manufacturable.
- *Boxing the design into a corner.* Don't make so many other arbitrary decisions that the design becomes unnecessarily complicated.

When faced with too many open choices, seek out additional constraints. Not doing so may result in arbitrary decisions that will preclude adding them later. When stuck, seek out additional constraints. It may be that a solution cannot be justified to solve *one* problem, but it can if it solves *multiple* problems.

1.9 DFM AND DESIGN TIME

Some designers may be tempted to think that considering all these constraints will mean more time needed to complete the design. However, it really takes no more time (maybe even less time), because thinking about all the constraints at once will steer the designer more quickly to the optimal design.

Theoretically, the ideal number of goals and constraints would lead the designer directly to the single optimal design. Too many constraints would result in no solution. However, too few constraints would result in multiple solutions, which must be systematically evaluated. When this is not done, it is unlikely that the chosen solution will optimize manufacturability and all the other constraints.

The net result of not considering manufacturability early is a design that will not easily incorporate DFM principles later. In order to make such a design manufacturable, it may be necessary to make later changes in the design, as discussed next.

1.10 ENGINEERING CHANGE ORDERS

One of the biggest payoffs of do-it-right-the-first-time product development is avoiding expensive and time-consuming engineering change orders (ECOs).

Early and thorough inclusion of all the design considerations can do a lot to minimize the need for change orders. Methodically defining the product to satisfy customer needs (Section 2.11) will avoid changes to "satisfy the customer"—a common, but illusory complaint. Changing the design to satisfy the customer really means the product definition phase did not thoroughly gather the voice of the customer in the first place.

One of the most insidious aspects of change orders is that a change to fix one problem may induce new problems. And changes to fix those may induce even more problems. Toyota says that late design changes are "expensive, suboptimal, and *always degrade both product and process performance.*"[31]

Further, change orders will probably negate any qualifications or certifications the product received while in its original form.

1.11 DO IT RIGHT THE FIRST TIME

Do it right the first time, because you will not have the chance to do it again, especially in the following cases:

- Fast paced projects, which you do not have time to do over
- Expensive development projects, which you cannot afford to do over
- Complex projects, where each change may induce other changes
- Regulations contracts, or change restrictions, that discourage change or force requalification of the product after any changes

And this is the setting in most companies today.

Everyone who practices DFM should adopt the motto: *Do it right the first time.* This advice seems obvious because no designer would begin a design *expecting* to redesign any part of it later. However, it is distressing how many companies routinely tolerate change orders, maybe because they always have. Some companies camouflage the process by euphemistically calling them "revisions" or even "updates." But, regardless of the label, changes can have a severe impact on a product's cost and time schedule, not to mention employee morale and corporate reputation.

On the other hand, if designers use these DFM techniques to actually do it right the first time, the new product will sail through product introduction into stable, trouble-free production with the best cost, delivery, and quality to provide early customer satisfaction.

Companies' goals should be that each product is designed *once* and that the initial design can be manufactured easily as a high-quality product and, of course, one that works properly. In many companies, the DFM program evolves from some "never again" trauma to a do-it-right-the-first-time thrust, including the strategies and methodologies in the following section.

1.12 STRATEGY TO DO IT RIGHT THE FIRST TIME

The strategy starts with multifunctional teams with all specialties present and active early, as well as good product definition to eliminate later engineering changes to satisfy the customer.

Throughout the product development process, the team must raise issues (the "what ifs" and the "what abouts") and resolve them early through simulations, experiments, research, and early models, mock-ups, solid models, and rapid prototypes, with statistical significance assured by the design of experiments. Critical applications may need failure modes and effects analyses (FMEAs).

The team must understand the lessons learned from previous projects through summaries of lessons-learned databases, investigations of lessons learned from previous projects, and presentations and feedback from previous projects, vendors, and in-house production.[32] This is discussed further in Section 3.3.4.

The team needs to formulate "Plan B" contingency plans to deal with the most likely changes, setbacks, delays, shortages, or other problems regarding technology, processing, customers, markets, regulation, and so forth. For instance, products can be designed to readily accept the Plan B part if the Plan A part does not work out or is not available in time.

The team needs to achieve concurrence before proceeding. Another MIT study, *Made in America: Regaining the Productive Edge,* offers additional insight into Japanese product development project management:

> "A key task of the manager is to make sure that all disagreements [issues] are aired and resolved at the outset. *Achieving consensus takes a great deal of effort,* but by skillful management at this point it is possible to gain the full commitment of all members of the program team so that *subsequent progress is very rapid.*"[33] (emphasis added)

The design of the product can be done right the first time by simplifying the concept and optimizing product architecture, thoroughly designing the product, focusing on new and pivotal aspects with optimal use of previous engineering, peer review and design checking, and good documentation management.

The design needs to be conveyed to production unambiguously so the product is built right the first time. And all changes, updates, and revisions must be implemented promptly and accurately so that subsequent designs will be done right the first time.

Finally, fast development has the least vulnerability to changing environments with respect to changing customer preferences, markets, competitors, regulation, and trends.

1.13 COMPANY BENEFITS OF DFM

The following slogan sums up the importance of DFM:

> *Functionality gets us into the game;*
> *Quality and reliability keep us in the game;*
> *Manufacturability determines the profit.*

And yet most engineers and managers focus primarily on functionality. In order to stay in the game, products need to be produced with high quality and reliability. And what about profits? Unless the product has a formidable head start or incredibly strong patents, the product will have to be priced competitively, which then means that profits will be determined by the cost. And, as pointed out earlier and in Section 6.1, it is very difficult to reduce cost by "cost reduction" efforts after the product is designed. Therefore, profits are determined by how well low cost can be assured *by design*—that is, *designed for manufacturability*.

The benefits of DFM range from the obvious cost, quality, and delivery, to some important subtle benefits:

- *Lower production cost.* Designing for simplicity, fewer parts, and easier assembly results in lower assembly cost. Lower cost of quality (Section 6.9) results from fewer parts and foolproof assembly. Smoother product introduction means less time spent on costly change orders and firefighting to deal with product introduction problems.
- *Higher quality.* Higher quality results from more robust designs, fewer parts, foolproof assembly (Section 10.9), optimal process selection and design, the use of more standardized parts with known good quality, and designing around proven engineering, parts, modules, and processes.
- *Quicker time to market.* DFM products fit better into existing processes and are less likely to require special equipment and procedures. The use of standard parts means most will be on hand or be easy to procure. Better DFM means fewer product introduction problems, leading to a quick and smooth introduction.
- *Lower capital equipment cost.* Designs that assemble easily need less time on assembly machinery. Less need for special equipment saves equipment capital. Designing to minimize setup (Section 4.6) and

the use of standardized parts result in fewer setup changes, thus leading to greater machinery utilization.

- *Greater automation potential.* Designing for automatic assembly maximizes the potential for automation, with all its cost and quality advantages.
- *Production up to speed sooner.* Faster development, fewer introduction problems, and less need for special equipment or procedures result in production that will be up to speed sooner.
- *Fewer engineering changes.* Early adjustments are much easier to make than later changes that are under change control procedures. If the original design satisfies all the goals and constraints (Section 3.5), it will not have to be changed or redesigned for manufacturability or any of the other design considerations.
- *Fewer parts to purchase from fewer vendors.* Having fewer parts to purchase saves purchasing expense, especially for standard parts. Dealing with fewer suppliers strengthens relations with those suppliers and results in less cost and effort to qualify parts and deal with quality problems.
- *Factory availability.* Fewer production problems and greater machine tool utilization make factories more available for other products.

1.14 PERSONAL BENEFITS OF DFM

The previous discussion cites many *corporate* justifications for DFM. The following points are used in the author's seminars[34] to motivate engineers at the *personal* level.

Why bother to do anything differently? Many engineers think that their companies are doing fine and they are really too busy to do anything differently, especially if company success is somehow based on technology. However, there are many compelling reasons to apply these methodologies:

1. Ensure the health of the company and job security. Very few companies are free of competitive pressures, even those with an apparent technological lead. Very few can coast along in such a turbulent and fast-changing environment.
2. World-class new product development increases the fortunes of the company *and* its employees and offers many career opportunities.

3. Work will be more interesting because, by using these principles, engineers will spend a higher proportion of their time doing fun design work and less time on change orders and firefighting.
4. The next project will be a better experience because you will not be distracted by change orders or firefights on past projects. Further, new projects can build on successful new practices and get better each time.
5. Thrive easily during growth. Without DFM, production workers will be under pressure to meet demand. Engineers will have to help solve production problems that may have been tolerable in lower volumes but are show stoppers at peak demands. New opportunities will probably be more challenging, thus compounding new problems if not done right the first time. Since new engineers will be less adept at solving problems on existing products, they will probably be assigned to new designs, leaving current engineers to continue to be the troubleshooters on existing products.
6. Enhance company success, and your career, with receptivity. Enthusiastically accept, support, and implement new product development methodologies. This will maximize your chances of being selected for the first teams to apply these principles. Work cooperatively with others applying new methodologies. Be openly receptive to new ways of designing and manufacturing parts and products. Enthusiastically accept new, different, and innovative assignments. Suggest design and processing innovations. Volunteer to work on or lead innovative endeavors.

1.15 CONCLUSIONS

DFM alone may make the difference between being competitive and not succeeding in the marketplace. Most markets are highly competitive, so slight competitive advantages (or disadvantages) can have a significant impact.

DFM may make the difference between a competitive product line and, in the extreme, products that are not manufacturable at all. Products fail and go out of production because costs are too high, quality is too low, the introduction was too late, or production could not keep up with demand. These are all manufacturability issues and therefore very much affected by DFM.

NOTES

1. David M. Anderson, *Build-to-Order & Mass Customization: The Ultimate Supply Chain Management and Lean Manufacturing Strategy for Low-Cost On-Demand Production without Forecasts or Inventory* (2008, CIM Press). See book description in Appendix D.
2. See also Section 6.1 ("How *Not* to Lower Cost") and Section 11.5 ("Stop Counterproductive Policies").
3. Anderson, *Build-to-Order & Mass Customization*.
4. This data was generated by DataQuest and presented in the landmark article that started the concurrent engineering movement: "A Smarter Way to Manufacture: How 'Concurrent Engineering' Can Invigorate American Industry," *Business Week*, April 30, 1990, p. 110. In the author's in-house seminars, he presents similar data from Motorola, Ford, General Motors, Westinghouse, Rolls Royce, British Aerospace, the Allison Division of Detroit Diesel, Draper Labs, Rensselear Polytechnic Institute, and several other published sources.
5. Satoshi Hino, *Inside the Mind of Toyota: Management Principles for Enduring Growth* (2006, Productivity Press), Chapter 3, "Toyota's System of Management Functions," p. 133.
6. Joachim Ebert, Shiv Shivaraman, and Paul Carrannanto, http://www.industryweek.com/companies-amp-executives/driving-growth-through-ultra-low-cost-product-development, February 23, 2010.
7. Ibid.
8. Robin Cooper, *When Lean Enterprises Collide* (1995, Harvard Business School Press), Part Three, "Managing the Costs of Future Products," p. 131.
9. Ibid., Chapter 7, "Target Costing."
10. Yasuhiro Monden, *Cost Reduction Systems: Target Costing and Kaizen Costing* (1995, Productivity Press).
11. See the offshoring article at www.HalfCostProducts.com/offshore_manufacturing.htm.
12. "A Smarter Way to Manufacture: How 'Concurrent Engineering' Can Reinvigorate American Industry," *Business Week*, April 30, 1990, p. 110.
13. Robert G. Atkins and Adrian J. Slywotzky, "You Can Profit From a Recession," *Wall Street Journal*, February 5, 2001, p. A22.
14. A.S.K. Gupta and D.L. Wileman, "Accelerating the Development of Technology-Based New Products," *California Management Review*, Winter 1990.
15. Jim Brown, *The Product Portfolio Management Benchmark Report, Achieving Maximum Product Value*, August 2006, the Aberdeen Group, http://www.aberdeen.com/link/sponsor.asp?spid=30410396&cid=3359.
16. In the Aberdeen Group study, *product failure* was defined as "products that are not launched or launched products that significantly fall below revenue, market share, or profit targets."
17. Configurators can automate the processes of determining the feasibility, cost, and time to do customizations, instead of the time-consuming and less accurate manual estimating. See Anderson, *Build-to-Order & Mass Customization*, Chapter 8, "On-Demand Lean Production."

18. Jordan D. Lewis, *The Connected Corporation: How Leading Companies Win Through Customer–Supplier Alliances* (1995, Free Press), p. 38.

19. David Pringle, "How Nokia Thrives by Breaking the Rules," *The Wall Street Journal,* January 3, 2003.

20. Shu Shin Luh, *Business the Sony Way* (2003, John Wiley), p. 195.

21. Patrick Lencioni, *Overcoming the Five Dysfunctions of a Team: A Field Guide for Leaders, Managers, and Facilitators* (2005, Josey-Bass).

22. Dan Steinbock, *The Nokia Revolution: The Story of an Extraordinary Company That Transformed an Industry* (2001, AMACOM), p. 186.

23. James Womack, Daniel Jones, and Daniel Roos, *The Machine that Changed the World: The Story of Lean Production* (1990, Rawson Associates; 1991, paperback edition, Harper Perennial), pp. 129–130.

24. Lee Dongyoup, *Samsung Electronics: The Global Inc.* (2006, YSM Inc., Seoul, Korea), Chapter 5, "Research & Development," p. 97.

25. Bill George, *Authentic Leadership* (2003, Jossey-Bass), Chapter 12, "Innovations from the Heart," pp. 133–134.

26. "Out of the Dusty Labs," *The Economist,* March 3–9, 2007, pp. 74–76.

27. Dan Steinbock, *The Nokia Revolution: The Story of an Extraordinary Company That Transformed an Industry* (2001, AMACOM), p. 185.

28. "The New Heat on Ford," *Business Week,* June 4, 2007, pp. 32–38.

29. "Smart Partners," *Business Week;* review of *The Connected Corporation* by Jordan D. Lewis. December 10, 1995. http://www.businessweek.com/stories/1995-12-10/smart-partners.

30. Bryan Bunch with Alexander Hellemans, *The History of Science and Technology* (2004, Easton Press), Appendix D.

31. James Morgan and Jeffrey K. Liker, *The Toyota Product Development System* (2006, Productivity Press), Chapter 4, p. 40, "Front-Load the PD Process to Explore Alternatives Thoroughly."

32. The feedback forms in Appendix C can be used to solicit valuable feedback from factories, vendors, and field service.

33. Dertouzos, Lester, and Solow, *Made in America: Regaining the Productive Edge,* from the MIT Commission on Industrial Productivity (1989, Harper Perennial), p. 71.

34. For more information on customized in-house DFM seminars, see Appendix D or www.design4manufacturability.com/seminars.htm.

2

Concurrent Engineering

Concurrent engineering is the practice of concurrently developing products and their manufacturing processes in multifunctional teams, with all specialties working together from the earliest stages.

The most critical factor in the success of concurrent engineering is the availability of resources to form multifunctional teams with all specialties present and active early.

The consequences of inadequate resources at the beginning are
significant delays and wasted resources,
which in turn will delay other projects and deplete their resources,
while expending more development cost for all projects.

2.1 RESOURCES

The most distinct contrast between advanced and primitive product development methodologies is the concept of ensuring that all the specialties are present and active early. This concept was presented in the landmark $5 million MIT study, *The Machine That Changed the World: The Story of Lean Production.*[1] In Chapter 5 on product development, authors Womack, Jones, and Roos conclude:

"In the best Japanese 'Lean' projects, the numbers of people involved are highest at the very outset. All the relevant specialties are present, and the project leader's job is to force the group to confront all the difficult tradeoffs they'll have to make to agree on the project."

The authors also conclude:

> "By contrast, in many mass-produced design exercises, the number of people involved is very small at the outset but grows to a peak very close to the time of launch, as hundreds or even thousands of extra bodies are brought in to resolve problems that should have been cleared up in the beginning."

This concept supports the philosophy of this book, which emphasizes thorough up-front work. Figure 2.1 graphically shows this front-loading in the advanced model (lower graph) as compared to the traditional team participation (upper graph).

The traditional product development gets off to a bad start, with a vague understanding of customers' needs and the production definition based on technological advancements, whims, or previous or competitive products. Typically, only a few people are involved at the beginning, either because of resource availability problems or by choice because project management does not appreciate the value of complete teams. In some cases, product development begins with a small clique because of downright exclusivity or because some elite people think that "DFM starts after they are finished" (which was actually said by one physicist).

Whether or not deficiencies in the team composition are acknowledged, schedule pressures will force the "team" to make some "progress." And a key part of making progress is making decisions. However, without the benefit of a complete team, the decisions will probably not address all the considerations discussed in Chapter 3. The problem will be made worse if there is no diversity among the people involved; for instance, if everyone works in the same department and has the same education and experiences.

Unfortunately, without a complete team, many early decisions will be arbitrary, which is especially problematic because these arbitrary decisions then become the basis for subsequent decisions, which in turn, will allow for even fewer open options. After several levels of subsequent arbitrary decision making, the product architecture becomes cast in concrete, which makes it very hard to optimize or correct later.

Continuing to follow the sequence in the top graph of Figure 2.1, what is perceived to be a "complete" team eventually forms, but it is not as complete as recommended herein. The team may proceed for a while in a state of naïve contentment, but eventually there will have to be some form of redirection because of the inadequate product definition or because of the arbitrary decisions made by an incomplete team. So then more effort is expended, possibly with more people added, because the project is starting to get into trouble.

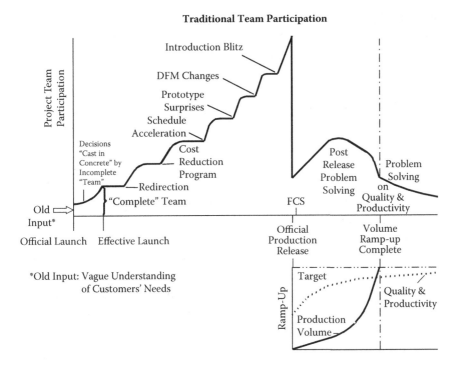

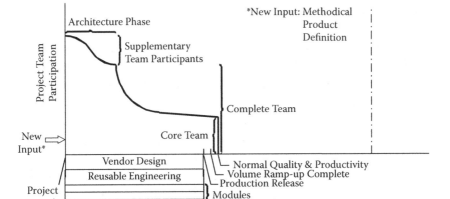

FIGURE 2.1

Team participation: traditional versus advanced models.

By the time cost estimates are generated, word gets out that the cost is too high, so then a cost reduction program is initiated. However, it will be difficult to reduce cost at this stage, because 60% of cumulative cost is committed in the architecture stage, as shown in Figure 1.1.

After the above redirections and delays, the project is now behind, so the schedule needs to be accelerated. This is so common that one product development book even has a chapter titled, "Throw Money at It," based on the thinking that time is more valuable than money at this point.

Then come the prototype surprises, which are the inevitable consequences of an incomplete team, cumulative arbitrary decisions, and failure to address all the design considerations detailed in Chapter 3. Work then proceeds, after many fire drills to try to correct problems and get the prototype to work. Of course, one prototype is not a statistically significant sample, so real-life production problems could be worse than indicated by the prototype.

Then, the typical project starts to consider DFM only as production ramps approach. If DFM did not occur early, it will probably be very difficult to make the product manufacturable through changes at this late a date (Murphy's law of product development). Faced with the formidable scope of implementing DFM by change order under intense time pressures, the team pursues only the easy changes and production soon begins on a product with questionable manufacturability.

As the product goes into production, manufacturability shortcomings manifest as painfully slow ramps, sometimes taking months to reach the volume production target. Manufacturability problems also show up as poor quality and disappointing productivity, which may take even longer to attain acceptable levels. Not only do these delays and shortcomings disappoint customers, but they also consume a great deal of resources—resources that should have been utilized more wisely and proactively at the beginning rather than at the inefficient, reactive end of the project. This, of course, emphasizes the importance of measuring time-to-market as the time of full stabilized production, instead of first-customer-ship (labeled "FCS" in Figure 2.1), which is meaningless as a measure of time-to-market—the factory could build three and ship the one that works!

Does this scenario sound familiar? In fact, most of the attendees queried in the author's seminars[2] admit that many elements in this sequence are quite familiar, many painfully so.

In the advanced model in Figure 2.1, all the relevant specialties are present and active early. If individual team members have versatile backgrounds and can represent multiple specialties, then the team may be smaller and easier to manage.

The complete team is formed at the beginning to simplify concepts and optimize product architecture (Section 3.3). In addition to the full-time core team are vendors, consultants, and part-time specialists for specific tasks such as various analyses and regulatory compliance.

The activities start with a methodical product definition, as discussed in Section 2.11. After the architecture phase is thoroughly optimized, the remaining workload actually can drop off because: (1) many tasks may be completed; (2) the off-the-shelf parts selected avoid the associated design efforts; (3) vendors help design parts or actually design parts entirely; and (4) previous modules can be utilized or the design of new modules can be shared with other projects.

The result is that the volume ramp is completed quickly. Similarly, normal quality and productivity targets are reached rapidly. One important result is the ability to cut in half the real time-to-market as measured to stable production. The other equally important result is that the cost of engineering resources (the areas under either curve) is half compared to the traditional model.

2.1.1 Front-Loading at Toyota

"Because front-loading solves problems at a root cause level early in the process, it nearly eliminates the traditional product development problem of late design changes, which are expensive, suboptimal, and always degrade both product and process performance."[3]

2.2 ENSURING RESOURCE AVAILABILITY

Schedules or workloads should never exceed the capability of design teams to develop good products. If this is a problem, the solutions are to (1) prioritize product development projects to focus on the highest return for effort; (2) hire more people; or, (3) use more efficient product development methodologies, as presented in this book.

The success of product development will be dependent on how complete—how multifunctional—the teams are and how early the complete team is active. Most companies understand the value of complete teams but do not form them early because of resource availability problems. The resource situation is even more difficult when a company is converting from a back-loaded model to a front-loaded model, in which case it is difficult to find enough people for a complete team when many of the people are still busy fixing the problems of the last product developments.

Resource availability issues can be solved with good prioritization of product development efforts and efficient use of team members' time. In a recent *Harvard Business Review* issue with a spotlight on recession recovery, one of the focus articles recommended: "allocating resources according to their articulated strategic priorities."[4] The strategic priorities get articulated by good product portfolio planning, which is discussed in Section 2.3.

Good product development methodologies can solve resource availability problems and save development cost and time. There are dozens of techniques to solve resource availability problems.

2.2.1 Prioritization

Rather than competing head-to-head on price for similar commodity products, focus on *differentiating* superior product designs, which have inherent advantages such as lower cost, better performance, higher quality, and better delivery. Rather than copying or "specing" competitors' products, leap past them with better designs.

Jim Collins, writing in *Good to Great*,[5] asserts that companies gradually build up momentum, like a giant flywheel. Great companies build up this momentum over time, providing many competitive advantages. By the time competitors notice news or ads about the great company's new product introductions, they are years behind and, thus, have little chance of catching up.

2.2.2 Prioritizing Product Portfolios

Prioritize product development efforts to get the highest return from given resources, especially if you don't have enough engineers, managers, and workers with experience, talent, or skill.

> *Prioritization is necessary to grow revenue or profits without adding proportionate resources.*

An Aberdeen Group report provided a blunt assessment on common portfolio shortcomings: "Companies frequently develop portfolios of products that are more comfortable than profitable, develop product strategies based more on wishful thinking than on sound reasoning, and create project and resource plans that are not designed for success."[6]

UCLA strategy professor Richard Rumelt, writing in *Good Strategy, Bad Strategy*, notes that, "Most complex organizations spread rather than concentrate resources."[7]

2.2.3 Prioritizing Product Development Projects

The most effective prioritization of resources comes from prioritizing product development projects and only embarking on high-return development efforts that have the greatest synergy with other development projects.

The book, *Fast Innovation*,[8] presents a case study at Motorola that clearly shows how too many projects diminish the chances of project success. In 2002, Motorola's Computer Group tried to develop 120 products, but resources were spread so thin that *no products were introduced at all!* The next year they cut the development load to 22 projects and were able to introduce eight products. In 2004, as they became more focused (with only 20 projects), they were able to successfully launch almost twice as many products in half the time of the previous year!

Further, the Motorola case study also correlated project success with the total number of legacy products in the portfolio: The year when they introduced no products, this division had 3,500 products in its portfolio. After this dropped to 2,000, they launched eight products, and after it dropped to 500, they were able to launch 14 products.

During this span, manufacturing productivity tripled, early life failures decreased by 38 times, customer satisfaction rose from 27% to 90%, revenue increased by 2.4 times, and operational earnings increased from −6% to +7%.

2.2.4 Prioritization at Leading Companies

2.2.4.1 Prioritization at Apple

When Steve Jobs returned to Apple in 1997, "he found a demoralized company that had spread its resources across no fewer than fifteen product platforms. Those teams were competing with one another for survival. ... Jobs slashed the company's offerings from fifteen to four. ... Every employee understood that the project that he or she was working on represented fully one-quarter of Apple's business and there was no possibility that it would be killed by an accountant scrutinizing the balance sheet. Optimism soared, morale turned 180 degrees, and the rest, as the saying goes, is history."[9]

2.2.4.2 Product Development Prioritization at HP

"Of the $3.5 billion that goes into HP Labs, the company allocates the money based on where it expects the biggest payoffs."[10]

2.2.4.3 Prioritization at Toyota

Toyota invests in research and module development, which can benefit a broad stream of future products. At Toyota, "managers might spend over half their time on a portfolio of ideas and projects."[11]

2.2.4.4 Product Prioritization for Truck Bodies

The largest manufacturer of truck bodies in the US, Knapheide, "pays strict attention the product pipeline or 'funnel.' The funnel now has only five to seven projects in it. In order to 'do fewer projects really well instead of fifty things poorly and late'."[12]

2.2.5 Prioritizing Resources for Custom Orders, Low-Volume Builds, Legacy Products, and Spare Parts

Avoid blanket policies to take all orders or accept all customizations and throw them over the wall thinking that "our guys can do anything and make money at everything!"

Don't take hard-to-build orders for revenue, because the total cost may exceed revenue thus losing money, while draining resources from designing and building more profitable products.

Don't be misled by obsolete "profit margins" that were computed when a product was at its peak. Very low-volume legacy orders will lose money if setup costs exceed revenue regardless of how good the "margin" looks.

Instead, prioritize all orders by *real* profitability, based on total cost numbers. Prioritize all low-volume orders by potential: the top priority would be those orders that lead to more higher volume sales. For those that don't lead to more high volume sales, consider the following approaches:

- Steer customers to newer versions, emphasizing quicker delivery, better quality, and low enough cost to pay for some adaptation effort on their part. Overcome inertia and resistance to change.
- If considering reviving a legacy design, do the trade-off for the minimum total cost (to all) of reviving legacy production and upgrading designs and supply chains compared to adapting the newer version to the customers' needs.
- Consider outsourcing legacy products and spare parts, realizing that the total cost charged by the outsourcer will have to be fairly traded off against *all* the internal costs, including the cost of overhead people (who are *not* "free") and legacy and spare parts fire drills that could drain away NPD (new product development) people and compromise NPD opportunities.
- Consider reselling another company's subsystems, or even entire products that may be rebranded if necessary. Mutual arrangements can be made, even among competitors, to cover each other's weakest offerings that are hard to build.
- For synergistic spare parts and legacy subassemblies, incorporate them into flexible cells that can build any part on demand,[13] supplied by the *spontaneous resupply* of standard materials.[14] If not possible, outsource them to a part supplier whose product line is synergistic.
- If accepting these orders for in-house production, *do not just throw them into "the system."* If there is any chance these activities will drain resources from new product development or any other improvement programs, then it is imperative to have all this work done in a self-funding profit and loss center that:
  - Is *fully staffed* with a complete set of its own people to do everything, including documentation retrieval, tooling retrieval, relearning the "learning curve," all setup, all manufacturing steps, all purchasing, all upgrades, all conversions, all change orders, and all firefighting without needing anyone working on new product development and improvement programs. Staffing would include resourceful technicians and experienced purchasing agents.
  - *Charges enough* to pay for this self-supporting operation. If not, then prices need to be raised accordingly, to cover all costs. If this is not possible, just say no.

- For other products, consider *rationalizing*[15] product lines to scrutinize low-volume, hard-to-build, money-losing products for elimination, improvement, or outsourcing, or to incorporate into new platform development (Appendix A).

If some of these opportunities are potentially promising, then proactively develop product families that are easy to customize.

2.2.6 Develop Acceptance Criteria for Unusual Orders

Consider the above options first. Find out the actual resources that will be needed for each contemplated customization and configuration. Ask people for time and cost data before accepting orders. Get approvals from engineering, manufacturing, purchasing, etc., for each custom order. Create databases containing enough time and resource data to generate accurate estimates, which can then be built into software configurators.[16]

Develop profiles that provide general characteristics of acceptable orders and unacceptable orders. Approvals should be needed for anything not in the profiles. For low-volume orders, an important profile criteria would be the ratio of setup to run time and if the total cost for all setups exceeds revenue.

If the original profile is too limiting, develop *mass customization*[17] capabilities to expand the range of quick and efficient customizations and configurations with parametric CAD, quick program generation, or CNC machine tools, standard parts, and versatile quick-loading fixtures.

If customizations cannot be built quickly and efficiently in the production factory without draining resources from product development, create a profit and loss center that has enough dedicated resources to do all tasks for all customizations and configurations. If that cannot make a profit, then raise prices, make the operation more flexible, or narrow the focus.

2.2.7 Make Customizations and Configurations More Efficient

Develop the ability to make customizations and configurations more efficient with modular design, mass customization,[18] and configurators.[19] Only take custom orders that can quickly and efficiently be customized and designed, without draining resources from new product development.

Proactively ensure quick and easy configuration by concurrently engineering hardware, software, and tooling.[20] Pay for new or special configuration development either through investments or by charging customers and committing resources to do that rather than tapping "overhead" resources who should be helping develop new products.

Utilize a *postponement* strategy[21] and develop versatile product platforms that can be configured by adding preplanned options. Develop modules that can be used for many product variations or subsequent product developments.

Practice "flexible design,"[22] which recommends designing products "cleverly to deal with *future eventualities*." This includes designing in *versatility* to enable adaptability to changing market conditions. For instance, designers should endow the baseline product with enough utility capacity for many evolution scenarios, along with extra ports, connectors, mounting holes, and space for future add-ons.

2.2.8 The Package Deal

If low-volume, hard-to-build, or custom products are perceived to be necessary to get "the big sale," consider the following alternatives:

- Before making any offers or commitments, quantify all the costs, including all the resource demands and setup costs at the build quantity. Then adjust your selling price and bid price accordingly.
- If customers are just asking for hard-to-build products as *deal sweeteners,* resist that and realize (and convey) that it will distract your company from the main order. Given that, a better deal sweetener would be a lower cost or faster delivery, *which should be offered only if the company has learned how to do that efficiently.*
- If the original design cannot be built as designed and needs requalification, steer the customer to a newer version, offering them help with integrating your product into their system or adapting your product. This may cost less and require fewer resources than a difficult build.
- Push back—the customer may not really go to a competitor over this.

If this still *must* be done and it will lose money:

1. Don't dilute resources in operations, supply chain management, and engineering, and don't let the costs of the lower volume order be added to the overhead burdens of the main order. This will make it

less competitive or earn less money. Rather, outsource the production or do it in your own self-supporting profit and loss center.

2. Consider the loss a sales and marketing expense, and don't burden good products and cash cows to pay for their losses.

Not doing the above may drain resources from new product development, burden cash cow products with excessive overhead to pay for "the losers," and may delay the time-to-revenue for profitable products.

2.2.9 Rationalize Products

Rationalize products to eliminate or outsource demanding products, legacy products, and spare parts currently in production that consume too many hours of product development resources to set up production and find tooling, instructions, unusual parts, and people who remember how to build it. It also takes resources to deal with the problems typically encountered with unrefined, seldom-built products.

A proliferation of too many products in the portfolio uses up valuable resources from manufacturing, engineering, and purchasing. Instead of helping teams develop new products, those employees must spend time building a multitude of low-volume and unusual products, with the usual fire drills to set up production, deal with the problems typically encountered with unrefined products, and find unusual parts.

Because these oddball products have such high overhead costs, they are probably losing money (or making significantly less than desired). To pay for that loss, the *good* products will have to subsidize them. Think of it as a "loser tax" on all good products.

Portfolio proliferation places a double whammy on new product development. Not only are resources drained away to build oddball products, but when the new product is launched, it too will have to pay this loser tax to subsidize unprofitable products, thus raising the new product's selling price, which, in turn, will make it less competitive.

The solution is to *rationalize* product lines to eliminate or outsource "loser" products to free up valuable resources to help develop "winner" products. Rationalization not only improves product development immediately (because resources are freed up whenever a company turns down a high-overhead product), but it also will improve profits immediately by eliminating the money-losing products. Effective methodologies for product line rationalization are presented in Appendix A.

Rationalize away or outsource legacy products and spare parts production and management, unless synergistic with current production processes and supply chains. At a minimum, segregate legacies and spares into a separate profit and loss center that has enough dedicated resources to do all tasks. If that cannot make a profit, narrow the focus (through rationalization), raise prices, or make the operation more efficient. This can free up valuable resources who could now help develop new products, instead of the tremendous fire drills to gear up to build unusual products or products "revived from the dead." If you *must* build legacy products and spares, be sure to charge enough to cover all the costs and use that money to hire more people to build them.

Outsourcing hard-to-build legacy products has the following benefits: (1) It stops *overhead demands* for local support people so they can help teams develop new products; and (2) it segregates *overhead charges* for legacy products that are then applied (charged) to these legacy products, so that new products will not have to subsidize them, which would raise the new products' cost unfairly.

Outsourcing legacy products will also probably appear to "raise" their cost because then customers will be paying for their full overhead costs, which may not have been the case when internal overhead costs were averaged. If their total cost exceeds their sales price, then this would encourage rationalizing them away, hopefully before expending the cost to transfer them. Outsourced products will also need to have good documentation, without the need for undocumented "tribal lore."

When outsourcing to your own plant, especially to offshore plants, all transfer costs should be paid for by the *destination plant* or, ideally, by the product itself. Transfer efforts should not be performed by corporate support people, who should be helping product development teams develop products.

If legacy products are truly necessary to support the sales of cash cow products, be sure to:

- Quantify all overhead costs, including the cost of all people who should be involved in product development. For each legacy sale contemplated, understand the man-hours that would be taken away from all the people who should be designing products or supporting product development teams.
- Charge enough for legacy products to pay for all their costs.

- Use this income to pay for all costs, especially personnel who may need to be committed to these tasks, using contractors for temporary spikes.
- If necessary, create a separate profit and loss center to ensure that legacy and spare part activities are fully staffed and supported without taking people away from product development.

For products sold as "loss-leaders," any shortfall between revenue and total cost would have to be reimbursed as a sales expense or paid for by the products that benefit. Failure to do this may drain resources and funds away from new product development and other improvement programs.

Don't offer customized legacy products, unless they have been designed for mass customization[23] and can be built quickly and cost-effectively by flexible operations (see Chapter 4).

Outsource parts and subassemblies that are hard to build, are too different, or require special skill, talent, and equipment.

2.2.10 Maximize Design Efficiency of Existing Resources

Ensure complete and stable product definitions and requirements documents that avoid expending valuable resources to redo previous work to accommodate new or different requirements. Changing previous work is much less efficient than getting the specs right the first time, takes more calendar time, risks introducing other problems, and compromises thorough up-front work on the next project.

Preserve, document, and learn from lessons learned to avoid wasting resources and time repeating past mistakes and missing out on more efficient techniques and methodologies.

Ensure fewer engineering changes to write when teams *do it right the first time*. This will ensure that team members are not lost to finishing incomplete designs, which lowers the overall effectiveness of product development because it is so inefficient to finish designs under change control or in the field.

Focus on new and pivotal aspects of the product design, not boilerplate, and parts that could be reused or bought off-the-shelf. Buy standard off-the-shelf parts instead of designing redundant versions. Reuse previously designed details, parts, and subassemblies and software code. Use modular design to incorporate existing modules or share engineering with multiple projects.

Maximize procurement's contributions by avoiding looking for cheaper parts (unless this provides a substantial net cost reduction without risking time or quality). And shift resources from managing bidding, especially on custom parts, to helping teams do concurrent engineering, which will save more money than bidding.

Arrange for vendors to help design what they build, which will save more money than any "saving" from low bidding (Section 2.6). This will result in: (1) more resources on the design team without increasing the development budget; (2) quicker part development and less chance of delays from part design or tooling shortcomings; and, (3) fewer team resources wasted dealing with manufacturability problems that come from less experienced designers.

Document well to avoid wasting resources to *buy* the wrong thing and *build* the wrong thing, and then spending more effort to *figure out* what's wrong, *correct* the documentation, *expedite ordering* the right thing, *build* the right thing, *reintegrate* the right thing, and various other damage control activities.

Avoid the drain of NPD resources for other improvement projects. Make sure other programs and initiatives (such as big IT implementations) are fully staffed and do not consume NPD resources or even distract them. The funding for all improvement programs should include enough to pay for all the resources necessary for implementation without expecting departments or NPD projects to chip in "free" resources. Focus "change energy" on one major change for each person, for instance: Each design engineer should focus change energy only on product development improvement initiatives. Each manufacturing engineer should be able to help teams design products and get them into production while focusing change energy on manufacturing or quality improvement initiatives. Part of the project leader's responsibility at Motorola is to be "able to shield their team from distractions and people who are not actively contributing to the completion of the project."[24]

Avoid the practice of raiding internal resources from critical functions. This just shifts the hiring and training burden to the critical functions, which will be weakened by both the loss of experience and the need to find and train new hires. Poorly documented companies that rely on "tribal lore" must minimize all forms of turnover for groups that have people that remember or know valuable knowledge and can convey that to new product development projects. *Maximize overall effectiveness by hiring for growth from the outside, unless the internal transfer results in a net gain for the company.*

Avoid problems with factory ramp-up and productivity targets with more thorough up-front work. Level production to avoid draining resources away from NPD for artificially created emergencies, such as big shipments or burning the midnight oil to meet periodic targets. This pulls key resources from NPD, possibly at critical times. It wastes the time of purchasing and operations people, who will then have less time to contribute to NPD teams. Womack and Jones, authors of *Lean Thinking*, note:

> "Raising awareness of the tight connection between sales and production also helps guard against one of the great evils of traditional selling and order-taking systems, namely the resort to bonus systems to motivate a sales force working with no real knowledge of or concern about the capabilities of the production system."[25]

Maximize the chances of project completion and minimize the chances of major redirection or outright cancellation by following the most effective product development methodologies to avoid big delays, which minimizes obsolescence risks. These methods include good and stable product definition, complete multifunctional teamwork with all specialties present and active early, drawing on lessons learned about what works and what to avoid, and thorough up-front work for the fastest ramps.

2.2.11 Avoid Product Development Failures

An Aberdeen Group survey[26] cited the top reasons for product development failures, which are defined as "products that are not launched or launched products that significantly fall below revenue, market share, or profit targets." Product development failures can be avoided with a good and stable product definition to satisfy the voice of the customer, complete multifunctional teams with all specialties available and active early, learning from past lessons, thorough up-front work that includes raising and resolving issues and optimizing product architecture, and fast enough development to avoid obsolescence or changing markets when launched.

2.2.12 Avoid Supply Chain Distractions

Avoid supply chain distractions that keep manufacturing engineers and purchasing people from making significant contributions to product development team participation because they are distracted by:

- Finding, qualifying, evaluating, and choosing among multiple vendor bids. As Womack, Jones, and Roos noted in the book that started the Lean Production movement in the US, in the best companies studied, vendors "are not selected on the basis of bids, but rather on the basis of past relationships and a proven record of performance."[27]
- Finding, qualifying, or evaluating for new low bidders to replace current vendors.
- Working with new low bidders to help them up the learning curve.
- Writing new contracts and change orders to accommodate new vendors with respect to material changes, different processes, translations, converting CAD drawings, and generating new machine tool programming.
- Dealing with new low-bidder problems regarding quality, performance, delivery, etc.

2.2.13 Optimize Product Development Project Scheduling

Optimize product development scheduling and planning to ensure each multifunctional team can form early with a complete mix of talent *throughout the project*, including *specialized or scarce talent*. The count and mix of resources are allocated proportional to the challenge. Ambitious projects or bids should receive adequate resource commitments before they are started or a binding proposal is submitted.

New product development resource commitments should be less than 80% of capacity to allow for statistical variation and ensure that all teams have the right mix of talent at the right time, so complete multifunctional teams can form early for each project. (See Figure 5.3, "Effect of Overburdening Capacity on Development Lead Time," in *The Toyota Product Development System*.[28])

2.2.14 Ensure Availability of Manufacturing Engineers

To protect product development support from firefighting distractions and maximize their participation in product development teams, restructure manufacturing engineering into three subgroups that are financed and staffed differently:

- *New product development* support (paid for by the development project; staffed with manufacturing engineers most experienced in new product introduction)
- *Process improvements* (on overhead, but expected to pay back investment in process cost savings; staffed by the manufacturing engineers most competent in process improvement)
- *Firefighting and engineering change orders* (ideally, paid for by the product involved; staffed by lower cost personnel who are experienced in the change order and firefighting process—not necessarily degreed engineers)

2.2.15 Correct Critical Resource Shortages

Correct critical resource shortages by selectively hiring permanent or temporary employees whose absence has created gaps in the multifunctional teams. Typically, the most common gaps are in the following positions:

- *Manufacturing engineers*, who generally spend most of their efforts on change orders and firefighting instead of showing design teams how to *eliminate* the firefighting by design
- *Quality and test engineers*, who can help ensure quality in the design stage, instead of expending many times the effort to fix quality problems later
- *DFM engineers* with specific expertise (e.g., circuit boards or plastics), who can eliminate subsequent problems that cost many times the cost of designing for manufacturability

At Toyota, "Skillful improvements at the planning and design stages are ten times more effective than at the manufacturing stage."[29] Given this ten times leverage, hiring people to fill these gaps will represent a ten times net savings for the company.

Groups that support product development (manufacturing, testing, quality engineering) should be billed to the development project, to avoid resistance to increasing overhead costs.

2.2.16 Invest in Product Development Resources

View product development as an *investment*, not a "cost." A Battelle R&D report on product development drew the conclusion that "the support of research and development runs the risk of being viewed as an expense and a luxury, rather than an investment, and one that can be shelved until more funds are available."[30] And a Deloitte study of 650 companies in North America and Europe revealed that "while manufacturers cite launching new products and services as the No. 1 driver of revenue growth, they also view supporting product innovation as one of the *least important* priorities."[31]

2.2.16.1 R&D Investment at Medtronic

According to Bill George, chairman and CEO at Medtronic: "Companies must rigorously reinvest a significant portion of their increased profits in R&D, market development, and future growth opportunities, and not let it all go to the bottom line."[32]

Medtronic increased R&D spending from 9% of revenues to 12%, knowing that these investments would not produce any bottom line return for 5 to 10 years.[33] Medtronic's shareholder value increased *150 times* over a period of 18 years.[34]

2.2.16.2 R&D Investment at General Electric and Siemens

The Economist reported in September of 2010 that "Jeff Immelt, GE's chief executive, has increased its spending on research and development to 5.6% of its industrial revenue, a ratio he calls 'pretty world class'… Siemens spends about the same proportion."[35] In April of 2012, the *Financial Post* reported that GE announced "industrial orders had risen 20% in the quarter and selling prices had improved in most of its businesses."[36]

2.2.16.3 R&D Investment at Apple

One of the key principles of innovation at Apple is, "If you believe in the future, and your future lies in R&D, don't starve R&D."[37]

2.2.16.4 R&D Investment at Samsung

"Samsung believes that constant advances are the only way it can reach a sustainable competitive advantage" and "has consistently been a first-to-market player with pioneering products—thanks in large part to its continuing investment in research and development."[38]

2.3 PRODUCT PORTFOLIO PLANNING

The very first step in product development is deciding what to develop. Product portfolio planning is the *proactive* determination of what products to develop.

Prioritize product development efforts to focus resources on the *most profitable products* to maximize the ratio of total gain over total cost. Total gain includes the potential gains for all variations and derivatives over time. Total cost includes all overhead costs to be incurred by the candidates. Toyota uses "multiproject management to optimize the sharing of resources across multiple, concurrent projects."[39]

Prioritize customers and customer segments and focus on the ones with the best current and future opportunities for profit and growth.[40]

Allocate the optimal resources (count and mix) proportional to the challenge for new technology, risk, cost, time, and so forth.

Know the true profitability of all product variations to deliver the highest return from given resources.

Develop *profiles* that describe the characteristics of the most profitable customers and customer segments. Use these profiles to help prioritize product development opportunities, scrutinize unusual or low-volume sales, and rationalize away high-overhead, low-profit products.

Focus on *product families* that can benefit from synergies in product development, operations, and supply chain management. Identify opportunities where products could benefit from build-to-order of standard products and the mass customization of specials.[41] Toyota produces an average of seven different vehicles on each platform, which maximizes reliability across vehicle types. Platforms are designed to be a basis for these vehicles for up to 15 years.[42]

Plan the portfolio around versatile products and flexible processes that (1) can easily satisfy the anticipated range of product breadth and customization, (2) have the potential to satisfy even broader ranges of customer needs, and (3) can easily adapt to evolving trends and upgrade possibilities.

Plan the portfolio to avoid the commodity trap, with its inherently low profits, and evolve to innovative products designed for low cost and sales at high profit. Set aside resources and budget for breakthroughs[43] in *blue oceans*, where innovative products would create uncontested markets. Based on analysis of 150 strategic moves spanning 100 years and

30 industries, the *blue ocean strategy* recommends using *value innovation* to pursue differentiation and low cost simultaneously,[44] a goal that is also supported by mass customization, which can achieve premium prices at lower cost.[45]

Use all the above techniques to prioritize the portfolio to maximize the return from available resources. Make sure all approved projects will have enough resources with the right mix of talent for thorough up-front work and consistently methodical work throughout the project.

Make all product portfolio decisions rationally and objectively. Avoid temptations to base portfolio decisions on automatic upgrades, enticing market opportunities, exciting technology, whims, pet projects, competitive precedents, and so forth. Do not allow portfolio decisions to be influenced by politics, powerful backers, attachments, inertia, fears, unrealistic expectations, folklore (e.g., "I heard somewhere that…"), and so forth.

Don't limit the portfolio to only products. Invest in research and module development that can benefit a broad stream of future products.

Make total cost numbers the cost basis of all product portfolio decisions, as discussed in Chapter 7. Relieve resource demands caused by existing products by planning versatile new products that can also replace hard-to-build or low-profit products that the company may not be able to rationalize away.

2.4 PARALLEL AND FUTURE PROJECTS

Each product development team should be coordinated with other *parallel* product development teams to simultaneously work together to set compatible design strategy, share engineering effort on common design features, determine common parts for use on all projects, and design modules for use on multiple projects. Further, *future* projects should be considered to set design strategy for current and future projects, establish an upgrade path for current products, isolate areas most likely to change to minimize engineering on future projects, and design modules on which future projects can be based.

Management guru Peter Drucker has presented a powerful version of this principle.[46] He recommends the following procedure, in which

"A single team of engineers, scientists, marketers, and manufacturers works simultaneously at *three* levels of innovation:

1) At the lowest level, they seek incremental improvement of an existing product;
2) At the second, they try for a significant jump;
3) The third is true innovation.

The idea is to produce three new products to replace each present product, *with the same investment of time and money*—with one of the three then becoming the new market leader."

Along a similar vein, *Megatrends 2010* reports that at Medtronic, "for each product launched, the company is working on four generations of upgrades."[47]

At Philips, "At any one time, Philips Consumer Electronics is actively working on three product generations: one in production, one in final development (for which major changes are not acceptable), and one at the concept generation stage (which is where new ideas enter)."[48]

At Sony, "from the moment the first machine was made, and before it even hit the market, Sony's engineers were already back at the design table, refining and tweaking the original blueprints to get better sound and better quality tapes to the consumers."[49]

At Crown, "When the company started development of its three-wheel sit-down counterbalanced lift truck, the design team all had in mind the future development of a four-wheel model. This allowed them to consider design challenges associated with both applications and essentially address them for both models at the same time."[50]

Knapheide, the largest truck body manufacturer in the United States, "develops digital models of their products and shows them to potential customers in advance. This allows them to get feedback early in the design process. ... Then, they continue to share virtual prototypes with customers throughout development. The result is greater interest by customers when the product is first released resulting in new fleet business and sales ramping up faster because the customers get what they want."[51]

In all these examples, many parts and design aspects are the same while others are similar. At the first level, designers can leave connectors, extra capacity, and space for future features. The customers and marketing channels may be the same. While purchasing agents are getting quotes on the current models or volumes, they can easily get quotes on alternative parts and higher volumes for subsequent endeavors.

2.5 DESIGNING PRODUCTS AS A TEAM

The following discussions apply both to formal multifunctional teams and to individuals, who should proactively engage people in operations, supply chain management, quality, customer liaison, etc.

The team should be designing the product concurrently *as a team*. In multifunctional design teams, *all members are expected to jointly design the product.*

The team should set the intermediate milestones between concept/architecture and the design phase. These are not assigned unilaterally.

Team members should *not* discuss important issues only on email; instead, use the rapid dialogue of face-to-face communication.

Team members should: *not* react to drawings or prototypes; *not* give their first "input" after something has been designed or is in design reviews; and *not* interact only in weekly meetings.

The team should work well as a *team* rather than just a group of *individuals.* Team members should communicate, discuss, and resolve all issues early. A popular saying at IDEO is "all of us are smarter than any of us."[52] Don't do concurrent engineering primarily in period meetings.

2.5.1 The Problems with Phases, Gates, Reviews, and Periodic Meetings

Consulting firm A. T. Kearney wrote in *Industry Week* about the typical rigid processes for new product development:

> "Most companies have well-established product-development processes that are highly rigorous with fixed tollgates, multiple interim milestones, and reviews to ensure that product development is on track and on budget. This approach, while inherently sound, contributes to a lot of inefficiency and lack of speed while adding additional cost elements."[53]

The context of the article was *ultra-low-cost product development,* which has been used to develop the $2,200 Tata Nano, $100 computers (for the One Laptop per Child Foundation), $35 cell phones, and low-cost medical products, such as the Siemens Essenze, which was designed for small clinics and rural hospitals to provide access to quality healthcare services at a fraction of the cost of standard MRI equipment.

The article summarized how to achieve ambitious cost and development time targets:

> "To develop an ultra-low-cost product while meeting aggressive time-to-market goals, companies should develop and deploy a flexible product development process that eliminates much of the intermediate reviews and all but the major tollgates.
>
> The process should instead require more frequent cross-functional engineering reviews. This ensures timely cross-functional engineering input into the development process while minimizing the amount of post-design-freeze changes."[54]

This corresponds to the principles of DFM, which advocate not only replacing these intermediate reviews but also recommend replacing scheduled *periodic* meetings with *on-demand huddles* (discussed next). Tim Brown, the CEO of IDEO, agrees: "Good ideas rarely come on schedule and may wither and die in the interludes between weekly meetings."[55] Instead, utilize on-demand meetings.

2.5.2 Huddles

Rather than trying to do concurrent engineering in scheduled periodic meetings and many formal reviews, team members should be continuously working together and calling *huddles* to make decisions and resolve issues as they need to be addressed. Thus, the design is continuously "reviewed." People inside and outside the team should huddle on demand for "peer review," rather than accumulating issues for an event.

The team leader calls huddles whenever appropriate and invites relevant participants. In addition to continuously talking with other team members, any team member can ask the team leader to call a huddle. When appropriate, the team leader calls huddles to present progress to management. And management or staff can ask the team leader to call a huddle to update the team on new developments in the company, markets, technologies, regulations, and so forth.

At Ford's successful 2009 skunkworks project (skunkworks are discussed in Section 11.7.2), which quickly developed the new "Scorpion" diesel engine: "We saved months by knowing hourly what the other guys were thinking and what their problems were."[56]

2.5.3 Building Many Models and Doing Early Experiments

IDEO emphasizes the value of building models, which they call "prototypes":

> "Prototyping allows exploration of many ideas in parallel. Early prototypes should be fast, rough, and cheap. The greater the investment in an idea, the more committed one becomes to it."

> "Product designers can use cheap and easy-to-manipulate materials: cardboard, surfboard foam, wood, and even objects and materials they find lying around – anything they can glue or tape or staple together to create a physical approximation of ideas."[57]

IDEO avoids arbitrary decisions by trying many experiments: *"Most problems worth worrying about are complex, and a series of early experiments is often the best way to decide among competing directions."*[58]

2.5.4 Manufacturing Participation

One of the most effective ways to ensure manufacturability is early and active participation from manufacturing, including manufacturing engineers, tool designers, and whoever is experienced with problems and change orders regarding assembly, throughput, quality, testing, repair, and ramping into production. It is *much* easier to prevent these problems in the design phase instead of trying to deal with them when the product is going into production.

The following should be early priorities for manufacturing personnel on product development teams:

- Fully convey the difficulties encountered, and their consequences, when products are not designed for manufacturability.
- Turn this experience into actionable proactive design recommendations and manufacturing strategies, including:
  - Processing strategy, including process selection, the flow of parts and products, and flexible cell and line design for product families
  - Investigating the optimal use of automation and CNC operations and, if necessary, making them flexible enough for high-mix operations (mass customization)
  - Optimizing outsourcing and internal integration decisions.[59]
  - Helping the team choose and find off-the-shelf parts

- Resupply strategy for parts and raw materials
- Identifying the supplier base for purchased parts and materials
- Vendor strategy for custom parts; possibly part design by vendor
- Identifying potential vendor partnerships early and arranging for early participation on the design team
- Strategies for quality and reliability assurance and test
- Mistake-proofing (poka-yoke) strategies
- Manufacturing strategy for customization, configurations, product variety, extensions, and derivatives.

- Help the team actually design the product.
- Push back on any distracting activities that contribute less to the company's real profitability than new product development, such as cost reduction on existing products, building low-volume or oddball products, accepting unusual customizations, building legacy products, spare parts production, implementing big IT programs, qualifying new low bidders and getting them up the learning curve, dealing with quality problems from low bidders, and so forth.
- Be an early and active team member, thoroughly raising and resolving all issues related to manufacturability. Be proactive and forceful to ensure the product is designed for manufacturability.

The Toyota perspective:

"Manufacturing and production engineers are now involved very early in the design process—working with design engineers at the concept development stage, to give input on manufacturing issues."[60]

2.5.5 Role of Procurement

The *procurement* function needs to shift from just *purchasing* "the" part that engineers want to searching for broad *ranges* of parts to maximize part availability throughout the life of the product. Broad searches are discussed further in Section 5.19.1.

Be prepared to pay for this availability. But any increase in the purchase price will be saved many times over by avoiding change orders to solve availability problems.

Procurement people should look for suppliers whose other customers have similar challenges, quality demands, and life spans. They should also qualify suppliers and vendors for quality, ability to deliver, and stability.

Potential new parts should be prequalified so they will be ready for engineers to incorporate into new designs. Long lead times can be avoided by selecting standard parts that are readily available. In emergencies, projects can borrow those standard parts from each other.

2.5.6 Team Composition

The multifunctional aspect of teamwork requires representation of all relevant specialties, as discussed in greater detail in Section 2.9. Key team members may be full time, while some members may be part time. The team should have a strong team leader, as will be discussed in Section 2.7.

In the concept/architecture phase, team staffing should consist of:

- System engineers (system architects), whose focus is simplifying concepts, optimizing product (system) architecture, system integration, wiring, off-the-shelf part decisions, and so forth
- All the functions and specialists that are needed to optimize the product architecture (Section 3.3)
- Designers of *critical* parts; they should be early and active participants in system engineering of the product or subassembly to help optimize the systems architecture concept that determines the part's requirements

In the design phase, team staffing should consist of:

- Enough system engineering resources to ensure optimal system architecture and integration
- A well-coordinated group of part and subsystem designers and manufacturing people concurrently designing the parts and processing
- Manufacturing engineers working with part and subassembly designers
- Purchasing agents to help select parts and subassemblies
- Tests engineers to develop and implement tests, fixtures, and equipment
- Software engineers to write code

At Apple, they believe that: "having all the experts in the same place— the mechanical, electrical, software, and industrial engineers, as well as the product designers—leads to a more holistic perspective on product development."[61]

2.5.7 Team Continuity

Project success depends on staff continuity and consistent responsibilities. The team should be assigned a complete mix of talents that will have adequate bandwidth throughout the project. Corporate resource planning should ensure that this is so by following the dozens of methods presented in Section 2.2.

Here is the team continuity policy at Philips:

> "[The team] continues from planning through design and interactive problem solving to final products being shipped. The constant makeup of the team supports easy raising of issues, shared understandings, and fast decisions. While new people occasionally enter the process, total continuity is essential during the final design iteration."[62]

2.5.8 Part-Time Participation

Smaller companies may not have enough resources for several complete product development teams staffed exclusively with full-time personnel. In these companies, specialists (e.g., regulatory compliance, heat flow analysis, stress analysis, tolerance analysis, design of experiments, etc.) may be assigned to multiple products on a part-time basis. It is especially important that each team have a complete set of expertise early, when fundamental decisions are being made. However, these part-time people must not be spread so thin that they cannot make meaningful contributions to all projects.

Xomed (a division of Medtronic) makes sure part-time team members have enough focus on all their projects:

> "To assure that engineers pay close attention to projects, Xomed limits them to no more than two projects concurrently. Xomed can't afford to have time conflicts delay moving a hot new product to market."[63]

2.5.9 Using Outside Expertise

Consider bringing in consultants, experts, and contractors to help smooth out peak demands, enhance diversity, and add specific expertise when needed. Be careful to ensure the availability of these workers when the team needs them and ensure security of critical or proprietary knowledge.

> "To avoid compromising either quality or time-to-market with not-invented-here issues, Xomed contracts expertise when needed."[64]

2.5.10 The Value of Diversity

One important benefit of having multidisciplinary membership is to provide a variety of experience and viewpoints. This diversity will help the team perform some of its most important tasks of raising issues, resolving issues, solving problems, and generating ideas. When Bill George was chairman and CEO of Medtronic, he said:

> "It is diversity, and the intense debates it generates, that leads to the best decisions. By calling upon the broad experiences of team members, you can avoid pitfalls and make better decisions."[65]

2.5.11 Encouraging Honest Feedback

Encourage feedback and be receptive to all news about product developments. Create an open culture where issues can be raised and discussed early, with the focus on issue resolution.

When the current Ford Americas President Mark Fields came to Ford from IBM, he was discouraged from airing problems at meetings unless his boss approved first![66]

In *How the Mighty Fall*, Jim Collins relates that when leaders start to fall in Stage 3, denial of risk and peril, "leaders discount negative data, amplify positive data, and put a positive spin on ambiguous data."[67] He describes teams on the way down as follows: "People shield those in power from unpleasant facts, fearful of penalties and criticism for shining light on the rough realities." Teams on the way up are described this way: "People bring forth grim facts—'Come here and look, man this is ugly'—to be discussed; leaders never criticize those who bring forth harsh realities."

2.6 VENDOR PARTNERSHIPS

2.6.1 The Value of Vendor/Partnerships

One of the main strengths of concurrent engineering is early and active participation of vendors (which is defined here as a supplier who builds your custom parts). The only way to get this is to preselect the vendor or partner on the basis of past relationships and a proven record of performance. They will not participate early unless they are reasonably sure paid work will follow. Implying that certain vendors will get your business and then going out for bids will alienate those vendors and ruin previous relationships.

Industry Week's "Best Plant" survey of the 25 top-performing candidate companies found that 92% emphasize early supplier involvement in product development.[68]

> "Toyota selects suppliers early in the product development program, guarantees the business, and incorporates them as part of the extended product development team."[69]

> "Teamwork at Motorola is imbedded in the firm's culture, and this is one reason for its success with supply alliances. ... [The teams] focus on quality (as broadly defined), speed (in terms of removing non-value-added steps), and cost reduction. ... [The ultra-low-cost Tata] Nano development also was characterized by extensive supplier collaboration from the early stages of product design with suppliers being given extensive design flexibility."[70]

> "The ultra-low-cost Tata Nano development also was characterized by extensive supplier collaboration from the early stages of product design with suppliers being given extensive design flexibility."[71]

2.6.2 Vendor/Partnerships Lead to Lower Net Cost

Vendor/partnershps will result in a *lower net cost* because:

Having the vendor help design the part will greatly improve the manufacturability, quality, and lead time, thus resulting in lower manufacturing and quality costs, because vendors thoroughly understand the DFM rules and guidelines for their process, in general, and for their equipment, in particular.

Vendor partners who work with their customers from the beginning will be able to charge less because they (1) understand the part requirement better, due to more thorough interactions, (2) are able work with their customers to minimize cost, and (3) won't have to add a "cushion" to deal with an unknown customer. A leading expert on supplier relations said:

> "Suppliers often add a risk premium to their pricing (thus raising the customer's cost) to cover nondisclosed or unexplored customer requirements or design flaws that may require later adjustments."[72]

Vendors can help avoid arbitrary decisions, which unnecessarily raise cost, delay delivery, and compromise quality. One of the worst causes of arbitrary decisions is styling, especially when the designer throws a pretty shape over the wall to engineering, which then throws it over another wall to manufacturing, which then throws it over yet another wall to the vendor—so the tool maker is three walls away from the designer! When designers work directly with tooling engineers, the results are designs that both look good *and* are easily manufacturable.

Vendor partners will provide the lowest *total cost* because interacting with the customer's team results in vendors thoroughly understanding the challenges and issues, sharing their experiences and their lessons learned with similar applications, making "what if" suggestions early on that will maximize manufacturability, and working with customers early on to minimize total cost. Vendor partnerships benefit from *learning relationships* in which the customer and vendor learn from each other, thus making each job better and faster.

After preselecting vendors, the team can then benefit from more participation from their own purchasing people, who will now be able to help the team make the best off-the-shelf decisions; find the best balance of cost, quality, and delivery; optimize availability for the life of the product; and so forth. When Motorola introduced concurrent engineering, it found that bringing vendors onto the team proved to be a major contributor to a project's success.[73]

Without vendor partnerships, the design would be thrown over the wall to the vendor. And unless the customer's engineers thoroughly understand the processes, which is rare, the design will not be optimized for manufacturability or, worse, it will be hard for the vendor to make changes to make it more manufacturable because (1) there is usually no calendar time for changes, (2) there is usually no budget for changes at either the customer's or the vendor's end, and (3) by this point, most changes will be difficult because of the reasons discussed in Section 6.11.

2.6.3 Vendor Partner Selection

Thoroughly evaluate potential vendors, and select them on the basis of:

- Capabilities, past relationships, and a proven record of performance—not low bidding.[74]
- Financial stability: get Dunn & Bradstreet or similar reports.

- *Your* business being an important share of *theirs,* especially if there are any unusual requirements or variations from typical operating procedures.
- Proximity: local vendors are preferred for contact and delivery.
- Similarities and synergies with vendors' other work, especially with respect to experience, parts used, fabrication/assembly machinery, and test equipment. This will minimize learning curves, delays, changeovers, part changes, and program rewriting, and, thus, minimize cost, quality problems, and delivery times.
- Willingness to work early with their customers to help design their parts and convey their process capabilities and constraints.
- Ability to work well together, contribute ideas, and provide honest, candid feedback.

"Toyota wants the suppliers to think for themselves, challenge the requirements, and provide value-added ideas to the process."[75]

Honda's criterion for selecting suppliers is the attitude of their management.[76] As a philosophy-driven company, Honda feels it is easier to teach product and process knowledge than to find a technically capable supplier with the right attitudes, motivation, responsiveness, and overall competence.[77]

2.6.4 Working with Vendor Partners

Vendor partnerships should be developed. Kiichiro Toyoda, founder of Toyota's automotive business, said: "First tier suppliers, in particular, must be partners in research. We don't just buy things from them. We have them make things for us."[78]

Don't just throw a spec at a vendor and ask for a quote; work with the vendor to optimize the design for manufacturability. Explore "what if" scenarios. Understand the vendor's processes, sensitivities, and process capabilities. Direct interaction and visits are preferred.

Team members should interact directly with the appropriate people in the vendor's factory (being sure to include the team's purchasing member), not just the procurement department or the vendor's reps, who may not understand your products or even their processing.

Vendors should be willing and able to do the following (HP's criteria): (1) help design the product, (2) build quick-turn prototype parts and parts for short-run projects, and (3) build production units.

Don't change vendors as volumes rise because this adds an additional ramp and new sources of statistical variation at the worst time for unexpected problems to occur. Similarly, don't change vendors for a "lower cost" on parts because the total cost including the cost of the change will most likely be higher. Don't dump a vendor at the first hint of disappointment; work with them to improve. In general, don't change vendors at all, so as to preserve the learning relationship in which every job improves the rapport, cooperation, dialogue, and feedback.

Don't beat up vendors to lower cost. If they don't know how to lower cost, they will either cut corners or cut margins, neither of which is good for the OEM (original equipment manufacturer). Rather, work together with vendor partners early to proactively minimize total cost. Taiichi Ohno, the father of the Toyota Production System, said:

> "Achievement of business performance by the parent company through bullying suppliers is totally alien to the spirit of the Toyota Production System."[79]

Here is how Toyota treats suppliers. A survey of suppliers found that although Toyota is rated as "their most demanding customer," it also received the most favorable ratings overall: 415 out of 500 compared to GM at 114 out of 500. Specifically, Toyota:

- Works with new or struggling suppliers to get up to speed
- Makes commitments to suppliers early in the product development process and makes good on promises
- Constructs contracts that are simple and for the life of the product
- Is the best at balancing a focus on cost with a focus on quality
- Honors the contracts—does not renege on them
- Treats suppliers respectfully and respects the integrity of intellectual property
- Works with suppliers to achieve price targets[80]

2.7 THE TEAM LEADER

The product development team leader is key to success. There should be a *single team leader* from the earliest stages through stable production, responsible for all goals, activities, schedule, and deliverables.

An MIT study provides insight into the most successful product development project managers, who serve as both a leader and a project champion:

> "In the Japanese auto companies, each new product is assigned a program manager who: acts as the product's champion, carries great authority with the firm, and along with his staff, stays with the product from conception until well past the production launch."[81]

In addition, a good product development team leader:

- Has the respect of the team and management and is more than an administrator; in fact, excess focus on budgets and schedules will stifle creativity and the crucial architecture phase optimization
- Provides the proper broad focus and resists the natural temptation to think only about functionality and design parts prematurely
- Understands the importance of thorough concept/architecture, pursues optimization, and resists temptations and pressures from the team or management to do otherwise
- Makes sure team members work together as a team, instead of just retreating to their cubicles, and leads the team to do all the thorough up-front work discussed in Chapter 3
- Forces the team to confront and resolve all issues and encourages team members to bring up issues and pounce on the *issue*, not the messenger
- Ensures that the tasks in each phase are completed before moving on

Finally, if all the principles are not implemented company-wide, the team leader creates a *microclimate* where the team can follow these principles right away.

2.7.1 The Team Leader at Toyota

The characteristics of a good team leader at Toyota[82] (called "chief engineer") are

- An instinctive feel for what customers want
- Exceptional engineering skills
- Intuitive yet grounded in facts
- Innovative yet skeptical of unproven technology

- Visionary yet practical
- A hard-driven teacher, motivator, and disciplinarian, yet a patient listener
- A no-compromise attitude to achieving breakthrough targets
- An exceptional communicator
- Always ready to get his or her hands dirty

"Key decisions, mentoring, lobbying for resources, building a shared vision, pushing the product to higher levels, and achieving quality, safety, cost, and timing targets all start with the chief engineer."[83]

2.7.2 The Team Leader at Motorola

"A good project leader's worst enemies are a chair and desk. Instead of sitting in an office, a project leader should visit the project team members at least once a day. … Project leaders have to be tenacious and want to make things happen. … They must be willing to make noise at the top and ask embarrassing questions when obstacles arise. … [Project leaders] make sure that communication among team members happens on an ongoing basis."[84]

2.7.3 Team Leaders and Sponsors at Motorola

"Essentially, a cross-functional team is selected, dedicated, co-located, and put under the direction of a general manager who serves as a full-time project leader for the duration of the effort."[85]

Motorola teams have an *executive sponsor*, who is "an early champion and supporter, as well as direct supervisor in selecting the project leader and helping get the team underway." In some cases, the executive sponsor makes the investment proposal to the Board of Directors.[86]

2.8 CO-LOCATION

Concurrent engineering works best when the product development team is in close proximity with manufacturing operations, so that the whole team can meet frequently and, as a team, do all the tasks recommended by this book. Separating manufacturing people geographically from the product development team will impair the team's ability to design for manufacturability, sometimes seriously. The main issues are

- *Awareness.* Design engineers will not be able to frequently observe production operations firsthand and get direct face-to-face interactions with production personnel, and thus will be less able to concurrently engineer products and processes.
- *Teamwork.* The manufacturing team members' contributions to the teamwork will be compromised if distance decreases the number of meetings and interactions. Greater distance makes face-to-face contact even less likely, thus diminishing the effectiveness of teamwork and discouraging spontaneous interactions.

2.8.1 Effect of Onshoring on Concurrent Engineering

Outsourcing manufacturing far away compromises concurrent engineering even more, because employees of other companies are less accessible in addition to the challenges of distance, time zones, languages, and cultural differences (between both companies and countries). The effect of this separate engineering is discussed in Section 4.8 and in the articles on outsourcing[87] and offshoring[88] at the author's website: www.HalfCostProducts.com.

2.8.2 The Project Room (The "Great Room" or *Obeya*)

Each multifunctional team should have a dedicated project room (*Obeya* in Japanese) for each project to accommodate spontaneous huddles and display the team's charts, graphs, drawings, experiments, samples, models, prototypes, and so forth. Not having this would discourage spontaneous discussions simply because of the lack of availability of somewhere to meet. Here is how the Obeya works in Toyota culture:

> "The Obeya integrates various product development participants throughout the life of a program [facilitating meetings several times a week, which] enable fast decision making and information sharing."[84]

The team and the team leader (chief engineer) meet almost daily in the Obeya to make decisions in real time, not waiting for periodic meetings: "Usually, once every two days at least the whole team assembles there."[90]

At IDEO, "we have dedicated rooms for our brainstorming sessions, and the rules are literally written on the walls."[91] "The simultaneous visibility of these project materials helps us identify patterns and encourages creative synthesis to occur much more readily than when these resources are hidden away in file folders, notebooks, or PowerPoint decks."[92]

If project room space is not readily available, fully understand that:

> *The business model should determine the facilities planning, not the other way around.*

2.9 TEAM MEMBERSHIP AND ROLES

Design teams should consist of design engineers, manufacturing engineers, service representatives, marketing managers, customers, dealers, finance representatives, industrial designers, quality and testing personnel, purchasing representatives, suppliers, regulation compliance experts, factory workers, specialized talent, and representatives from other projects. First, this helps to ensure that all the design considerations will be covered. Second, such diversity can lead to a better design because of contributions from many perspectives. This synergy results in a better design than could result from a homogeneous team consisting only of design engineers or scientists.

Key tenets of the Honda product development process are *trust*, which comes from teamwork and shared knowledge, and *equality*, which means to "recognize, respect and benefit from individual differences."[93] Kaj Linden, the research director at Nokia, says, "The common denominator of Nokia's R&D stems from concurrent engineering efforts in which product development, sales, and production units cooperate significantly."[94]

It is very important that all team members are present and active early so that they will make meaningful contributions to the design team. Everyone on the team should work well together and be receptive to everyone else's contributions. Team members need not all be full time, but they should not be so preoccupied with other tasks that they cannot make meaningful contributions. All team members should actively participate in the product development, not waiting for "designers" to design something and then reacting to their designs. Some of the key team members are described here.

2.9.1 Manufacturing and Service

The participation of manufacturing and service personnel is crucial to ensure that the product development team designs manufacturability and serviceability into the product. Manufacturing engineers have the responsibility of making sure that products are being designed for stable processes that are already in use or for new processes that will be concurrently designed as the product is designed. Manufacturing and service representatives must not wait until the stage where there are drawings to mark up. Their role is to help design the product and constantly influence the design to ensure manufacturability and serviceability. Manufacturing representatives must be isolated from the daily emergencies and firefighting that occur in manufacturing, because *urgent* matters usually take precedence over *important* matters.

Problems can arise when manufacturing engineers view team participation as a career opportunity to migrate into engineering, designing one portion of the product very well for manufacturability, while the remainder of the product has manufacturability ignored.

Valuable knowledge can be obtained from all manufacturing and service people by asking them to fill out survey forms for factory feedback and field service feedback (see Appendix C).

2.9.2 Tooling Engineers

When new or custom processes and tooling are required, they should be designed concurrently with the products. If current manufacturing procedures are inadequate for cost, time, or quality, then new processes may need to be concurrently developed.

Look for opportunities to create innovative tooling that replaces slow, costly, and poor-quality processing. Concurrently develop tooling and fixtures that are faster, more efficient, quick loading, more accurate, partially mechanized, or automated.

2.9.3 Purchasing and Vendors

Purchasing agents should help design teams select parts for the best balance of cost, quality, and availability. Instead of just throwing a single spec over the wall and telling purchasing to "just buy it," engineers should provide a performance range so purchasing agents can look for the best price and availability within the entire range. Sometimes, a higher performance part may have a lower price and better availability if it is mass-produced and is in widespread use, as shown in Figure 5.6.

Purchasing's other role should be to set up and manage vendor partnerships for custom parts, as discussed in Section 2.6. These partnerships should be used instead of bidding, for reasons discussed in Section 6.11.

2.9.4 Marketing

Marketing is the link to the customer and must help the team define the product so that it listens to the voice of the customer. Product definition will be discussed in Section 2.11.

Cooperation between engineering and marketing is a key determinant of product development success. A study of 289 projects found that when there was "harmony" between engineering and marketing, there were only 13% failures. On the other hand, when there was "severe disharmony," the results were the opposite: only 11% of projects succeeded![95]

2.9.5 Customers

Product development teams should be close to customers and understand how they use products. Toyota engineers spend months talking to customers and dealers to understand what customers want in new product designs.[96]

A key aspect of Guidant's product development process for surgical tools is to have engineers observe surgical procedures and gain understanding and get feedback directly from customers.

When Hewlett-Packard's (now Philips) Medical Products Group develops ultrasound imaging systems, they construct full-size nonfunctional models for doctors to evaluate usability in hospitals. This gives the product development team valuable early feedback on all aspects of the user interface, including how well a doctor wearing gloves can grasp probes and switches, extend and retract cords, read displays, move the system from room to room, and so forth.

One reason for Rubbermaid's early success was that its product development teams were so close to customers and products that they could minimize market testing, which dramatically reduced the time-to-market and made it harder for competitors to copy their products as clones or knock-offs.[97]

It is becoming more common to have customers themselves actually participating on product development teams. When Boeing developed the 777, they invited representatives of their customers, the airlines, to help design the product. At first, Boeing engineers were apprehensive, but they soon learned the value of hearing detailed customer input in the design stage.[98]

Xerox does brainstorming or "dreaming with the customer" on its product development efforts. The goal is "involving experts who know the technology with customers who know the pain points."[99]

Toymaker LEGO Group "went straight to its customers—namely robot enthusiasts—several years ago to help guide product development improvements to its Mindstorm product, creating a users' panel and challenging them to improve the company's robotics kit."[100]

Half a century ago, the Swiss–Swedish corporation ABB (formerly ASEA Brown Boveri) bought a minor US company that made electrical meters. Their product development team brought in seven customers (utilities) representing various types of rural and urban markets. The customers signed nondisclosure agreements and these utilities got deals on the forthcoming product they helped develop. Because of this customer input, the new meter took over the market and became the standard electrical meter.

In 1929, a consortium of 28 streetcar operators (customers) and 25 manufacturers spent 5 years to jointly develop the next-generation "PCC" car for rapid acceleration, a quiet ride, aerodynamic styling, and regenerative braking. It was an overwhelming success, with 5,000 sold in the United States and 20,000 more in Europe. Hundreds still run today in cities such as San Francisco, Philadelphia, and Boston.[101]

An added value to involving customers in the design process is that it *bonds* them to the product and, thus, tends to make them more loyal customers when the new product comes out.

Industry Week's "Best Plant" survey of the 25 top-performing candidate companies indicated that 96% of the companies surveyed used customer participation in product development efforts.[102]

2.9.6 Industrial Designers

These creative people need to be part of the design team so that product styling is not thrown over the wall to engineering, who must then fit everything into a pretty enclosure. It is an encouraging trend that the leading industrial design firms are evolving away from a styling emphasis to include engineering, manufacturability, and usability.[103]

2.9.7 Quality and Test

The need for diagnostic testing is dependent on the "quality culture" of the company. If quality is designed into the product and then built in by processes that are in control, then the "fall out" will be so low that diagnostic tests may not be needed. At IBM, products that were expected to have higher than a 98.5% first-pass-accept rate could avoid diagnostic test development and the expensive ATE (automatic test equipment) "bed-of-nails" testers. Above this threshold, it was more cost effective to discard defective printed circuit boards than to pay for the testers and test development. ATE testers cost millions of dollars, and for some printed circuit boards, test development can exceed the cost and the calendar time of product development!

2.9.8 Finance

Finance representatives can help decision making by providing relevant cost data, which does not automatically come from most accounting systems. Implementing activity-based cost management, as discussed in Chapter 7, can provide data based on *total cost* considerations, which will lead to much more rational decision making (e.g., for tradeoff analysis of quality/diagnostics, make/buy decisions, off-the-shelf parts, quantifying quality costs, and quantifying overhead cost savings resulting from standardization and modularity).

2.9.9 Regulatory Compliance

Every design team needs representatives who can ensure that all applicable regulations are satisfied by the *initial product design*, not by costly and time-consuming changes. Future regulations must also be considered, because regulations sometimes change faster than manufacturers can respond with another product development cycle. Some companies have legislative or environmental lawyers on the design teams to anticipate the impact of future regulations. Efficiently producing regional product variations, for instance, for many countries, is one promising application of mass customization (Section 4.3). This requires that the design team incorporate the regulations of every customer country into the design process.

2.9.10 Factory Workers

Factory workers are a valuable source of input, either from actual participation on the design team or from surveys like the factory feedback form in Appendix C. Factory workers usually have no feedback channel for their vast amount of knowledge on past manufacturability issues. Factory worker participation in the design process may have the added benefit of improving labor relations and making the new product more easily accepted as it is launched into manufacturing.

2.9.11 Specialized Talent

Design teams may need help from specialized talent for automation, simulation, stress analysis, heat flow analysis, solid modeling, rapid prototyping, design of experiments, "robust" tolerancing, lab testing, safety, product liability, patent law, and so forth.

2.9.12 Other Projects

Coordinating multiple product development projects is important to maximize synergies, share work on common module design, and standardize parts, modules, tooling, and processes.

The success of product development projects is determined by *how well* concurrent engineering is practiced: *how complete* the multifunctional team is; *how early* the entire team is active; and *how well* the team is led. Locating people very close together (co-location) also helps ensure the success of product development teams.

The largest study of corporate failures, *Why Smart Executives Fail: And What You Can Learn from Their Mistakes*, emphasized the importance of multifunctional teams:

> "Create crossfunctional teams and diverse work groups whose members will see things differently. Such heterogenous groups have been shown to be much better than homogeneous groups when it comes to developing new knowledge."[104]

2.10 OUTSOURCING ENGINEERING

Many managers are intrigued with the prospect of outsourcing engineering as a way to "save cost" on product development with foreign engineers whose wages are a fraction of domestic engineers. Further, there is the allure of speeding up product development by keeping engineering working three shifts a day, with work passed off around the world every day.

This might work with independent activities such as call centers or insurance or loan processing, but product development should be a highly interactive and integrated team activity. Product development is most effective and efficient when products are designed by complete multifunctional teams that are co-located with manufacturing operations and close to customers and vendors, as discussed earlier.

Any proposals to save cost must be based on minimizing total cost. Chapter 6 emphasizes the many costs that make up the total, of which product development labor cost is just one element. Some of the consequences of chasing cheap manufacturing labor—lower labor efficiency and poorer quality—might cancel out the labor rate savings; adding in other costs could push results into a net loss.[105]

A *Business Week* article titled, "The Hidden Costs of IT Outsourcing," confirmed this phenomenon, saying that it applies even more so for software outsourcing: "Offshoring—sending work overseas—isn't always all it's made out to be. Particularly with information technology, which can be a lot more complicated than moving traditional manufacturing operations overseas."[106] The article also pointed out that IT quality is harder to ascertain because it requires more communication and management. When total cost was taken into account, the cost was generally comparable to domestic work. Further, some experts claim outsourced software has 35% to 40% more bugs, and some outsourcers have experienced missing code and unconnected pages, which made updating a nightmare. According to the article, the problems encountered are a "dirty little secret" because companies don't publicize the problems they have run into.

The best way to lower product development expenses is to *maximize the efficiency of the whole process* through concurrent engineering (with co-located teams) and design for manufacturability (with continuous interactions with manufacturing people), as discussed throughout this chapter.

A more efficient product development process consumes fewer labor hours to (1) develop new products and (2) solve problems during product launches, ramping up to volume targets, achieving quality and productivity targets, implementing engineering change orders, and all the firefighting associated with these activities, which are really avoidable with an efficient product development process, as discussed throughout this book.

Optimal product development practices not only minimize *engineering costs* but also can substantially reduce *product costs*, which is one of product development's primary goals. Spreading out teams geographically—with people who may not even be working at the same time—compromises the teamwork that is necessary to optimize product architecture, which determines 60% of the product's cost (Figure 1.1). Outsourcing subsystem and part design to remote engineers can produce products that may be hard to integrate, have part interaction problems, or miss out on synergistic system opportunities, as discussed in Chapter 3.

Physically integrated teamwork is even more important when products need to be designed for Lean Production and build-to-order. Design teams need to work closely with manufacturing people to design around standard materials and modules, design for no setup, design for CNC, and concurrently design versatile product families and flexible processes, as discussed in Chapter 4. For mass customization, engineers need to work closely with manufacturing, marketing, and the sales force to establish parametric CAD templates that can automatically generate CNC programs (see Chapter 4).

The transition to outsourced engineering incurs a certain cost and may cause significant consequences. It will cost money to find engineers, arrange to hire and pay them, and prepare and transfer documents, drawings, and CAD files. The effort to implement these changes will probably consume most of the available change energy and preclude or compromise other more valuable improvement efforts, such as DFM, robust design, better CAD tools for domestic engineers, Six Sigma, build-to-order, and so forth.

The consequences can be even more severe if the "cost savings" from outsourcing engineering are to come from layoffs. In addition to the general consequences of layoffs,[107] there are special consequences for product development. Once word spreads, morale and efficiency will drop and the best engineers may leave, thus weakening engineering efforts and making product development even less efficient. Engineers who quit or are laid off will probably take with them valuable skills and knowledge, some of which may be unique, and thus weaken product development. If documentation is less than complete, which is usually the case, departures may create serious documentation gaps. Continuity may suffer for ongoing projects. The remaining engineering capability will be less efficient with such losses. Further, teamwork will be less effective because of missing talent, knowledge, and diversity.

2.10.1 Which Engineering Could Be Outsourced?

Certain engineering tasks could be outsourced, as long as they help rather than hurt the overall product development effort. *Research* may be more suitable than *development*. Outsourcing certain tasks could help new product development efforts. Outsourcing certain other tasks could relieve domestic engineers from distracting tasks and thus indirectly benefit new product development efforts.

Tasks that support domestic new product development include: computational intensive research tasks, as long as they do not specify or imply product architecture or part design; computational intensive analysis of performance, strength, dynamics, heat flow, reliability, design of experiments, robust tolerances, etc.; materials research; structuring and adapting parametric CAD templates and CAD/CAM programs;[108] and literature or web searches in general.

Tasks that usually distract new product development efforts include the following activities on current products: engineering change order processing; changing drawings; analyses of problems on current products; cleaning up or converting documentation, drawings, or part numbers for standardization efforts or absorbing acquired products; maintenance engineering and conversions from obsolete materials for older legacy products and spare parts; cleaning up documentation for outsourcing the manufacturing of legacy, spare parts, and oddball products;[109] conversion of drawings into geometric dimensioning and tolerancing (GD&T) format; and conversion of drawings to metric or dual dimensioning.

Pfizer created a program to handle these types of tasks. Pfizer Works "permits some 4,000 employees to pass off parts of their job to outsiders," including web searches, number crunching, market research, presentation preparations, graphic design, data mining, and so forth. "They write up what they need on an online form, which is sent to one of two Indian service-outsourcing firms." Pfizer estimates that it has freed up 66,500 employee hours in its first year.[110]

Section 4.8 discusses the offshoring of manufacturing, why it will not save money on a total cost basis, and its detrimental effects on product development, Lean Production, and quality improvements.

2.11 PRODUCT DEFINITION

Before product development can begin, the product must be methodically *defined* with clear and realistic goals that will satisfy customers. Sometimes products are defined as a hodgepodge of unrealistic ambitions.

2.11.1 Understanding Customer Needs

Understand customer needs and develop products to satisfy customer needs rather than buying into the philosophy of "build it and they will come." Offer customers *solutions* instead of just *products*.

How do you understand customer needs? Ask customers what they want with respect to functionality, cost, cost-of-ownership, reliability, and so forth. Ask customers to rank what is most important and how well your company does compared to the competition.

Understand customer root needs well enough to predict what customers may need but don't realize. HP LaserJet Printer division has an "imaginative understanding of user needs." Procter and Gamble recommends "understanding your customers so well that you can predict what they want but don't know that they want."

Understand what customers will need in the future; study trends and develop versatile products that can adapt to future environments. Understand *new potential customer* wants, which may be the basis for new markets. Understand what customers would want if they knew all the things that were possible. For example, the Sony Walkman "just wasn't an option that consumers fathomed would even be possible."[111] Another example from Sony: "Sony's dealers in the US and some of its engineers balked at the idea of miniaturizing the transistor radio."[112]

Henry Ford knew that if he asked people what they wanted, they would have asked for faster horses! So he wrote to *The Automobile* magazine in 1906 with his own product definition: "The greatest need today is a light, low priced car with an up-to-date engine with ample horsepower, and built of the very best material. ... It must be powerful enough for

American roads and capable of carrying its passengers anywhere that a horse-drawn vehicle will go."[113]

2.11.2 Writing Product Requirements

Make sure all customer/market needs are specified in the product requirements document, *never adding to it or changing it after the product development has commenced.* Avoid late changes by proactively bringing up early all the requirements that historically were omitted in previous requirement documents and came in later as changes:

- Ask the customer: "What about this? ... What about that?" Then make sure all of the requirements are incorporated into the initial requirements document.
- Base your questions on your experiences with previous changes with this customer and this type of product, plotted in Pareto order, ranked by the most difficult, most costly, most responsible for delays, and most draining of resources.
- If probing questions are inconclusive or noncommittal, identify possible changes (with separate "trend research") and proactively create versatile designs that can easily handle anticipated changes.

Product requirements should be noted *generically* without specifying or implying how they should be done. When providing requirements to customers or regulators, state them generically without constraining product architecture, to keep it open for optimization. State "what will be done," not the "how we will do it."

Write rational, objective, and appropriate product requirements. Avoid temptations to base product requirements on enticing market opportunities, exciting technology, whims, attachments, aversions, prejudices, competitive precedents, or the "next bench syndrome" (designing products for colleagues). Make sure the requirements are thoroughly written for the current products. Avoid the temptation to base the requirements on previous products or form-letter templates. Make sure all goals, targets, specs, and metrics are clearly delineated deliverables and are *relevant to customers.*

Make sure the system architecture is optimized *before* writing component and subsystem-level specs. Having vendors on the team early to help do this will (1) ensure this is done fast and correctly and (2), if this is not done well, avoid taking more calendar time to clear up an over-the-wall spec and fix incorrect or suboptimal specs later.

Avoid feature creep and unnecessary complication. Philips Electronics found that "at least half of returned products had nothing wrong with them. Consumers just couldn't figure out how to use them."[114]

Think in terms of synergistic product families (platforms) and their evolution over time.

2.11.3 Consequences of Poor Product Definition

Time, money, resources, and opportunities are wasted when products are developed that customers don't want. Product development costs and valuable resources are wasted and the project is delayed when product design requirements have to change to reflect what the customer really wanted in the first place.

2.11.4 Customer Input

First, create a list of factors that would be important to customers. A baseline list is shown in the first column of Figure 2.2. Second, ask customers for the relative importance of their preferences, from a low of "1" to a high of "10" (there may be more than one 10, etc.). Ask them to grade your product (in the second column) compared to the leading competitor(s) in the third

Rating of Importance	Grade	Compared to:
_____Functionality	_____	_____
_____Purchase cost	_____	_____
_____Quality	_____	_____
_____Reliability/Durability	_____	_____
_____Delivery/Availability	_____	_____
_____Ergonomics; ease of use	_____	_____
_____Appearance/Aestetics	_____	_____
_____Service, repair, maintenace	_____	_____
_____Cost of ownership	_____	_____
_____Technical support	_____	_____
_____Customizability/Options	_____	_____
_____Safety	_____	_____
_____Environmental	_____	_____
_____Other_____	_____	_____

FIGURE 2.2
Customer input form.

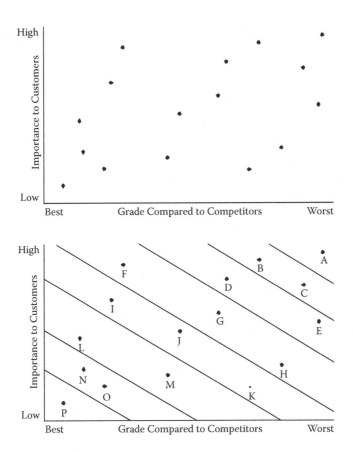

FIGURE 2.3
Customer importance versus competitive grade.

column. Use the academic scale of A (best), B (above average), C (average), D (below average), and F (worst). For plotting, use F = 0 through A = 5.

Next, prioritize customer preferences, which is valuable information in itself. This will help neutralize misunderstandings or internal biases toward existing technology, markets, or "pet" features. Finally, label each at its appropriate numerical position on the vertical scale in Figure 2.3, from a low of 1 to a high of 10. Then plot their positions against the competitive grade on the horizontal axis, as shown in the upper graph in Figure 2.3. The graph can then be prioritized with diagonal zones (lower graph), which quickly show where to place the most effort. The first zone, with point "A," presents the most opportunity because it is most important to customers but our company is ranked worst. The next zone, with points "B" and "C," is the next priority, and so forth through the zones to point "P," at which our competitive grade is best on a factor that is least important to customers.

2.11.5 Quality Function Deployment

Quality function deployment (QFD) is a tool for systematically translating the voice of the customer into product design specifications and resource prioritizations.[115] Its strength is to translate *objective and subjective* customer wants and needs into *objective* specifications that engineers can use to design products.

In the most general sense, the input to QFD is a set of customer preferences, and the outputs are product specifications and resource prioritization (Figure 2.4). The values in the "design specifications" rows are the actual values that engineers will use to design products. The "resource prioritization" row is the percent of the design team's effort that should be spent on each aspect of the design. Using this prioritization will ensure that the design team devotes its efforts toward features that customers

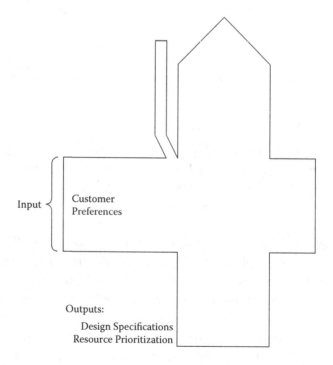

FIGURE 2.4
QFD executive overview.

want the most. And, thus, they will not waste effort "polishing the ruby" more than the customer wants it polished. This can be a temptation when engineers are personally excited about certain new technologies that are being used in the product.

2.11.6 How QFD Works

Figure 2.5 shows the complete QFD "house of quality" chart, with each area labeled according to its function in the methodology. The "customer preferences" are listed, one per row, in words that are meaningful to the customer, not in "specsmanship" jargon.

Two of the most valuable aspects of QFD are obtaining customer preferences and competitive grades that can be documented in the form shown in Figure 2.2 and graphed in the format of Figure 2.3. The value of having this information can be quickly demonstrated by surveying an internal

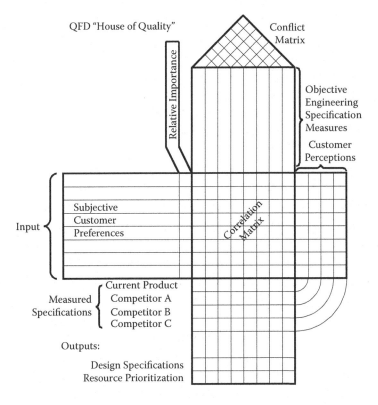

FIGURE 2.5
QFD "house of quality" chart.

group; for instance, a newly formed product development team. It is truly amazing how much the results can vary—some answers for estimated customer importance vary the entire range from 1 to 10! In one workshop, the team reaction was, "How can we design a product when even *we* can't agree on what's most important?" Of course, the real input will have to come from real customers themselves. The customer values (from 1 to 10) are entered under the "chimney" of the house of quality and competitive rankings are entered on the right of Figure 2.5.

The full QFD procedure uses these two sets of inputs to calculate the optimal development budget allotments for various tasks. If a competitor was ranked best about something important to the customer, the company would be wise to analyze that product, using the data in the "measured specifications" section, and try to understand how it achieved customer satisfaction in this area. The things to measure would be the specifications that appear under the columns labeled "objective engineering specification measures." These contain the objective measurements that will eventually be the target numbers for discrete products or ranges for mass-customized products.[116] The objective measures would be in engineering units that quantify dimensions, force, torque, energy, decibels, etc. All of those objective measurements are then measured for the current product and several competitors and entered in the chart in the "measured specifications."

The "correlation matrix" correlates which customer preferences are affected by which engineering specifications. Symbols are placed in the square to indicate the type of correlation. Usually rankings are "positive" or "negative" correlation; "strong," "some," or "possible" correlation; or "strong positive," "medium positive," "medium negative," and "strong negative" correlation. Roughly half the boxes should be checked. Results become less valuable as one approaches either extreme of all boxes checked or no boxes checked.

The "conflict matrix" tabulates any specifications that might be inherently in conflict with others, to aid in making tradeoffs of one feature versus another. An example would be a more powerful car engine that may adversely affect mileage or handling because of its extra weight.

Various cells in a QFD chart are used to make calculations and normalize them to useful percentages. One of the bottom lines is the design target row, labeled "design specifications." The other is the "resource prioritization," which can be given as a percentage of the engineering budget or hours spent achieving the various design targets.

Defining products to satisfy the voice of the customer is one way to ensure customer satisfaction, which has an enormous effect on sales, profits, and shareholder value.[117]

NOTES

1. James P. Womack, Daniel T. Jones, and Daniel Roos, *The Machine That Changed the World: The Story of Lean Production* (1991, Harper Perennial), Chapter 5, pp. 115–116.
2. For more information on customized in-house DFM seminars, see Appendix D or www.design4manufacturability.com/seminars.htm.
3. James Morgan and Jeffrey K. Liker, *The Toyota Product Development System* (2006, Productivity Press), Chapter 4, "Front-Load the PD Process to Explore Alternatives Thoroughly."
4. Pankaj Ghemawat, "Finding Your Strategy in the New Landscape: The Postcrisis World Demands a Much More Flexible Approach to Global Strategy and Organization," *Harvard Business Review*, March 2010, pp. 54–60.
5. Jim Collins, *Good to Great: Why Some Companies Make the Leap ... and Others Don't* (2001, Harper Business), Chapter 8, "The Flywheel and the Doom Loop."
6. Aberdeen Group report, "Product Portfolio Management: Targeting and Realizing Product Lifecycle Value."
7. Richard Rumelt, *Good Strategy, Bad Strategy: The Difference and Why it Matters* (July 2011, Crown Business).
8. Michael L. George, et al., *Fast Innovation* (2005, McGraw-Hill), Chapter 7, "Spotlight on Conquering the Cost of Complexity," p. 167.
9. Tim Brown, *Change by Design* (2009, Harper Business), p. 77.
10. Cliff Edwards, "The Return on Research: HP's R&D Productivity Index Shows Which Projects Have the Biggest Payoff," *BusinessWeek*, March 23 & 30, 2009, p. 45.
11. Matthew E. May, *The Elegant Solution* (2007, Free Press), p. 41.
12. Aberdeen Research, "NPD—The 2011 Growth Imperative" (2010), Chapter 2.
13. David M. Anderson, *Build-to-Order & Mass Customization: The Ultimate Supply Chain Management and Lean Manufacturing Strategy for Low-Cost On-Demand Production without Forecasts or Inventory* (2008, CIM Press), Chapter 8, "On-Demand Lean Production." See description in Appendix D.
14. Ibid., Chapter 7, "Spontaneous Supply Chains."
15. See http://www.design4manufacturability.com/product_line_rationalization.htm or Chapter 3 in *Build-to-Order & Mass Customization*.
16. Configurators can automate the processes of determining the feasibility, cost, and time to do customizations, instead of the time-consuming and less accurate manual estimating. See Anderson, *Build-to-Order & Mass Customization*, Chapter 8, "On-Demand Lean Production."
17. Anderson, *Build-to-Order & Mass Customization*.
18. Ibid.
19. Anderson, *Build-to-Order & Mass Customization*, Chapter 8, "On-Demand Lean Production."

20. Anderson, *Build-to-Order & Mass Customization*. Chapter 10 "Product Development for BTO&MC."
21. Ibid., Chapter 9, "Mass Customization," p. 293.
22. Richard de Neufville and Stefan Scholtes, *Flexibility in Engineering Design* (2011, MIT Press).
23. Ibid.
24. Michael McGrath, Michael Anthony, and Amram Shapiro, *Product Development: Success Through Product and Cycle-Time Excellence* (1992, Butterworth-Heinemann), Chapter 11, "Project Team Leadership."
25. James P. Womack and Daniel T. Jones, *Lean Thinking: Banish Waste and Create Wealth in Your Corporation* (1996, Simon & Schuster), Chapter 3, "Flow."
26. From a study of 153 companies by Jim Brown, *The Product Portfolio Management Benchmark Report: Achieving Maximum Product Value*, August 2006, the Aberdeen Group, available at http://www.aberdeen.com/link/sponsor.asp?spid=30410396&cid=3359.
27. Womack et al., *The Machine That Changed the World*, Chapter 6, "Coordinating the Supply Chain."
28. Morgan and Liker, *The Toyota Product Development System*, Figure 5.3, "Effect of Overburdening Capacity on Development Lead Time."
29. Satoshi Hino, *Inside the Mind of Toyota: Management Principles for Enduring Growth* (2006, Productivity Press), Chapter 3, "Toyota's System of Management Functions," p. 133.
30. John Teresko, "Recapturing R&D Leadership," *Industry Week*, August 2006, p. 29.
31. From a 2005 Deloitte study, cited in an *Industry Week* article on management strategies, May 2005, p. 46.
32. Bill George, *Authentic Leadership: Rediscovering the Secrets of Creating Lasting Value* (2003, Jossey-Bass), Chapter 4, "Missions Motivate, Dollars Don't," p. 65.
33. Ibid., Chapter 12, "Innovations from the Heart," p. 137.
34. Ibid., p. 63.
35. "A Giant Awakens," *The Economist*, September 11–17, 2010, pp. 81–83.
36. *Financial Post*, April 20, 2012, http://business.financialpost.com/2012/04/20/ge-shares-rise-after-profit-revenue-top-wall-street-forecasts/
37. Jeffrey L. Cruikshank, *The Apple Way* (2006, McGraw-Hill), p. 26.
38. Lee Dongyoup, *Samsung Electronics: The Global Inc.* (2006, YSM Inc., Seoul, Korea), Chapter 4, "Research & Development," p. 83.
39. Morgan and Liker, *The Toyota Product Development System*, Chapter 4, "Front-Load the PD Process to Explore Alternatives Thoroughly."
40. Richard Koch, *The 80/20 Principle: The Secret of Achieving More with Less* (1998, Currency/Doubleday), Chapter 4, "Why Your Strategy is Wrong."
41. Anderson, *Build-to-Order & Mass Customization*.
42. Morgan and Liker, *The Toyota Product Development System*, Chapter 4, "Front-Load the PD Process to Explore Alternatives Thoroughly."
43. Mark Stefik and Barbara Stefik, *Breakthrough: Stories and Strategies for Radical Innovation* (2004, MIT Press).
44. W. Chan Kim and Renee Mauborgne, *Blue Ocean Strategy: How to Create Uncontested Market Space and Make the Competition Irrelevant* (2005, Harvard Business School Press).
45. Anderson, *Build-to-Order & Mass Customization*, Chapter 9, "Mass Customization."

46. Peter Drucker, quoted in the cover story, "The Innovation Gap," *Fortune,* December 2, 1991, p. 58. http://money.cnn.com/magazines/fortune/fortune_archive/1991/12/02/75823/index.htm

47. Patricia Aburdene, *Megatrends 2010* (2005, Hampton Roads Publishing), p. xv.

48. Jordan D. Lewis, *The Connected Corporation: How Leading Companies Win Through Customer–Supplier Alliances* (1995, Free Press), Chapter 5, "Cooperating for More Value," p. 92.

49. Shu Shin Luh, *Business the Sony Way* (2003, John Wiley & Sons), Chapter 3, "Stay Ahead: Feeding the Innovation Engine," p. 94.

50. Patricia Panchak, "The Virtues of Vertical Integration," *Industry Week*, September 2003, pp. 50–52.

51. Aberdeen Research, "NPD—The 2011 Growth Imperative."

52. Brown, *Change by Design*, p. 26.

53. Joachim Ebert, Shiv Shivaraman, and Paul Carrannanto, http://www.industryweek.com/companies-amp-executives/driving-growth-through-ultra-low-cost-product-development, February 23, 2010.

54. Ibid.

55. Brown, *Change by Design*, p. 31.

56. "Putting Ford on Fast-Forward," *Business Week*, October 26, 2009, pp. 56–57.

57. Brown, *Change by Design,* p. 90.

58. Ibid., p. 89.

59. Ibid., Chapter 6, "Outsourcing vs. Integration."

60. Jeffrey Liker, *The Toyota Way* (2004, McGraw-Hill), p. 62.

61. Cruikshank, *The Apple Way*, p. 38.

62. Lewis, *The Connected Corporation,* Chapter 5, "Cooperating for More Value," p. 93.

63. Robert W. Hall, "Medtronic Xomed: Change at 'People Speed'," *Target*, 2004, Vol. 20, p. 14, http://www.ame.org/sites/default/files/target_articles/04-20-1-Medtronic_Xomed.pdf.

64. Ibid.

65. George, *Authentic Leadership*, Chapter 7, "It's Not Just the CEO," p. 97.

66. "The New Heat on Ford," *Business Week*, June 4, 2007, pp. 32–38.

67. Jim Collins, *How the Mighty Fall, and Why Some Companies Never Give In* (2009, HarperCollins).

68. "The Complete Guide to America's Best Plants," *Industry Week* (1995, Penton Publishing), p. 12.

69. Morgan and Liker, *The Toyota Product Development System*, p. 193.

70. Lewis, *The Connected Corporation*, Chapter 13, "Successful Alliance Practitioners," p. 273.

71. Joachim Ebert, Shiv Shivaraman, and Paul Carrannanto, http://www.industryweek.com/companies-amp-executives/driving-growth-through-ultra-low-cost-product-development,February 23, 2010.

72. Ibid., Chapter 4, "Practices for Joint Creativity," p. 74.

73. Kim B. Clark and Takahiro Fujimoto, *Product Development Performance* (1991, Harvard Business School Press), p. 349.

74. Womack et al., *The Machine that Changed the World*, Chapter 6, "Coordinating the Supply Chain."

75. Morgan and Liker, *The Toyota Production Development System*, p. 185.

76. Jeffrey Pfeffer and Robert I. Sutton, *The Knowing–Doing Gap: How Smart Companies Turn Knowledge into Action* (2000, Harvard Business School Press), p. 23.
77. John Paul MacDuffie and Susan Helper, "Creating Lean Suppliers: Diffusing Lean Production Through the Supply Chain," *California Management Review,* Summer 1997, pp. 118–150.
78. Satoshi Hino, *Inside the Mind of Toyota: Management Principles for Enduring Growth,* (2006, Productivity Press), Chapter 1, "Toyota's Genes and DNA."
79. Morgan and Liker, *The Toyota Product Development System,* Chapter 10, "Fully Integrate Suppliers into the Product Development System," opening quote.
80. Ibid., p. 181.
81. Michael L. Dertouzos, Richard K. Lester, and Robert M. Solow, *Made in America, Regaining the Productive Edge,* 1989, HarperPerennial, p. 71.
82. Morgan and Liker, Chapter 7, *The Toyota Product Development System,* "Create a Chief Engineer System to Lead Development from Start to Finish."
83. Ibid., p. 137.
84. McGrath et al., *Product Development.*
85. Kim B. Clark and Steven C. Wheelwright, *Managing New Product Development and Process Development: Text and Cases* (1993, The Free Press), p. 391.
86. Ibid., p. 534.
87. See the outsourcing article at the author's website: www.HalfCostProducts.com/outsourcing.htm.
88. See the offshoring article at the author's website: www.HalfCostProducts.com/offshore_manufacturing.htm.
89. Morgan and Liker, *The Toyota Product Development System,* p. 308.
90. Liker, *The Toyota Way,* p. 62.
91. Ibid.
92. Ibid., pp. 33–34.
93. Micheline Maynard, *The End of Detroit: How the Big Three Lost Their Grip on the American Car Market* (2003, Currency/Doubleday), Chapter 2 on Toyota and Honda, p. 74.
94. Dan Steinbock, *The Nokia Revolution: The Story of an Extraordinary Company That Transformed an Industry* (2001, AMACOM), Chapter 8, "Nokia's R&D: Focusing and Globalizing," p. 209.
95. William Souder, *Managing New Products* (1987, Lexington-MacMillan).
96. Maynard, *The End of Detroit,* Chapter 2 on Toyota and Honda, p. 67.
97. Sydney Finkelstein, *Why Smart Executives Fail, and What You Can Learn from Their Mistakes* (2003, Portfolio/Penguin Group), Chapter 3, "Innovation and Change," p. 60.
98. From a speech by Boeing President, Philip M. Condit, presented May 7, 1993, at the Haas Graduate School of Business at the University of California at Berkeley.
99. Nanette Byrnes, "Xerox' New Design Team: Customers," *Business Week,* May 7, 2007, p. 72.
100. Jill Jusko, "Customer Created: Open Innovation, Internet Help Put Customers in the Product Development Drivers Seat," *Industry Week,* March 2007, p. 44.
101. John Westwood and Ian Wood, *The Historical Atlas of North American Railroads* (2007, Chartwell Books), "The Last Great Streetcar," pp. 322–325.
102. "The Complete Guide to America's Best Plants," *Industry Week* (1995, Penton Publishing), p. 12.

103. Artemis March, "Usability: The New Dimension of Product Design," *Harvard Business Review*, September–October 1994, p. 144.

104. Sydney Finkelstein, *Why Smart Executives Fail: And What You Can Learn from Their Mistakes* (2003, Portfolio/Penguin Group), Chapter 7.

105. Anderson, *Build-to-Order & Mass Customization*, Chapter 6, "Outsourcing vs. Integration." Also see the article on outsourcing at www.HalfCostProducts.com/outsourcing.htm and a discussion of offshoring at www.HalfCostProducts.com/offshore_manufacturing.htm.

106. Olga Kharif, *Business Week Online*, "The Hidden Costs of IT Outsourcing," October 26, 2003, http://www.businessweek.com/technology/content/oct2003/tc20031027_9655_tc119.htm.

107. Anderson, *Build-to-Order & Mass Customization*, Chapter 13. See the section, "Don't Lay Off People," pp. 442–445.

108. Parametric CAD is a key tool of mass customization that allows dimensions to "float," with customized data plugged in for various product variations. Once inserted, CAD drawings can generate CNC machine tool programs through CAD/CAM software. See Chapters 8 and 9 in *Build-to-Order & Mass Customization*.

109. Anderson, *Build-to-Order & Mass Customization*, Chapter 6, "Outsourcing vs. Integration." Also see the article on outsourcing at www.HalfCostProducts.com/outsourcing.htmandoffshoringatwww.HalfCostProducts.com/offshore_manufacturing.htm. Also see recommendations about outsourcing unusual products to free up resources in Appendix A.

110. Jenna McGregor, "The Chore Goes Offshore," *Business Week*, March 23 & 30, 2009, pp. 50–51.

111. Shu Shin Luh, *Business the Sony Way* (2003, John Wiley & Sons), Chapter 3, "Stay Ahead: Feeding the Innovation Engine," p. 91.

112. Ibid., p. 95.

113. Robert H. Casey, "The Model T Turns 100! Henry Ford's Innovated Design Suited the Nation to a T," *American Heritage's Invention and Technology*, Winter 2009, pp. 36–41.

114. James Surowiecki, "Feature Presentation," *The New Yorker*, 28 May 2007, p. 28.

115. John Hauser and Don Clausing, "House of Quality," *Harvard Business Review*, May–June 1988; reprint number 88307.

116. QFD for mass customization is presented in Chapter 10 of *Build-to-Order & Mass Customization*, pp. 324–328.

117. Chris Denove and J. D. Power, IV, *Satisfaction: How Every Great Company Listens to the Voice of the Customer* (Portfolio, 2006); interesting quotes and results summarized at the end of Chapter 10.

3

Designing the Product

The primary focus of product development should be to *develop products*, not the project management concerns that typically dominate product development efforts, with:

- *Early deadlines*, which usually discourage thorough up-front work
- *Cost "targets,"* which can lead to much trouble if teams don't know how to meet the targets or if all costs are not included in the measured target
- *Development budgets*, which rarely include the costs of changes and firefighting downstream; further, budget pressures can encourage releasing unfinished or suboptimized designs to manufacturing
- *Time-to-market*, which is usually defined as the "release" (which encourages throwing designs over the wall "on time") or as "first-customer-ship," when the first unit that works gets shipped

If the actual product development is done right, it will naturally optimize the deadlines, total cost, development budget, and time to stable production.

No matter how many times Tom Cruise shouted "Show me the money!" in the title role of the movie, *Jerry Maguire,* he and his client weren't shown any money until they focused on the *activity* that achieved the goal instead of the *goal* itself. The activity of product development that achieves the goals is *designing the product.* This is a *design* activity, so the focus must be on optimizing the *design.*

As discussed in the last chapter, the team must have early and complete participation of all the specialties and an effective team leader who can focus and lead the team to design the product to satisfy the voice of the customer and for manufacturability, in the broadest sense, as defined on the opening page of Chapter 1. In order to design a successful product that satisfies the *voice of the customer,* the focus needs to be much greater than just designing something that "works" and then later dealing with cost, quality, service, regulations, supply chain management, and customer satisfaction.

3.1 DESIGN STRATEGY

In addition to the *design specification* targets generated by the product definition phase (Section 2.11), design teams should establish optimal *design strategies*. These usually focus on making big improvements in manufacturability, quality, delivery, etc.; starting early to solve major problems at the concept/architecture level; adherence to certain rules and practices and avoiding certain parts and practices altogether; and maximizing, minimizing, or otherwise optimizing certain practices.

Establishing optimal design strategies will maximize the chance that the design team will produce a clean, optimized design. If these are not proactively established early, undesirable practices may slip into the design because individual engineers (1) didn't understand the importance, especially as it affects the big picture, (2) didn't understand how their work affects other subassemblies and the product as a whole, or (3) let themselves get boxed into a corner so that the only way out was to fall back on some suboptimal practice that compromises the product. The following are design strategy examples.

3.1.1 Designing around Standard Parts

Standard parts lists should be established or adopted early with predefined goals and expectations for adherence. These standard parts should be common across many products and readily available over the life of the product (see Chapter 5).

Designing around standard raw materials can save a lot of money and improve availability in the plant, especially for expensive or bulky raw materials. The total cost will probably be less, even if some parts get better materials than needed.

3.1.1.1 Sheet Metal

If sheet metal can be standardized on one type, then heavy users can buy that grade in a coil, feed it through straightening rollers, and then cut each piece on demand on a programmable shear, thus minimizing shearing cost, material waste, handling/storing damage, and all the overhead costs to inventory, and distribute many sheet metal pieces.

3.1.1.2 Bar Stock

If bar stock can be standardized on one type, then machine tools can be more efficient by avoiding setup delays and costs to change bar stock. A proliferation of types of stock can incur large inventory carrying costs and waste valuable space. Many different remnants are hard to keep track of in inventory management, which can result in perceived shortages, unnecessary ordering of excess materials, and expensive expediting. In addition, remnants may not retain identifying grade marks (usually at the end of bar stock), which would either discourage use or risk use of the wrong material.

3.1.2 Consolidation

Consolidation of expensive parts/modules will raise order volumes, increase purchasing leverage, minimize setup changes, reduce inventory for multiple versions, and arrange steady flows or kanbans of the consolidated part that will be used one way or another. Even if simpler products appear to get "a more expensive part" than they need, there is great potential for a net cost savings from greater economies of scale, better build efficiencies, and less material overhead cost. Total cost measurements (Chapter 7) must be used to justify consolidation or it may appear to raise material cost.

Consolidation may get the combined order volume over the threshold that makes available more sophisticated processes, custom silicon, and dedicated cells or lines. (See the graphs in Figure 5.3, "Standardization of Expensive Parts.")

3.1.3 Off-the-Shelf Parts

The off-the-shelf part utilization strategy should be optimized early. One of the paradoxes of product development is that:

Designers may have to choose the off-the-shelf parts first and literally design the products around them, or else they will probably make arbitrary decisions that will preclude their use.

But incorporating off-the-shelf parts early into the design will greatly simplify the design and the design effort (see Section 5.19).

3.1.4 Proven Processing

Proven processing should be *designed for*. If this is not done properly, then proven processing cannot be used and special processes will have to be concurrently developed, with more cost, delays, and risk.

This may be necessary for leading-edge products—and be an element of a company's competitive advantage—but it is an unnecessary waste of resources if it was needed only because designers didn't know how to design for existing processes, didn't follow the design rules for the equipment, or exceeded the equipment's capabilities.

3.1.5 Proven Designs, Parts, and Modules

Proven designs, parts, and modules should be specified early as a key foundation of product architecture. A high percentage of complaints, field failures, recalls, and lawsuits do not involve new features or new technology. Rather, they involve boilerplate functions that should be based on proven designs, parts, and modules.

For instance, in the automobile industry, the most serious problems and consequences involve fuel systems, seat belts, steering, brakes, suspension, tires, and so forth—all subsystems for which wise companies reuse proven subassemblies. Ironically, these are not the parts that companies are advertising or customers are clamoring for, which are more likely to be things like styling, sound systems, navigation systems, safety systems, and hybrid drives.

The success of a reuse strategy will be maximized by making reuse a key design strategy; designing versatile parts, modules, and subsystems that can be used in many designs over time; encouraging receptivity to reuse and discouraging the not-invented-here syndrome; and avoiding arbitrary decisions that can exclude the use of proven design.

3.1.6 Arbitrary Decisions

Arbitrary decisions should be avoided because they will very likely preclude meeting goals, satisfying the design considerations, and taking advantage of opportunities such as off-the-shelf parts and the reuse of proven designs, parts, and modules. Avoid arbitrary part choice decisions that unnecessarily proliferate part variety and needlessly complicate operations and supply chain management. Avoid the practice of independently selecting parts for each situation, which leads to a crippling proliferation. Instead, for each situation, select the best *standard parts* that have widespread applicability.

3.1.7 Overconstraints

Make sure there are no more *constraints* than the minimum necessary; for instance, avoid situations where four points try to determine a plane, four linear bearings try to guide precise movements, or two parts are aligned with round pins/bolts in round holes. Overconstraints are costly and can cause quality problems and compromise functionality because the design will work only if all parts are fabricated to tight, maybe unrealistic, tolerances. Fortunately, overconstraints are easy to avoid in the architecture stage by specifying the exact number of constraints that will do the job (see Guideline A3 in Chapter 8).

One of the reasons for the phenomenal success of Henry Ford's Model-T was its ability to handle the uneven roads that existed a century ago. The solution was an engine that mounted to the chassis on three points and a chassis that was supported on three points: at both ends of the rear axle and at the center of the front axle.[1]

3.1.8 Tolerances

Optimize tolerances by design by:

- *Choosing* the optimal tolerance, not relying on block tolerances or arbitrary assumptions
- *Specifying* all tolerances to be the widest that will still ensure functionality, quality, and safety; don't use automatic block tolerances
- Analyzing worst-case tolerance situations
- Methodically specifying tolerances with Taguchi Methods™ for a robust design (see Section 10.2.5)

3.1.9 Minimizing Tolerance Demands

Eliminate the cost, quality, and performance problems of tight tolerances by identifying tolerance sensitivities and *creating designs that are not as sensitive to tolerances.*

Avoid design approaches that depend on tight tolerances, calibration, alignment, or parts that have to be matched, screened, or special ordered.

Avoid overconstraints (Guideline A3), which can increase tolerance demands.

Understand tolerance step functions for all contemplated processes to avoid unknowingly specifying processes that are more expensive than necessary (Guideline P23 in Chapter 9).

Combine parts (Section 9.5); design machined parts so all dimensions are machined in the same setup (Guideline P14).

Avoid getting boxed into a corner and "rescuing" the design with tight-tolerance parts.

Avoid cumulative tolerance stacks. When many stacked parts must mate with another part or another stack, the tolerances of all parts will be cumulative for the stack. Solutions include: (1) control the tolerances of all parts on both sides of the stack; (2) eliminate parts by simplifying product architecture and combining parts in the stack(s); or (3) drill, clamp, or spot-weld one set of parts at assembly, assuming that those parts do not need to be interchangeable; for instance, for structural assemblies.

3.1.10 System Integration

System integration should be optimized early, not after several subassemblies have been independently designed with little regard to how they integrate and interact in the product.

3.1.11 Optimizing All Design Strategies

Do it right the first time so the team will not have to make changes to correct things. You're most likely to do it right the first time when enough time is allocated for thorough up-front work.

Reinventing the wheel can be avoided by cooperatively designing modules and reusing or sharing previous, proven engineering.

Avoid past mistakes by understanding lessons learned from previous projects (Section 3.3.4). This will require an active effort to investigate, summarize, and disseminate this information.

Avoid troublesome practices entirely, such as assembling with adhesives or using liquid locking compound to retain fasteners instead of using easier and more consistent solutions. Instead, use screws or nuts coated with retention compound, fasteners with deformed threads, or optimal use of lockwashers. Be sure to keep in mind service needs and cycle limitations.

Develop a software debug strategy. The choices are (1) make embedded software upgradeable or patchable or (2) make software perfect using object-oriented programming and basing subsequent code on previously written and debugged modules (objects).

Vendor assistance is maximized by having preselected vendors on the design team early to help the team design the product, or, when appropriate, by having the vendors design the parts they will be building.

Variety and customization strategies should be developed and implemented early. Product developments will not achieve their potentials if different efforts focus only on developing many discrete mass-produced products.

Options and upgrades should be designed to be easy to add. Sometimes upgradeability can be enhanced by simple additions (that cost little more), such as extra mounting holes, signal ports, power ports, utility capacity, accessibility, and convenient mounting spaces.

Total cost should be used as the basis for measuring all costs and the basis for all decisions involving cost.

Establish design practices for procedures, documentation, standards, new design guidelines, and new design practices for designing for specific processes. Follow the procedures on an *ongoing basis*; don't wait until a review to start considering procedures.

3.1.12 Design Strategy for Electrical Systems

Complete electrical systems should be optimized at the architecture level.

Wiring and cables are key elements of that architecture and should not be left as an afterthought.

Optimize architecture for electrical systems by first considering higher levels of integration (VLSI, custom silicon, FPGAs) to minimize connections between components, speed signal paths, reduce space, and thus reduce the number of circuit boards.

3.1.13 Electrical Connections: Best to Worst

Maximize the best means to convey signals and power with the highest reliability and lowest total cost:

- *Best*: components are automatically soldered directly to *traces* on printed circuit boards (PCBs), which are routinely produced at Six Sigma quality levels. Examples:
  - Components are automatically soldered to a single PCB.
  - Multiple PCBs are connected by *flex layers,* which are actually layers on all connected boards.
- *Second best*: PCBs are connected by a standard off-the-shelf card cage, which also opens options for standard off-the-shelf boards for processing, control, I/O, memory, etc. These have one mechanical connection per pin, which could be gold plated for better reliability.
- *Third best*: Connect PCBs and devices with off-the-shelf cables plugging into standard connectors automatically soldered onto PCBs. These have two mechanical connections per wire. All connectors should be polarized to prevent mistakes.
  - Standardize all cable connectors throughout the product: one cable type for each voltage and one type for each signal type, each with its own unique pin configuration.
  - Provide strain relief to protect cables and limit stress on wires/contacts.
- *Fourth best.* Ribbon cables, which have four mechanical connections per wire and more failure modes. Ribbon cables should not be plugged into even-pin-count headers on PCBs, unless a polarizing feature is included.
- *Next to worst.* Lugs/pressed-in wires and terminal blocks, which are time-consuming to wire and prone to manufacturing and installation errors. Further, lugs can loosen over time or cause short circuits during service. If used, specify color-coded wires or print on wires to prevent mistakes.
- *Worst.* Hand soldering. Quality is orders of magnitude worse than the best.

Other innovative routing techniques include *LAN cables* for multiplexed signals or a power buss, like a *buss bar,* power layer on a back plane, or a shared power cable instead of many individual routings from the power supply to every power user. Interconnection reliability is even more important when products are expected to perform with less local service support.

Minimize or eliminate extra connections for modularity, shipping, and service. Try to align necessary connections with module boundaries when modules are shipped separately. If not possible, specify unbroken cables and coil them for shipping, either shipped separately or connected at one end. Concurrently engineer the product and its installation, service, and repair to minimize the need for extra connections.

All decisions must be based on the *total cost for the system;* do not present or review innovative parts and subassemblies outside the system context. Optimize the trade-offs of serviceability and one-time ease of installation versus total cost and long-term reliability.

3.1.14 Optimizing Use of Flex Layers

Flex layers consist of many traces in a flexible Kapton® encasement that is also a layer on all connected boards with components automatically soldered to the traces in the flex layer. This is different than *flex cables* that link connectors. Here is how to take full advantage of flex layers.

Early in the architecture stage, develop an optimal strategy to connect boards to other boards and off-board components and other devices. Thoroughly explore various interconnect concepts. Thoroughly evaluate vendors that have experience in your regulatory environment. Preselect a capable vendor who will work early with your team to optimize the interconnect design; *do not try to design this alone and send out for bids.*

Opportunities to implement flex layers are more limited after the product is designed. Do not compromise the attempt with inadequate effort, resources, materials, samples, or vendors. Do not be discouraged by "tried it; didn't work" folklore; investigate previous attempts in the context of the above commitment.

To justify, use total cost numbers for the entire assembly; until total costs are available, use *total cost thinking.* Ensure team consensus and management support, including enough calendar time, resources, and budget. If this is beyond the scope of a single project, spin it off as a research project.

3.1.15 Voltage Standardization

As a system architecture strategy, the number of different voltages should be minimized and standardized to minimize power supply design complications and allow the use of reliable off-the-shelf power supplies or modules.

3.1.16 DFM for Printed Circuit Boards

Obey all design rules for printed circuit boards for component placement, spacing, and layout, and for proper geometries/spacing for pads, holes, vias, traces, probes, test access, test fixtures, locating features, and stay-out zones. Violating PC board design rules either induces quality/rework problems for automatic processing or prevents automatic placement/insertion of components, which then requires hand placement or soldering, which should not even be considered an option for surface-mounted components.

Avoid overcrowding PC boards by identifying functional density challenges early and proactively pursing solutions at the system architecture level, such as higher levels of integration (VLSI, ASICs, FPGAs, etc.), more space-efficient circuitry, more compact components, removing circuitry that does not have widespread use or value to customers, and so forth.

Ensure automatic placement/insertion of all components possible to maximize throughput, minimize assembly cost, and minimize the chances of assembling the wrong component, wrong orientation, or bad insertions/placements. No hand placement should be allowed for any surface-mount components. Automatic assembly should be a criteria for component selection, even if auto-placeable components cost more to purchase.

No hand soldering should be allowed because automatic soldering (reflow or wave soldering) is probably the most refined process in industry, with many operations routinely achieving Six Sigma performance (around three defects per million). Hand soldering quality is inferior by orders of magnitude; further, some hand-soldered joints can pass tests in the plant but fail in the field. No hand soldering should be allowed at all for surface-mount components.

Component selection criteria should include the ability to be automatically soldered and survive the heat of wave or reflow soldering and cleaning processes, even if these parts cost more to purchase.

Standardize PCB parts to prevent proliferation, because if part variety exceeds circuit board equipment part capacity, then boards will have to go through the machines twice with setup changes in between.

Avoid components that need to be screened, matched, calibrated, or adjusted. These steps add labor cost, increase the manufacturing time, and create supply chain challenges.

Avoid hand soldering on circuit boards for off-board wiring, bottom-sided leaded connectors, unusual components, etc. Instead, find auto-solderable components.

3.2 IMPORTANCE OF THOROUGH UP-FRONT WORK

The process of optimizing product architecture starts with an early balance of all design considerations by a complete multifunctional team. Optimizing product architecture is the highest leverage activity in product development and has the greatest potential for ensuring success. But, as with product definition, the importance of this stage is often ignored by merely assuming that the product architecture will be the same as previous or competitive products.

One of the biggest causes of suboptimal product architecture starts with the seemingly innocuous step of building a breadboard "just to see if it works." Breadboards are designed to prove functional feasibility and are usually built with the materials on hand (not widely available production-grade materials) in the most expedient way (not the most manufacturable way). Further, breadboards are built by prototype technicians, who can usually make a single unmanufacturable unit work as a matter of pride. Product architecture optimization and manufacturability are rarely even considered at the breadboard phase, based on the assumption that those tasks will be done "later."

Unfortunately, once the breadboard "works" and is demonstrated to management or customers—you guessed it—there is a strong temptation to "draw it up and get it into production." The unfortunate result is the company ends up mass-producing breadboards forever! Basing production designs on breadboard architecture misses the biggest opportunities to make significant reductions in cost and development time.

As shown in Chapter 1, 60% of a product's lifetime cumulative cost is determined by the concept/architecture phase of a project (Figure 1.1). By the time design is completed, 80% of the lifetime cumulative cost is determined. By the time the product reaches production, only 5% of the total cost can be influenced. This is why cost reduction efforts can be so futile, because cost is really determined by the design itself and is very difficult to remove later.

Similarly, other important design goals, such as quality, reliability, serviceability, flexibility, customizability, and regulatory compliance are most easily achieved by optimizing the product's architecture.

Thorough up-front work greatly shortens the real time-to-market and avoids wasting time and resources on firefighting, change orders, and ramp problems, as shown in Figure 3.1.

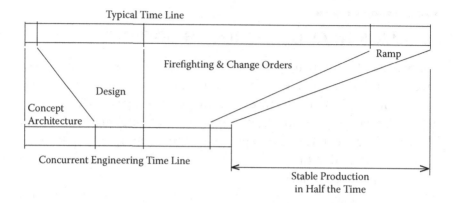

FIGURE 3.1
Traditional vs. front-loaded time lines.

The time to stable production can be cut in half because of the thorough up-front work, which minimizes the need for firefighting and change orders and makes the manufacturing ramp-up several times faster. Note that the concept/architecture phase went from a trivial amount in the traditional model to *an order of magnitude more* in the concurrent engineering model. More thorough up-front work decreases the post-design activities from almost three-fourths to less than one-half of the product development cycle. It is more efficient to incorporate a balance of design considerations early than to implement them later as changes.

Figure 3.1 emphasizes one of the most important principles to reduce the *real time-to-market*: thorough up-front work. This graphic, and its profound implications, generate much discussion in the author's in-house DFM trainings.[2] In fact, at one Fortune 50 company, we spent an hour discussing this graphic at all four seminars.

While developing the Prius, the Toyota team avoided the temptation to jump right into detailed design:

"With the extreme time pressure, the temptation would be to make a very fast decision on the hybrid technology and get to work on it immediately. Instead, the team reexamined all its options with painstaking thoroughness, ... considering 80 hybrid types and systematically ... narrowing it to 10 types. The team carefully considered the merits of each of these and then selected the best four. Each of these four types was then evaluated carefully through computer simulation. Based on these results, they were confident enough to propose one alternative, ... six months later."[3]

As engineers and managers realize the importance of thorough architecture optimization, they ask what more should be done in the order-of-magnitude increase in the concept/architecture phase shown in Figure 3.1 and how this can actually reduce the final timeline so much. The key elements of an optimal architecture phase are described here.

Product definition defines what the customer really wants and minimizes the chance that the product will be subject to change orders to reflect new customer needs that were not anticipated in the beginning.

Lessons learned should be thoroughly investigated and understood to learn what worked well and what caused problems in previous projects (see Section 3.3).

Issues should be raised and resolved before proceeding further, thus minimizing the chances that these issues will have to be resolved later when each change is harder to implement and when each change may, in turn, induce yet more changes.

Concepts should be simplified with clever, elegant designs, fewer parts, part combinations, higher levels of silicon integration, modular opportunities, and so forth. Design efforts with ambitious cost goals may need help from *design studies* to generate breakthrough ideas.[4]

The architecture should be optimized for the minimum total cost, for designed-in quality and reliability, for manufacturability and serviceability, and for flexibility and customizability.

Don't compromise the up-front work by rushing into design for the illusion of "early" progress because up-front work is unfamiliar, or because of temptations or pressures to launch the first working prototype into production.

3.2.1 Thorough Up-Front Work at Toyota

The 2006 book, *The Toyota Product Development System*, emphasized the importance of thorough up-front work at a company that some say is four times more efficient than the typical product development process.

> "The ability to influence the success of a product development program is never greater than at the start of a project. The further into the process, the greater the constraints on decision making. As the program progresses, the design space fills, investments are made, and changing course becomes increasingly more expensive, time consuming, and detrimental to product integrity."[5]

The book provides many details about "bringing together your brightest, most experienced engineers from all functional disciplines to work collaboratively, thoroughly thinking through all of the critical project details, anticipating problems, applying lessons learned, creating precise plans, and designing countermeasures from a total systems perspective... ."

3.2.2 Thorough Up-Front Work at Motorola

Motorola's most effective product development projects "invested relatively large amounts of effort early in the initial design phase so that most, if not all, of the problems that appeared later in the implementation phases had already been considered."[6]

3.2.3 Thorough Up-Front Work at IDEO

Contrary to the apparent paradox, IDEO firmly believes that if they build many models early, it will "slow us down to speed us up. By taking the time to [model] our ideas, we avoid costly mistakes such as becoming too complex too early and sticking with a weak idea for too long."[7]

3.2.4 Avoid Compromising Up-Front Work

Make sure the following problems do not compromise thorough up-front work.

3.2.4.1 Slow Processes for Sales and Contracts

Make sure that a slow sales and contract process does not delay the start of product development, especially for fixed delivery deadlines. Encourage customers to decide quickly and order early. Streamline the sales and contract process. Avoid contractual delays because of cumbersome or onerous terms and conditions. Plan ahead to start NPD as early as possible, even before the formal signing of a likely contract or award.

3.2.4.2 Rushing NPD for Long-Lead-Time Parts

When designs are based on parts with long lead times, the product's up-front work may be shortened to make time for their delivery before launch. The solution would be to eliminate long-lead-time parts with the techniques described in Section 5.19.2.

3.2.4.3 Rushing NPD for Early Evaluation Units

Similarly, pressures for early evaluation units could shorten the early up-front work. See solutions in the next section.

3.2.5 Early Evaluation Units

Early evaluation units are sometimes needed for qualifications, alpha tests, beta evaluations, trade shows, contractual obligations, or test markets, but hopefully not to get late customer feedback.

First, do all the thorough up-front work for the good of the overall project, but add the goal that some form of evaluation units can be delivered early. *Make sure there are enough resources available.*

Avoid the temptation to compromise the up-front work just to get units out early. Be careful that the structure of the early units does not specify or imply suboptimal product architecture of the production units.

Prove the feasibility of the functionality with a physical evaluation unit without necessarily achieving all the space, shape, interface, or weight constraints while, for contracts, presenting plans to achieve these goals in the production units

Maximize the proportion of time spent on the concept/architecture phase by accelerating early the unit design and building process with parametric CAD, modular design, rapid prototyping, additive manufacturing,[8] short-run vendors/partners, expedient tooling, or extensive "hogging" on CNC machines that would be replaced with more cost-effective processes in high-volume production. Two especially effective techniques are

- *The SWAT team.* Use skilled technicians to build, test, and validate evaluation units quickly, making sure they feed back lessons learned promptly to the design team.
- *The parts store.* If part delivery time for the evaluation units is so long that it threatens to shorten thorough up-front work, order all the possible part candidates before the BOM finalizes. These may appear to add cost or may not have an obvious funding source. However, getting these units out earlier will generate much more money in

increased profits. Actually, many of the parts in the "store" might be obtained at no cost as samples, which suppliers would gladly provide for free just to have their parts considered.

Keep track of the total cost and resource demands and make sure these provide the basis for planning and estimating, bidding, and negotiations on subsequent projects.

3.3 OPTIMIZING ARCHITECTURE AND SYSTEM DESIGN

Figure 1.1 shows that 60% of the product's lifetime cumulative cost is committed by the concept/architecture phase. Similarly, this phase has the most significant effect on quality, reliability, serviceability, flexibility, customizability, etc. The graphs in Figure 2.1 show that thoroughly optimizing the architecture phase results in faster ramps and eliminates post-release problem solving for volume, quality, and productivity. Similarly, Figure 3.1 shows that the real time to market can be cut almost in half with thorough up-front work, by spending a third of the timeline on optimizing architecture instead of rushing through that phase.

The following subsections detail important steps to take to optimize product architecture.

3.3.1 Generic Product Definition

Thoroughly understand what customers want—the "voice of the customer"—with optimal design specifications and resource allocations *before* starting the design process. Make sure the product definition (requirements document) is generic and does not specify, imply, or limit product architecture. Be sure to address and satisfy all aspects of the product requirements.

3.3.2 Team Composition and Availability

Make sure the product development team is complete, with all specialties available for immediate and sustained deployment.

3.3.3 Product Development Approach

The team must understand and agree on the new product development approach (presented herein), especially those aspects that are different from the way things were done in the past.

3.3.4 Lessons Learned

Avoid past mistakes by understanding lessons learned from previous projects and formulate appropriate action plans. Without this, the only "lessons" considered may be from the memories of individuals who are assigned to the design team.

The consequences of ignoring lessons learned are significant delays and wasted resources, which in turn will delay other projects and deplete their resources while expending more development cost for all projects.

The effort should focus on all relevant lessons, not limited to reactions to high-profile issues. Thoroughly investigate and understand what worked well and what caused problems in previous projects with respect to development time and effort, functionality, quality, ramps, and so forth.

3.3.4.1 Categories of Lessons Learned

- *Product development lessons*, such as causes of change orders, delays, budget overruns, requalifications, and the quantification of all cost categories (the *total cost* of the product)
- *Build lessons*, such as difficulties in fabrication, assembly, test, quality assurance, and delivery
- *Performance and reliability lessons*

Quality and reliability lessons can be visually prioritized by plotting issue frequency versus severity using the format presented in Figure 10.1.

3.3.4.2 Methodologies for Lessons Learned

There are three methodologies to understand lessons learned: databases, investigations, and presentations.

Databases: The company should compile, manage, and distribute summaries of lessons learned in databases. Until then, a database expert or a team member can investigate the data in company databases, summarize the lessons, and present them to the team.

> Toyota uses one-page summaries and engineering checklists to accumulate "the knowledge base reflecting what a company has learned over time about good and bad design practices, performance requirements, critical design interfaces, critical to quality characteristics, manufacturing requirements, as well as standards that commonize design."[9]

Investigations: Team members can investigate lessons learned about previous projects by interviewing people and acquiring data from past projects. As follows, the team should (1) investigate, (2) present in Pareto order, and then (3) *formulate and implement prevention strategies:*

> What worked; what did not.
>
> Causes of firefighting or change orders on the traditional timeline in Figure 3.1. For each cause, ask if it was preventable.
>
> Causes of delays.
>
> Summaries of cost of quality: defects, scrap, rework, field failures, etc.

Presentations: Presentations from in-house people and vendors involved with previous projects[10] (arranged by the team to be presented to the team) from the following groups, who should show parts that are examples of good and bad design practice:

> Manufacturing engineering
>
> Procurement and key off-the-shelf part suppliers
>
> Vendors/partners, who are especially effective if they have built your custom parts before
>
> Quality/reliability
>
> Field service/installation
>
> Causes of change orders and firefighting

Based on all this, action plans are developed to leverage what was good and avoid what was bad. Keep learning and improving all throughout the project.

3.3.5 Raising and Resolving Issues Early

The MIT study, *The Machine That Changed the World,*[11] summarized that in the best Lean projects: "… the project leader's job is to force the group to confront all the difficult [issues] they'll have to make to agree on the project." According to team-building expert Patrick Lencioni, team members must overcome any *fear of conflict*[12] to openly and effectively raise and resolve issues. And a *Harvard Business Review* article cited a survey that stated that, "Surprisingly, the most common reason for withholding input is a sense of futility rather than a fear of retribution."[13]

3.3.5.1 Project Issues

Each design team should candidly ask the following questions:

- Is there a reasonable chance of achieving the project goals in the scheduled time, given the allocated budget?
- Will the project receive enough resources to accomplish its goals? Will they be allocated early enough?
- Are there any red flags? For instance, is the product definition stable or does it keep changing?
- Are cost and time goals based on bottom-line metrics such as total cost and time to stable production? If goals are based on the wrong metrics, the design team may be encouraged to specify cheap purchased parts, go for low-bid vendors who won't help teams design their parts, let labor cost dominate sourcing and plant location decisions, or not optimize product architecture and throw the design over the wall to manufacturing "on time."

3.3.5.2 Team Issues

Each design team should address all the team issues. For instance, the team should discuss lessons learned from previous projects and formulate appropriate action plans. Team members should thoroughly understand the risks from the lessons learned, including:

- Risks that *were* mitigated early. How was that done?
- Risks that *could have been* mitigated early. How could that be done now?
- Risks that were *not* mitigated. What were all the consequences and associated costs?

Upon discovering unresolved issues, everyone should immediately raise them to the appropriate level.

3.3.5.3 *Mitigating Risk*

Discuss and mitigate risk issues, for instance:

- Are new product and process technologies proven enough and refined enough to incorporate into new designs and production?
- Are there multiple sources of risk regarding new product technologies or new manufacturing processes?
- How much does success, cost, and time depend on entities and developments not under the control of the development team, such as partners, suppliers, outsourcing, regulations, etc.?
- Are there supply chain risks such as bidding or multiple sources versus vendor partnerships?
- Will "cost reduction" directives compromise part quality, early vendor participation for custom parts, resource availability to ensure a complete team, or concurrent engineering itself if manufacturing is done far away?

3.3.5.4 *New Technologies*

Raise and thoroughly investigate new, unproven, or historically troublesome technologies, parts, processes, features, user interfaces, and so forth. Before deciding, thoroughly investigate all issues to ensure that all risk can be mitigated before pursuing. Make sure there is enough time and resources to do this.

Investigate and verify quality, reliability, manufacturability, usability, total cost, and functionality in all user environments by the expected range of users. Thoroughly investigate suppliers' application guidelines; if not well documented, contact suppliers' application engineers. Investigate the history, reputation, and financial strength of suppliers. Ensure availability of all parts throughout the life of the product.

3.3.5.5 *Techniques to Resolve Issues Early*

Some things you can do to resolve issues early include:

- Research. Make sure that technical feasibility studies do not specify or even suggest suboptimal product architecture.
- Experiments with statistical significance ensured by design of experiments (DOE), which should also be used for prototype testing, first article and beta test evaluations, pilot production, and tolerances.

- Enlist experts, such as those available from consulting associations like PATCA.org (which has 350 technical experts in Silicon Valley) or research services like Guideline.com (formerly TelTech), who can do research and provide access to 3,000 consultants that are experts on 30,000 specialties.
- Simulations, both for the product and the processing.
- Risk analysis and management.
- For critical applications, conduct failure modes and effects analyses (FMEAs).
- Make early models and rapid prototypes.

3.3.5.6 Contingency Plans

Formulate "Plan B" contingency plans to deal with the most likely changes, setbacks, delays, shortages, or other problems regarding technology, processing, customers, markets, regulation, and so forth. For instance, products can be designed to readily accept the "Plan B" part if the "Plan A" part doesn't work out or is not available in time.

3.3.5.7 Achieving Concurrence before Proceeding

Another MIT study, *Made in America: Regaining the Productive Edge,* gave additional insight into Japanese product development project management:

> "A key task of the manager is to make sure that all disagreements are aired and resolved at the outset. Achieving consensus takes a great deal of effort, but by skillful management at this point it is possible to gain the full commitment of all members of the program team so that subsequent progress is very rapid."[14]

At the Xomed division of Medtronic, "Xomed works harder validating process at each stage of a project, which forces them to ask the right questions earlier and earlier, and wastes less overall project time by reducing the number of bad assumptions made to compensate for incomplete information."[15]

3.3.6 Manual Tasks

High quality is hard to achieve by manual tasks, such as manual component placement and soldering, whereas automated processes, like PC board assembly and CNC machining, routinely reach Six Sigma quality levels.

Manual tasks are expensive for the assembly people and support people to train them, develop work and quality standards and procedures, fix quality problems, and implement change orders. These support demands keep manufacturing and quality people from helping multifunctional teams develop new products.

Manual tasks are hard to scale up quickly. People with the right skills would have to be found, hired, and then trained how to build company products. All these problems, costs, and delays *rise exponentially* with higher levels of skill and judgment required, for incomplete documentation. Documentation shortcomings rely on operators' memories, personal notes, or verbal queries to compensate.

Before trying to implement the use of robots to perform manual tasks, investigate ways to eliminate such tasks with standard CNC machine tools.

3.3.7 Skill and Judgment

There are two main risks with designs that need too much skill and judgment:

1. Needing skill and judgment presents the possibility for errors, so where that is possible, skill and judgment should be eliminated by design with poka-yoke (Section 10.7).
2. If the design requires skilled, experienced people, it will be (1) hard to ramp up and (2) hard to set up at new factories, especially in other countries.

To eliminate the need for skill and judgment:

- Minimize parts by using simplified design; combined, monolithic parts; and off-the-shelf assemblies that come assembled and ready to go.
- Ensure instructions are obvious, intuitive, graphical, animated, and, if text, in the language spoken on the floor.
- Ensure simple production machine–user interfaces that don't require computer skills; for instance, using bar codes to change machine settings.
- Avoid manual alignment, complicated procedures, calibration, tricky tasks, etc.
- Incorporate self-locating parts, preferably not needing fixtures. If necessary, concurrently design fixtures to ensure orientation, alignment, etc.

- Avoid liquid thread-lock; thoroughly pursue alternatives, such as self-locking fasteners.
- Avoid liquid sealant (see Guideline A10 in Chapter 8) and any need for tape, glue, etc.
- Avoid judgment needed on wire routing.
- Avoid blind assembly of fasteners.
- Incorporate poka-yoke to mistake-proof assembly (Section 10.7) with standard parts and processes that can't be confused, symmetrical parts that can't be put in backwards, polarized connectors (especially when headers with even numbers of pins are used as connectors on printed circuit boards), and readily available standard tools and standard torque settings.
- Avoid screening or matching parts.
- Eliminate the skill demands and quality problems of hand soldering and wiring lugs.

3.3.8 Technical or Functional Challenges

If there are significant challenges for new technology, functionality, or regulatory compliance, the easiest way engineers can ensure DFM is to avoid arbitrary decisions that might compromise manufacturability and then require changing the solution for DFM, which, in turn, may compromise the solution that solved the challenges.

Ensure optimal manufacturability by design the first time to remove that variable so that once the technical challenges are solved, the design can go into production and won't have to be redesigned for manufacturability, cost, quality, and so forth.

Given the value of solving challenges and the cost of troubleshooting, start out with the highest quality components, to eliminate those variables. *Do not try to minimize BOM costs while solving difficult challenges.* Even if those parts appear to cost more, the total cost will be less. Understand the challenges enough to commit enough money, resources, and time.

Thoroughly understand all ramifications of the technology for its intended use in the anticipated environment. Explore many ways to achieve the goals. Don't limit thinking to just extending conventional approaches. Big leaps forward may require research projects to develop next paradigm solutions for many subsequent projects, rather than trying to handle escalating challenges with each product development project. Rather, commission *research* to proactively generate solutions—and *commercialize* them (see next section). A "best of both worlds" approach to introducing innovation would be to develop new modules that can be introduced on existing products and then become the basis of next-generation products. When a breakthrough may come from new or innovative components or materials, commit the time and effort to research alternatives, evaluate samples, and qualify the components.

Thoroughly raise and resolve all relevant issues in the architecture phase. More resources can be applied early if they do not have to fight fires on other projects because *they* were not designed for manufacturability.

Don't accept potentially naïve customer expectations for functionality, cost, and deadlines. When appropriate, convey to the customer or prime contractor how minor changes in the specs could significantly lower risk, cost, and time. Look out for arbitrary decisions that would make your job more difficult.

For major challenges, be sure a reasonable return can be expected based on total cost numbers, and be sure to allot enough resources and calendar time.

3.3.9 Commercialization

Before anything can be launched into production, it may need to be *commercialized*[16] to preserve the "crown jewels" with the rest designed for manufacturability. Commercialization may need to be applied to new product designs, research, experiments, breadboards, prototypes, patents, and acquired technologies.

The *crown jewels* are the technology that is the basic premise of the innovation or the essence of what has been proven. Without changing the proven functionality, the parts and subassemblies surrounding the core technology and supporting systems are designed or redesigned for the best manufacturability, cost, quality, and time-to-market while being integrated into an optimal product architecture.

For key modules, commercialization should be done early, so remove that from the product development timeline.

With this in mind, write a description of the functions that capture the essence of the crown jewels, but are worded in *generic language* that leaves everything else open for optimizing the design for lower cost and better manufacturability, delivery, quality, and reliability.

3.3.10 Manufacturable Science

There is a temptation for scientists to specify *the* design or *the* process which will "work" or may have been "optimized" only for functional parameters and then throw it over the wall to the design engineers to build products or machinery that perform *the* specified process. Although this is common for experimental equipment at research labs or aerospace contractors (where equipment cost is buried in much larger research budgets), this approach will rarely result in the lowest cost and quickest development for cost-sensitive commercial products.

Instead, scientists should (1) understand all the issues regarding manufacturability and how problems with manufacturability affect functionality, quality, reliability, purchase cost, cost-of-ownership, and, ultimately, success; and (2) work closely with the multifunctional team members, from the earliest stages, to develop the practical science and robust processes that will work consistently in all anticipated environments and be easily manufacturable from readily available components and materials. When scientists specify parts, they should work with purchasing to look for a range of potential solutions and then narrow the search to the ones with the best cost, reliability, and availability.

3.3.11 Concept/Architecture Design Optimization

Concept/architecture design optimization is to be performed concurrently with manufacturing and supply chain strategies (Section 3.3.14). Establish design strategies with respect to design practices, such as standard parts, off-the-shelf parts, proven processing, reuse, avoiding arbitrary decisions, doing it right the first time, avoiding past mistakes, early vendor assistance, customization, and thorough concept/architecture optimization.

Thoroughly optimize the concept/architecture of the product, not just a collection of parts and subsystems. This phase provides the greatest opportunities for innovation and substantial cost reduction and ensures a quick ramp to volume production.

Keep focusing on the architecture, without designing a lot of detail, which may become obsolete as the architecture evolves. If necessary, keep a list of "loose ends" to finish after the architecture is optimized. Both engineers and managers need to avoid the trap of trying to "wrap up" certain parts or subsystems prematurely.

Focus on generating many concepts, not making only a few detailed designs or models. Document all ideas: save or print with meaningful labels, file names, and layer names. For sketches, keep dated copies of all ideas. If related ideas build on the same sketch, make a dark-line copy of the sketch for the records and then change the sketch to the next idea. Repeat as necessary.

Don't ignore obvious aspects of the design, since all aspects affect the system design. On the other hand, don't ignore vague aspects of the design, since again, all aspects affect the system design.

Build models, when applicable, to demonstrate or compare concepts. Companies such as McMaster-Carr have vast arrays of parts and materials available from stock for model building. McMaster-Carr's website (www.McMaster.com) has hundred of links, each of which lets you narrow down the search by many parameters. However, *never specify model building or lab parts for production products. Make sure unsuitable parts in prototypes never get into production bills-of-material.*

3.3.12 Optimizing the Use of CAD in the Concept/ Architecture Phase

Lay out drawings for easy editing: Leave entities in forms that are easy to edit. Defer tasks to join structural members (union, combine, etc.) until the concept has finalized; you may want to modify or replace one of the separate elements.

Optimize data and the drawing "origin" (0,0,0). Extend (extrude) cubes and cylinders from the optimal plane so that the ones mostly likely to change dimension can be easy to change later without redrawing. For instance, a centered shaft would be drawn as two adjacent half shafts so that each half shaft could then be lengthened or shortened and the overall shaft would always remain centered.

Don't be a purist about only using the most advanced technology; sometimes the fastest progress can be made by printing out what you have drawn in CAD (parts, grids, boundaries, or dimension lines) and sketching the next detail, and then repeating.

Omit details (bolts, etc.) until the concept has finalized; keep a list of what was omitted. Save "obsolete" details to an "extra parts" layer, which is normally turned off. Leave "ghosted lines" and construction lines in; concepts change often and you may need them for the next change.

3.3.13 Concept Simplification

Explore many potential concepts and approaches to optimize system architecture. Use creativity (Section 3.6) and brainstorming (Section 3.7) to generate many ideas.[17] Don't be limited by your current products or competitors' products. Don't latch onto the first idea to come along.

Strive for *design simplicity* by minimizing the number of parts and process steps and optimizing decisions on part combinations, off-the-shelf parts, and, in general, the simplest architecture.

Brainstorm for many ideas. Look for breakthrough ideas that would revolutionize the industry, even if that is not the project's stated goal. "Toyota considers a broad range of alternatives and systematically narrows the sets to a final, often superior, choice."[18] At Sony, "thirty engineers in Ibuka's [Sony co-founder's] team began exploring multiple approaches to color simultaneously," eventually converging on three electron beams from a single gun, which became the revolutionary *Trinitron*.[19]

Try to simultaneously optimize all the goals and satisfy all the constraints. Sometimes considering multiple goals and constraints can overcome design paralysis and avoid arbitrary decisions. This results in faster progress and better solutions than trying to solve one challenge at a time.

If a particular idea doesn't work out, look for more ideas to *make it work*. If that doesn't work, go back and make the original idea better or generate a better idea. Obstacles to individual problems might be overcome by expanding the scope of the idea to simultaneously solving several problems. For instance, a solution that appears to be too expensive to solve one problem may be justified if it solves three.

At Toyota, generating many ideas presents *patterns* and *possibilities*. This provides more opportunities to combine these ideas and multiply them into bigger ones.[20] And, contrary to popular belief, a steady stream of incremental innovation is what is most likely to lead to "the big idea." A *Fortune* article about 3M noted, "3M has never been about inventing the Next Big Thing. It is about inventing hundreds and hundreds of Next Small Things, year after year."[21]

Formulate the off-the-shelf strategy (Section 5.19) early because off-the-shelf parts need to be chosen before arbitrary decisions preclude their use. After the optimal set of off-the-shelf parts is selected, the rest of the design will be designed around them. Optimal off-the-shelf utilization allows the design team to focus on optimizing the architecture and designing the remaining parts. It also helps the design team meet its goals for the lowest total cost, quickest development, best quality, proven reliability, and lowest risk of problems or delays.

Formulate the strategy of what to leverage from previous designs and which designs, processes, and practices to avoid. This may require additional investigations and experiments.

Optimize integration of parts and their assembly, interfaces, wiring, cabling, and part/subsystem interactions. Decide on the level of standardization; create or adopt standard parts lists. Formulate strategies for part combinations and silicon integration (VSLI, ASICs) in support of strategies for product families and standardization.

Formulate the strategies for ensuring quality, reliability, mistake-proofing, repair, service, and test, including optimizing test points. Formulate the design strategy for variety, configurations, customization, derivatives, foreign versions, and subsequent products.

Optimize software configurability, with all hardware installed and easily accessible ports. Add flash memory into slots or build in enough memory for the most demanding version. Use standardized parts big enough to handle many product versions and variable parts or modules that can be plugged into universal connectors, with enough connectors provided for the most demanding version.

Structure architecture with variety provided by interchangeable modules. Adopt industry-standard interfaces or establish versatile interfaces. Developing versatile products, with extra parts and capacity, costs less than multiple versions. Ensure the product has enough space, connector openings, and utility capability.

Use parametric CAD (Section 4.6.5). Universal parametric templates can be created ahead of time for families of parts and structured so that, when the customized dimensions are plugged in, the *floating dimensions* are stretched and the drawing transforms into a customized assembly drawing, which automatically updates customized part drawings and documentation. Dimensional customization can be performed quickly and cost-effectively using a combination of CNC machine tools and parametric CAD. Another use of parametric CAD is to quickly show how changing a parameter impacts the system.

3.3.14 Manufacturing and Supply Chain Strategies

The following manufacturing and supply chain strategies are to be done concurrently with concept/architecture design optimization (see Section 3.3.11):

- Formulate processing strategies, including the process selection and the flows of parts and products, the manufacturing plan (what is manufactured where), and flexible cell/line design for product families.
- Formulate the resupply strategy for parts and raw materials: MRP-based purchase orders or steady flows, kanban, min/max, or breadtruck (free stock).[22]
- Optimize outsourcing and internal integration decisions[23] and identify the supplier base for parts and materials.
- Formulate the vendor strategy for nonstandard parts and outsourcing; identify potential vendor partnerships early and arrange for early participation on the design team.
- Formulate the strategies for quality and reliability assurance and test.
- Incorporate mistake-proofing (poka-yoke) into the design to prevent manufacturing errors by design features, and into the processing to make sure manufacturing errors don't happen in fabrication or assembly.
- Formulate the manufacturing strategy for customization, configurations, product variety, extensions, and derivatives.
- Arrange for relevant overhead allocations (if not automatic through total cost accounting) to (1) prevent new products from having to pay the high overhead charged to pay for less manufacturable products and (2) encourage behavior that further lowers overhead costs.

3.4 PART DESIGN STRATEGIES

Strategies for designing parts and subassemblies (hereafter called parts) include:

- Ensure product architecture and the conceptual product design have been optimized before designing any parts.
- Thoroughly pursue off-the-shelf (Section 5.19) and modularity (Section 4.7) opportunities as explored in the architecture phase before attempting to design any parts.
- Design the most pivotal and challenging parts first. Otherwise, the design of less-challenging parts may result in arbitrary decisions that may compromise the most challenging designs.

- Design all parts to support the system design and work well together *by design*. Don't just design a collection of parts that may be hard to integrate later. Don't structure the project management into independent part or subassembly design efforts.
- Allocate appropriate effort to the design of various parts and subassemblies according to customer input, as graphed in Figure 2.3 and calculated in one of the bottom lines of the QFD chart (Figure 2.5).
- Understand lessons learned about similar parts, including manufacturability issues and reliability track records.
- Understand the potential processes that will manufacture the parts. If multiple processes are candidates, investigate and understand all the processes; talk with appropriate vendors to help make the best decision.
- Collaboratively design parts with vendors who know the design rules for those processes, in general, and the process capabilities of their own equipment. For vendor-assisted design to be possible, vendors must be chosen first based on reputations and relationships. The primitive paradigm of part bidding precludes vendor-assisted design, by definition, because the part would have to be designed before potential vendors are asked to bid on it (see Section 6.11 on the shortcomings of low bidding).
- Cost strategy should be focused on minimizing the *total cost of the product or product family*. In some cases, seemingly more expensive parts may result in better quality and be more standardized, thus resulting in cost-saving synergies in supply chain management and operations. Don't compare new designs to previous designs on a part-to-part basis, because that may discourage part combination strategies where, for instance, two new parts replace five old parts (Section 9.5).

If the design of any part starts to run into trouble, immediately notify the rest of the team, who are still designing the rest of the product around that part. Then, the team can help get that part design back on track or quickly adjust the system design to accommodate changes in that part design. Toyota's product development process emphasizes *transparency*, as summarized by a senior executive: "If you have a problem, you'd better tell somebody, because eventually you will be found out."[24] Phil Condit, when he was chairman and CEO of Boeing, spoke on this problem of individual engineers running into trouble, not telling anyone, and stubbornly trying to fix it alone. The result, as Phil Condit postulated, might be that the airplane manufacturer would have to go to the airlines and say: "You know that plane you wanted to fly nonstop to Sydney—well, its range is going to come up about 500 miles short!"[25]

Complete all part designs and documentation. In the rush to get prototypes built, many companies use "expedient" documentation (e.g., sketches) or no documentation at all (e.g., verbal instructions), which can result in miscommunications and delay ramping into production where complete documentation is needed. Further, not fully completing and submitting CAD designs eliminates the opportunity to perform CAD assembly integration and error-checking, for instance, for interferences between parts. In extreme cases, designers may leave certain details "to be determined" by production workers! In one company, production line workers had to go to the local Lowe's hardware store to buy bolts for structural elements. Of course, this type of situation can lead to significant variations from the design intent, thus resulting in performance, quality, and even product liability ramifications.

Be sure to adhere to the following documentation principles:

- Ensure all documents are 100% complete. Consequences of incomplete documentation include: Local production people seeking more information from those who know, but information may be verbal and thus not correct documentation deficiencies. Many companies seem to get along with these documentation shortcomings *until* they outsource, which is worse when going offshore. Communications to offshore plants may be limited to only one email per day. Translations and misunderstandings can add even more delays. Production people may guess, interpret drawings, or make assumptions. If production is at a contract manufacturer, then, in order to get paid, the CM must "build to print" and their customer must pay, even if the parts don't work because of the customer's documentation problems.
- Ensure all documents are 100% correct and accurate. Consequences of incorrect or inaccurate documentation include: Astute local production people hopefully catching the errors and seeking correct information. Offshore contract manufacturers probably won't catch errors and will simply "build to print," ship, and expect payment. Problems will not be discovered until shipments arrive. If OEM rework isn't feasible, then another expedited procurement cycle begins (hopefully with correct documentation), followed by damage control.

- Make sure all changes are updated immediately into all documents to ensure subsequent work is based on the changes.
- For derivative products, implement procedures to ensure only relevant data carries through to subsequent designs. Structure data into modules that change and modules that don't change.
- Ensure drawings are complete and unambiguous and convey the design intent with geometric dimensioning and tolerancing (see Guideline Q13 in Section 10.1).

Implement effective product development methodologies, leadership, and a culture that encourages complete and correct documentation, and allow enough time for documentation, knowing that *do-it-right-the-first-time documentation results in a net savings in resources and time to stable production.*

Follow the part design guidelines in Chapter 9.

3.5 DESIGN FOR EVERYTHING (DFX)

Engineers are trained to design for functionality and their CAD tools predominantly design for functionality. However, really good product development comes from *designing for everything,* which is sometimes called *DFX.* This section details design considerations for DFX. The key is to consider all goals and constraints early.

3.5.1 Function

While the product has to work properly, it must be kept in mind that, although function is the most obvious consideration, it is far from being the only one. A redesign to correct a purely functional problem will result in another product development cycle, and another introduction. And those can introduce new, unknown manufacturability problems which can become an unexpected drain on manufacturing resources.

3.5.2 Cost

Cost has been the battleground of competition for decades. But the lowest product cost does not result from "cost reduction" measures per se. As pointed out in Chapter 1, design determines more than three-fourths of a product's cost.

For example, one high-tech company appointed a "cost reduction manager" for a critical new product line, who managed to reduce the *projected cost* to within the goal by buying the cheapest parts. However, the parts came from 16 different countries and it took 9 months to deliver first articles. And this was on a leading-edge product! Furthermore, when production began, the part quality was so poor that plant production actually ground to a halt, thus delaying delivery even further.

The subject of cost will be treated in more detail in Chapter 6.

3.5.3 Delivery

Delivery is greatly affected by the design because the design determines how difficult the product is to build and assemble. The choice of the parts determines how hard the parts will be to procure and how vulnerable production will be to supply glitches. Standardization (Chapter 5) will affect the effectiveness of Lean Production, which is the key to fast factory throughput (see Chapter 4).

3.5.4 Quality and Reliability

Like cost, quality and reliability are determined more by the design than is commonly realized. Designers specify the parts and, thus, the quality of the parts. Designers determine the number of parts and so determine the cumulative effect of *part quality* on *product quality*, which is especially important for complex products (see Section 10.3). Designers are responsible for the tolerance sensitivity. The processes specified by the designer determine the inherent quality of the parts. Designers are responsible for ensuring that parts are designed so that they cannot be assembled incorrectly, which in Japan is called poka-yoke, or what we would call mistake-proofing (see Section 10.7). These are very much manufacturability issues because quality problems must be consistently corrected in the plant before a product can be shipped. Quality and reliability are discussed further in Chapter 10.

3.5.5 Ease of Assembly

Ease of assembly is what comes to mind when most people think of DFM because much attention has been focused on design for assembly (DFA), later renamed DFMA, and software to analyze designs to look for opportunities to improve the assembly of high-volume products. The DFM techniques presented herein optimize the ease of assembly by design, independent of production volumes. Chapters 8 and 9 present general guidelines for designing products that can be easily assembled.

3.5.6 Ability to Test

Test strategy is very much affected by the company quality culture. At companies with a good quality culture, quality is everyone's responsibility, including designers! The TQM (total quality management) philosophy is that, instead of being *tested in*, quality should be *designed in* and then *built in* using process controls. Theoretically, products need not be tested if all processes are 100% in control. However, few factories are that confident in their processes, so they may elect to conduct at least a "go/no-go" functional test. Unsophisticated factories with higher fallout (failures) producing complex products may need tests to aid in diagnostics. Designers of these products are responsible for devising a way to not only *test* the product but also *diagnose* any problems, to show the Rework Department how to repair it. In complex products, test development costs can exceed product development costs and can even take more calendar time. Test guidelines are included in Chapter 8.

3.5.7 Ease of Service and Repair

Being able to repair a defective product is a manufacturability issue because any product failing any test will have to be repaired, thus consuming valuable manufacturing resources. Service and repair in the field can be more troublesome because field service centers usually have less sophisticated equipment than factories. In extreme cases, field failures may be sent back to the factory for repair, thus diluting manufacturing resources. Designers should design in ease of service and repair (see Section 8.9).

3.5.8 Supply Chain Management

Supply chain management can be greatly simplified by the standardization of parts and raw materials (Chapter 5), part selection based on adequate availability over time, and product line rationalization (see Appendix A) to eliminate or outsource the old, low-volume, unusual products that have the most unusual parts. In many cases, this simplification, performed in product portfolio planning and product development, will be essential to the success of supply chain management initiatives as well as programs to implement Lean Production, build-to-order, and mass customization.

3.5.9 Shipping and Distribution

The distribution of products will be revolutionized by *build-to-order,*[26] which is the ability to build products on demand and shipping them directly to customers, stores, or other factories instead of the mass production tradition of building large batches and then shipping them through warehouses and distribution centers. Selling products from inventory presents many problems: The whole system depends on forecasts, which rarely come even close to predicting customer demands, especially when markets are fast moving. Inventory costs money to carry, usually 25% of its value per year![27] If forecasts were too high, then inventory will have to be marked down to "move the merchandise." If forecasts were too low, then sales opportunities will be lost. Sometimes manufacturers try to compensate for inadequate forecasts by expediting production, but this can be expensive and can disrupt scheduled production.

3.5.10 Packaging

Packaging considerations should not be left until the first manufactured product reaches the shipping dock. Packaging variety and its logistics can be reduced with standard packaging that can be used for many products. Unique information can be added by printing on-demand labels or directly onto the boxes. Environmentally friendly packaging materials and recycled packages are becoming more important.

Designing inherent shock resistance into the product can reduce the size and cost of protective packaging. Another packaging implication is that returns due to shipping damage may come back to the factory, thus adding cost, upsetting customers, and depleting manufacturing resources.

3.5.11 Human Factors

Human factors and ergonomics are social considerations that should be considered at the very beginning, because ergonomic changes would be difficult to implement after the design is complete. Good human factors design (Chapter 10) of the product and process will reduce errors and accidents in use and during manufacture. In some industries (e.g., electronics), many service calls deal with usability issues. Philips Electronics found that "at least half of returned products had nothing wrong with them. Consumers just couldn't figure out how to use them."[28]

3.5.12 Appearance and Style

Appearance and style should be considered an integral part of the design, not something that is added later. Sometimes, the style is dictated by an early industrial design study. This can really hamper incorporating DFM principles if they were not considered in the styling design. All factors of a design, including styling, need to be considered simultaneously throughout the design.

3.5.13 Safety

Safety should not be considered only after a recall or lawsuit. Careful design and simulations should be utilized to prevent safety problems before they manifest. If a safety issue surfaces, the root of the problem must be determined and remedied immediately. Safety issues can create a major disruption to engineering, manufacturing, and sales, in addition to jeopardizing the product's and the company's reputation. Designers should make every effort to design safe products the first time as a moral and legal obligation.

3.5.14 Customers' Needs

The ultimate goal in designing a product is to satisfy customer needs. In order to do that, designers must thoroughly identify and understand customer needs (Section 2.11), and then systematically develop the product to satisfy those needs.

3.5.15 Breadth of Product Line

Using the principles of Lean Production and build-to-order, discussed in Chapter 4, products can be designed with standard parts and be produced on flexible manufacturing lines or cells. Common parts, standard design features, modular subassemblies, and flexible manufacturing can be combined to satisfy more customers.

3.5.16 Product Customization

Customized products can be built as quickly and efficiently as mass production if products and processes are designed for *mass customization*.[29] For more on mass customization, see Section 4.3.

3.5.17 Time-to-Market

Time-to-market is a major source of competitive advantage.[30] In fast-moving markets, being first to market can have major market share implications. Figure 3.2 shows the effect of an early product release on the revenue profile. The shaded area represents the extra *sales* due to the early introduction. However, because the product development and tooling costs were paid for by the baseline sales profile, the shaded area is really extra *profit*.

3.5.18 Expansion and Upgrading

Designers should design products so that they are easy to expand or upgrade by the plant or by the customer. This capability may allow the company

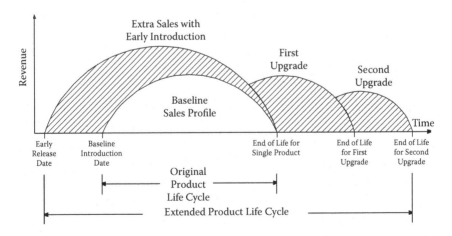

FIGURE 3.2
Increasing revenue with early introductions and upgrades.

to increase profits by extending the life of each product. Marketing and finance representatives should be involved early to help formulate the product upgrading strategy and calculate its value. Crunching the numbers will point out the high profit potential in the latter stages of the products' lifetimes after development costs are paid off, as shown in the upgrade extensions in Figure 3.2. Again, the product development and tooling costs were paid for by the baseline sales profile, so the shaded area is really extra profit. Planning to extend a product's lifetime with easy upgrades may be a very worthwhile goal to consider in the initial phases of the product design.

3.5.19 Future Designs

Similarly, current products should be designed so subsequent products can be based largely on current designs. This will save considerable time and cost in the next design if maximum use can be made of current engineering, parts, modules, and software.

3.5.20 Environmental Considerations

3.5.20.1 Product Pollution

Environmental design considerations should not be left to the first time the product or its process is fired up. Problems discovered at this stage may require major changes or a redesign to correct. Designers should anticipate environmental trends and design products clean enough for future environmental standards.

3.5.20.2 Processing Pollution

Product designers specify the process whether they realize it or not. Even specifying *the usual process* may continue a process that is causing pollution from solvents, combustion products, chemical waste, and so forth. Designers of new products have the opportunity to optimize the environmental cleanliness of the processes. This is much easier to do in the early stages of the design than later. Do not wait until environmental activists or a regulatory agency force your company to change your processes, which would result in disruptive changes in the factory, costly penalties, engineering change orders, and maybe a product redesign.

3M Corporation formulated an environmental strategy called the 3P program: "Pollution Prevention Pays." The theme is prevention of pollution at its source. The three elements of the program are recycling, redesigning products and equipment for less pollution, and creating products that do not pollute in the first place. Note that two out of the three methods depend on the *design* to reduce pollution.

3.5.20.3 Ease of Recycling Products

Similarly, companies should be concerned about what happens to a product after its useful life is over. Can it be recycled into new products? Can it be upgraded for extended life? The company may even be able to profit in some manner from the recycling or extension of its products. If the product must be disposable, it should degrade quickly and safely without aggravating solid waste disposal problems. These factors, like all the others, must be considered throughout the design phase.

3.5.21 Summary

All these factors should be emphasized early by product development teams because redesigns or major product design changes consume a great deal of design and manufacturing resources to implement. Remember that changes and redesigns consume engineering time and money that should be invested in new product development. This leads to one of the most important design principles for design in general, not just for DFM: *The further into a design, the harder it is to start satisfying additional needs.*

It is important to design with a balanced set of design considerations. Do not let any considerations dominate the design or others will suffer. Many people do not have a clear understanding of whose job is it to incorporate these considerations into the design. Some may think that ease of assembly comes from the manufacturing department. Some may think ease of service comes from the service department. Many think that quality is the responsibility of the quality department. The correct view of responsibility is a paraphrase of the motto of TQM programs:

> *It is* everyone's *responsibility to incorporate* all *considerations early into the design.*

The consequences of not considering everything early:

1. It is very difficult to integrate considerations later and results in delays, extra cost, and depleted resources.
2. The considerations are never incorporated into the design, resulting in a product that is (1) less competitive, (2) must be withdrawn from the market, or (3) needs to be redesigned.

3.6 CREATIVE PRODUCT DEVELOPMENT

Inventing requires *creativity*. And designing an invention for manufacturability requires even more creativity. At Toyota, "Invention without practical application is no more than a hobby. Invention becomes innovation only when stable duplication is achieved on a meaningful scale at a realistic cost."[31] This corresponds to when the stable target volume production is achieved, as shown in the center graph in Figure 2.1.

The prerequisite for creativity is an *open mind* and a *receptivity to ideas*. The chapter on "Innovation and Change" in the Dartmouth study on corporate success and failure stated: Innovation is not a "thing that just happens. It's a natural outgrowth of a culture of open-mindedness."[32]

3.6.1 Generating Creative Ideas

State the challenge *generically* without implying or excluding any solutions. For instance, in order to support electric cars and hybrids, the temptation would be to encourage development of better batteries. However, the broadest *generic* description would be "energy storage devices," which would also include ultra-caps (large capacitors), compressed air, flywheels, and hydrogen (which is an energy storage medium, not a fuel).

For new products, creative ideas should apply to all aspects of the design and manufacture. For redesigns and derivatives, the search for ideas could start with the most expensive part or subassembly or the hardest to build or any other challenges. Identify fundamental flaws in conventional designs and then think of new concepts that are inherently not flawed.

To get good ideas, ignore what has been done before and don't be deterred by obstacles. Ignore current designs and conventional thinking.

Rather, focus on what needs to be done. Pretend you are new to this industry. Ask: *What would revolutionize the industry?*

Don't limit idea generation to thinking only in one setting. Look for ideas in a wide range of situations, climates, and states of mind. Generate many ideas to accomplish all the challenges and possibly more. Find the best applications of various designs, in any industry. If possible, do teardowns and keep them available.

Consider *idea triggers*, even if they seem too expensive. If the concept technically would work, use it as an inspiration to find or develop lower cost versions.

Create an environment where people are encouraged to be creative and are not inhibited by fears or a perception of not succeeding. Don't limit thinking by excluding innovative solutions, especially when previous conventional thinking says, "it won't work" or "it's not possible" or it's "too risky." Tsuyoshi Kikukawa, former president of Olympus, said: "If you don't take a risk on a new idea, that in itself, becomes a risk."[33]

Don't limit the number of ideas by spending too much time modeling or drawing each one. At the idea stage, sketching or discussing many creative ideas is more important than presentation format.

Be aware of new developments that are relevant to the project, even if they are outside your industry. Look for many relevant precedents in the field that are lower cost or higher performance, as well as innovations outside the industry in product design, processing, etc. Read, do searches, and attend trade shows on anything that might benefit your project.

When applicable, create a visual model that contains only the most challenging elements. This avoids limiting creativity by temporarily omitting less challenging aspects, such as frames, enclosures, and so forth.

Continue to use creativity as the design progresses. *"The output from one creative process usually stimulates the need for more creativity."*[34]

3.6.2 Generating Ideas at Leading Companies

At Toyota, "The Toyota organization implements a million ideas a year."[35] Frank Nuovo, VP of Design at Nokia, said, "You cannot fall in love with your first idea. You have to be able to explore openly and accept input from lots of people. It is very important to be flexible."[36] And Sony co-founder Akio Morita asserted, "If you go through life convinced that your way is always best, all the new ideas in the world will pass you by."[37]

3.6.3 Encouraging innovation at Medtronic

In *Authentic Leadership*, Bill George, then-CEO of Medtronic, wrote:

> "A growing bureaucracy is a huge barrier to innovative ideas and dampens creativity, no matter how much it spends on research and development. Leaders committed to innovation have to work hard to offset these tendencies, giving preference to the mavericks and the innovators and protecting new business ventures while they are in the fragile, formative stage."[38]

He encourages "walking through the labs and learning about creative ideas before they get killed off" by the system.

3.6.4 Nine Keys to Creativity

From *The Manager's Pocket Guide to Creativity*,[39] here is the essence of the nine keys of creativity, with a quote from each:

1. *Ask a lot of questions*: "Most of our social institutions seem designed to limit, if not discourage, creative inquiry."
2. *Record all ideas*: "When you need new ideas, you can start by reexamining the old ones."
3. *Revisit ideas and assumptions*: "When revisiting, you often find that assumptions are more striking than ideas." IDEO culture emphasizes: "Refine or rethink our assumptions rather than press onward in adherence to an original plan."[40]
4. *Express ideas and follow through*: "Most ideas are cut short by our automatic self-censorship."
5. *Think in new ways*: "You don't get out of the box by doing what you've always done."
6. *Wish for more. Stretch*: "Creativity is nurtured by creative speculation."
7. *Everyone should try to be creative*: "Most people feel they're not creative and therefore don't try to be."
8. *Keep trying*: "The secret to incubation is revisiting the problem and doing so often."
9. *Encourage creativity and ask for creative behavior all the time*: Most managers think that "if you aren't visibly producing something tangible, then you're wasting the company's money."

3.6.5 Creativity in a Team

For innovation, Gary Hamel asserts, "the real returns come from harnessing the imagination of every single employee every single day."[41] Start with a creative, open-minded team that is stimulated by the challenge. The team should be diverse in knowledge as well as cultural and thinking styles.

Fire up the team.[42] Creativity is enhanced when people really want to invent something. Howard Schultz, CEO of Starbucks, said, "When you're surrounded by people who share a collective passion around a common purpose, there is no telling what you can accomplish."[43]

Allocate the time to be creative; hopefully, the project manager will realize the importance of the concept/architecture phase and allocate enough time to optimize this phase.

Put aside preconceived notions and conventional thinking and *think outside the box*. It may help to brainstorm, possibly with an experienced facilitator who comprehends the market and technical challenges but can also offer an outside perspective.

Do benchmarking and teardowns on successful products that use these parts. Don't limit analyses to just your industry or competitors.

To promote creative thinking, co-locate team workers close together and hold creative sessions in a permanent shared space or off site in a relaxed atmosphere away from phones and distractions. Consider a site that has some relevance to the project.

One of the goals of team building is to build up enough trust[44] so that people will open up and come forth with many creative ideas.

Build models, which IDEO says "is always inspirational—not in the sense of a perfected artwork but just the opposite: because it inspires new ideas. Model building should start early in the life of a project, and we expect them to be numerous, quickly executed, and pretty ugly. Each one is intended to develop an idea 'just enough' to allow the team to learn something and move on."[45]

Do not start creative product development with a discussion of administrative issues, especially deadlines and development budgets.

Backward compatibility may be important for minor improvements that can be quickly implemented, even retrofitted into products in the field. However, for major advances, innovation could be held back by pressures to be compatible with whatever it is replacing. In between, innovated advances can be implemented as new modules, and the modules can be introduced on current products and then be applied to new products.

Use creative solutions for follow-through and implementation activities.

3.6.6 The Ups and Downs of Creativity

The very nature of creativity and invention means they have many cycles of ups and downs, like a sinusoidal curve. Thinking of a good idea can be an exhilarating experience, which is the first "high" on the curve. Then you, or probably someone else, say, "What about ... ?" Then the curve hits the first "low," and the idea could easily die without a culture that encourages continued creativity and persistence. The down cycles are when ideas are most vulnerable to discouragement or lack of support. To proceed back up, new ideas and solutions will be required to solve problems encountered at each drop.

To get past obstacles, Toyota emphasizes, "the tenacity to bring ideas to fruition even in spite of initial setbacks."[46] Toyota "removes anything that stands in the way. That means looking at the target in a fundamentally different way. It means asking 'what's blocking perfection?' instead of what can we improve? That's what differentiates their brand of continuous improvement from all others."[47]

Determination is important. When developing the first high-speed train, Japan National Railroad used the Chinese motto, *Yu fa zi*, which means, "There is a way."

The book, *Breakthrough*, proposes *radical research* to get past obstacles, follow the problem to the root, and then look for solutions:

"In *applied research*, if there is an obstacle, the group tries to get around it. If stuck at an obstacle, the researchers look for a quick fix or give up."

"In *radical research*, however, obstacles focus the research. Typically, a multidisciplinary team is deployed to find perspectives on the obstacle. As the exploration deepens, more disciplines may be brought in as needed."[48]

At Toyota, engineers are expected to "break the problem down to its smallest definable elements and attack each one with ingenuity." And "big leaps forward are achieved not in one big swag but through the cumulative effect of a multitude of much smaller hits."[49]

3.7 BRAINSTORMING

Brainstorming is a technique, normally used in groups, that can generate many ideas. Stanford's Institute of Design emphasizes the right setting: "The power of group brainstorming comes from creating a safe place where people with different ideas can share, blend, and expand their diverse knowledge."[50]

The leader can organize a formal session or just steer a spontaneous discussion into brainstorming using the following rules:

- *Criticism and judgment are not allowed.* Criticism discourages the generation of new ideas and inhibits everyone's responses.
- *Praise all ideas.* Since judgment is not being applied, all ideas should be praised to encourage the generation of many ideas. IDEO encourages building on the ideas of others.[51]
- *Generate many ideas.* Nobel laureate Linus Pauling once said, "The secret to getting good ideas is to have a lot of ideas and some criterion for choosing." Further, one idea can lead to another.
- *Think wild.* What may seem to be a ridiculous idea might trigger the thought process that ultimately leads to a useful solution. Think of the mathematical process of using imaginary numbers (using the square root of -1) to derive real answers. There are many calculations using these imaginary numbers, but in the end, a real answer appears.

Then, after the brainstorming session:

- *Sort all the candidate ideas.* This is where judgment is applied. Try to prioritize the ideas to a list of leading candidates. Some ideas may seem promising but need more investigation.
- *Choose the final solution.* Choose the best solution based on all the objectives, constraints, and resources.

After design ideas have been chosen and survived the initial follow-through analysis, the team must perform a more thorough follow-through effort to reduce the ideas to practice. This must be pursued ambitiously until the design meets all its design goals. If the tentative design falls short, the designers should iterate the above process until the product meets all the goals. Many DFM problems are caused when the manufacturability goal is dropped because the design is having problems.

3.8 HALF-COST PRODUCT DEVELOPMENT

First, companies must ensure that their policies are not counterproductive to the best product development practices (Section 11.5). These counterproductive policies include outsourcing production away from engineering (Section 4.8), attempting cost reduction after design (Section 6.1), bidding custom parts (Section 6.11), and averaging overhead costs (Chapter 7).

3.8.1 Prerequisites for Half-Cost Development

3.8.1.1 Total Cost

The more important cost is, the more important it is to measure it properly. For ambitious cost goals, cost measurements *absolutely must quantify all costs* that contribute to the selling price (see Figure 6.3).

Until company-wide total cost measurements are implemented, the half-cost team needs to make cost decisions on the basis of *total cost thinking,* or, for important decisions, manually gather all the costs. Since a large portion of cost savings will be in overhead, the costing must ensure that new products are not burdened with the averaged overhead charges, but only the specific overhead charges that are appropriate.

3.8.1.2 Rationalization

Companies wanting to develop half-cost products will have to immediately rationalize their existing product lines to eliminate or outsource (1) demanding products that take resources away from new product development and (b2 "loser" products that must be subsidized by more profitable products (Appendix A). The effects of rationalization on a half-cost program are to ensure that (1) resources are always available for multifunctional teams and not pulled away to build fire-drill products or find unusual parts; (2) resources are available for the other half-cost strategies;

and, (3) the new half-cost product will not have to pay the "loser tax" to subsidize less-efficient products.

3.8.2 Designing Half-Cost Products

Half-cost products will not result just from setting an ambitious target (Section 1.4) or any other "show me the money" goal. If companies want to develop products at half the cost, *they must do everything right*, which includes implementing all the techniques presented in this book and implementing the strategies presented at www.HalfCostProducts.com, which presents a coordinated eight-point strategy to cut total cost in half or better (this strategy is summarized in Section 6.3). This book will describe how to accomplish the first point, which is product development, plus some of the other half-cost strategies. Companies should implement the other seven strategies to (1) help reduce the cost and (2) support product development by measuring total cost (Chapter 7), establishing vendor partnerships (Section 2.6), standardizing parts (Chapter 5), and rationalizing products to free up resources (Appendix A). Thus, a half-cost product development will consist of implementing all the principles of this book, with special emphasis on:

- *Product definition.* The product definition must reflect the voice of the customer (Section 2.11) and yet be worded generically (using words like *means to*) so as to maximize the design possibilities.
- *Lessons learned.* Thoroughly understand the total cost structure of previous or related products to identify sources of excessive waste, such as defects, rework, scrap, setups, warrantee costs, and other inefficiencies, as discussed in Section 3.3.
- *Breakthrough concept/architecture optimization.* Since 60% of cost is determined in this phase, half-cost product development must fully explore every possible way to simplify concepts with breakthrough ideas.[52] The multifunctional team should schedule enough time to creatively search for many ideas (Section 3.6) and conduct many brainstorming sessions (Section 3.7) to search for low-cost design approaches and manufacturing and sourcing strategies.
- *Tolerances.* A systematic approach to tolerancing may be a key element of a half-cost strategy. The architecture should be optimized

to eliminate the need for tolerances that have to be so tight that they add unnecessarily to the cost. Techniques to minimize tolerance costs include:

- Avoid overconstraints (as discussed in Guideline A3 in Chapter 8) and cumulative stacks (with techniques presented in Section 3.1).
- Use concurrently designed fixtures to precisely position parts and then bolt members together or drill, pin, or spot-weld at assembly.
- Understand tolerance step functions for all contemplated processes to avoid unknowingly specifying processes that are more expensive than necessary (see Figure 9.2).
- Use design of experiments and Taguchi Methods™ to methodically specify tolerances for a robust design to achieve "high quality at low cost," as practitioners say (see Guideline Q12 in Chapter 10).
- Combine parts (Section 9.5); design machined parts so all dimensions are machined in the same setup (Guideline P14, Section 9.2).

When products have dominant high-cost parts or subassemblies:

- Identify the most expensive potential parts or subassemblies.
- Look for a wide range of parts, as shown in Figure 5.6. In many cases, higher performing mass-produced parts may cost less but would not normally be considered if the engineer specifies only the one part that meets the exact need.
- Consolidate expensive parts into a few standard parts using the logic of Figure 5.3.

NOTES

1. Robert H. Casey, "The Model T Turns 100! Henry Ford's Innovated Design Suited the Nation to a T," *American Heritage's Invention and Technology*, Winter 2009, pp. 36–41.
2. For more information on customized in-house DFM seminars, see Appendix D of this book or go to the seminar page of the website: www.design4manufacturability.com/seminars.htm.
3. Jeffrey Liker, *The Toyota Way* (2004, McGraw-Hill), p. 57.
4. Dr. Anderson's design studies that generate breakthrough concept ideas are described in Appendix D.
5. James Morgan and Jeffrey K. Liker, *The Toyota Product Development System* (2006, Productivity Press), Chapter 4, "Front-Load the PD Process to Explore Alternatives Thoroughly."

6. Stephen B. Rosenthal, *Effective Product Design and Development* (1992, Irwin), Chapter 9, "The Quest for Six Sigma Quality at Motorola Corp."

7. Tim Brown, *Change by Design* (2009, Harper Business), p. 105.

8. Terry Walters and Tim Caffrey, "Additive Manufacturing Going Mainstream," *Manufacturing Engineering*, June 2013.

9. Morgan and Liker, *The Toyota Product Development System*, Chapter 6, "Utilizing Rigorous Standardization to Reduce Variation and Create Flexibility and Predictable Outcomes."

10. The feedback forms in Appendix C can be used to solicit valuable feedback from factories, vendors, and field service.

11. Based on the MIT $5 million dollar five-year study on the future of the automobile; James P. Womack, Daniel T. Jones, and Daniel Roos, *The Machine That Changed the World: The Story of Lean Production* (1991, Harper Perennial).

12. Patrick Lencioni, *Overcoming the Five Dysfunctions of a Team* (2005, Jossey-Bass); see the chapter on "Mastering Conflict," pp. 37–50.

13. John R. Detert, Ethan R. Burris, and David A. Harrison, "Debunking Four Myths About Employee Silence," *Harvard Business Review*, June 2010, p. 26.

14. Michael L. Dertouzos, Richard K. Lester, and Robert M. Solow, *Made in America: Regaining the Productive Edge,* from the MIT Commission on Industrial Productivity (1989, Harper Perennial).

15. Robert W. Hall, "Medtronic Xomed: Change at 'People Speed,'" *Target*, 2004, Vol. 20, p. 14, http://www.ame.org/sites/default/files/target_articles/04-20-1-Medtronic_Xomed.pdf.

16. For more on commercialization, see the author's web page: http://www.halfcostproducts.com/commercialization.htm.

17. Dr. Anderson's design studies that generate breakthrough concept ideas are described in Appendix D.

18. Morgan and Liker, *The Toyota Product Development System*, "Set-Based Concurrent Engineering," p. 47.

19. John Nathan, *Sony* (1999, Mariner Books), p. 46.

20. Matthew E. May, *The Elegant Solution* (2007, Free Press), p. 50.

21. "3M's Innovation Revival," *Fortune*, September 24, 2010, http://money.cnn.com/2010/09/23/news/companies/3m_innovation_revival.fortune/index.htm.

22. David M. Anderson, *Build-to-Order & Mass Customization: The Ultimate Supply Chain Management and Lean Manufacturing Strategy for Low-Cost On-Demand Production without Forecasts or Inventory* (2008, CIM Press), Chapter 7, "Spontaneous Supply Chains." See book description in Appendix D.

23. Ibid., Chapter 6, "Outsourcing vs. Integration."

24. Micheline Maynard, *The End of Detroit: How the Big Three Lost Their Grip on the American Car Market* (2003, Currency/Doubleday), Chapter 2 on Toyota and Honda, p. 62.

25. From a speech by Boeing Chairman and CEO, Philip M. Condit, presented May 7, 1993, at the Haas Graduate School of Business at the University of California at Berkeley.

26. Anderson, *Build-to-Order & Mass Customization*. See book description in Appendix D.

27. David M. Anderson, *Agile Product Development for Mass Customization* (1997, McGraw-Hill), Chapter 3, "Cost of Variety."

28. James Surowiecki, "Feature Presentation," *The New Yorker*, 28 May 2007, p. 28.

29. Anderson, *Build-to-Order & Mass Customization.*
30. George Stalk, Jr., and Thomas M. Hout, *Competing Against Time* (1990, Free Press).
31. Satoshi Hino, *Inside the Mind of Toyota* (2006, Productivity Press), Chapter 1, "Toyota's Genes and DNA," p. 3.
32. Sydney Finkelstein, *Why Smart Executives Fail, and What You Can Learn from Their Mistakes* (2003, Portfolio/Penguin Group), p. 73.
33. Norton Paley, *Mastering the Rules of Competitive Strategy: A Resource Guide for Managers* (2007, Auerbach Publications/Taylor & Francis), p. 2.
34. Dorothy Leonard and Walter Swap, *When Sparks Fly: Igniting Creativity in Groups* (1999, Harvard Business School Press), p. 15.
35. May, *The Elegant Solution*, p. xi.
36. Dan Steinbock, *The Nokia Revolution: The Story of an Extraordinary Company That Transformed an Industry* (2001, AMACOM), p. 273.
37. Shu Shin Luh, *Business the Sony Way* (2003, Wiley), p. 103.
38. Bill George, *Authentic Leadership: Rediscovering the Secrets of Creating Lasting Value* (2003, Jossey-Bass), Chapter 12, "Innovations from the Heart," pp. 133–134.
39. Alexander Hiam, *The Manager's Pocket Guide to Creativity* (1998, HRD Press). Excerpted in *The Futurist*, October 1998, pp. 30–34.
40. Brown, *Change by Design*, p. 16.
41. David Kirkpatrick and Gary Hamel, "Innovation Do's & Don'ts in Today's Economy: Good Ideas Aren't Just Good for Business, They're Essential for Growth." http://money.cnn.com/magazines/fortune/fortune_archive/2004/09/06/380347/index.htm
42. Leonard and Swap, *When Sparks Fly.*
43. Camine Gallo, "Inspire your Audience: 7 Keys to Influential Presentations," white paper, p. 3, www.caminegallo.com. Gallo also wrote, *Fire Them Up: 7 Simple Secrets to Inspire Colleagues, Customers, and Clients; Sell Yourself, Your Vision, and Your Values; Communicate with Charisma and Confidence* (2007, Wiley).
44. P. Lencioni, *Overcoming the Five Dysfunctions of a Team;* see the chapter on "Building Trust," pp. 13–35.
45. Brown, *Change by Design*, p. 106.
46. Satoshi Hino, *Inside the Mind of Toyota*, Chapter 1, "Toyota's Genes and DNA," p. 29.
47. May, *The Elegant Solution*, p. 42.
48. Mark Stefik and Barbara Stefik, *Breakthrough: Stories and Strategies for Radical Innovation* (2004, MIT Press), pp. 23 and 32.
49. May, *The Elegant Solution*, Chapter 2, "The Pursuit of Perfection," p. 49.
50. Robert I. Sutton, co-founder of the Hasso Plattner Institute of Design at Stanford University, "The Truth About Brainstorming," *Business Week*, September 2006, p. 17.
51. Brown, *Change by Design*, p. 78.
52. Dr. Anderson's design studies that generate breakthrough concept ideas are described in Appendix D.

Section II

Flexibility

4

Designing for Lean and Build-to-Order

In concurrent engineering, multifunctional product development teams design products for the production environment or concurrently design products and new processes. When companies embark on flexible manufacturing strategies, concurrent engineering is crucial to the success of Lean Production, build-to-order (BTO), and mass customization.

Products not concurrently engineered for flexible environments may impede implementations, diminish the payback, or even thwart success entirely. The product portfolio may have too many unrelated products that lack any synergy and, thus, too many different parts and processes. Even within a focused product portfolio, there may be a needless and crippling proliferation of parts and materials. The specified parts may be hard to get quickly. The products and processes may have too many setups designed in. Quality may not be designed into the product or process, which results in disruptions in the flow when failures loop back for correction. The product or process design may not make optimal use of CNC (computer numerically controlled) machine tools, because most CNC equipment is used in a batch mode, not flexibly. Before discussing how to design for these environments, they are briefly summarized below.

4.1 LEAN PRODUCTION

Lean Production accelerates production while eliminating many types of waste such as setup, excess inventory, unnecessary handling, waiting, low equipment utilization, defects, and rework. Former MIT researchers James P. Womack and Daniel T. Jones, authors of the definitive book on the subject, *Lean Thinking*, say that Lean Production "is *Lean* because it provides a way to do more and more with less and less—less human effort, less equipment, less time, and less space—while coming closer and closer to providing customers with exactly what they want." They summarize the corporate benefits of Lean Production as follows:

> "Based on years of benchmarking and observations in organizations around the globe, we have developed the following simple rules of thumb: converting a classic batch-and-queue production system to continuous flow with effective pull by the customer will:
>
> - double labor productivity all the way through the system (for direct, managerial, and technical workers, from raw materials to delivered product)
> - cut production throughput times by 90 percent
> - reduce inventories in the system by 90 percent
> - cut in half errors reaching the customer and scrap within the production process
> - cut in half job-related injuries
> - cut in half time-to-market for new products
> - offer a wider variety of products, within product families, at very modest additional cost
> - reduce capital investments required to very modest, even negative, levels if facilities and equipment can be freed up or sold.
>
> Firms having completed the radical realignment can typically double productivity again through incremental improvements within two to three years and halve again inventories, errors, and lead times during this period."[1]

4.1.1 Flow Manufacturing

A key element of Lean Production is *flow,* sometimes called "one-piece flow." This is especially important for build-to-order and mass customization where every piece may be different.

Flow manufacturing is achieved by reducing setup to the point where products can be efficiently built in a one-piece flow instead of the customary batches. A key element of flow is "dock-to-line" delivery,[2] in which parts are pulled directly to all points of use without the delays and cost of incoming inspection. To accomplish this, suppliers will need to be certified to ensure quality at the source.

Another aspect of flow manufacturing is the dedicated *cell* or line, which may be arranged to build any variation within a product family without any setup. Each cell has a compete set of "right-sized" inexpensive machines. Utilizing older machines that have been made obsolete by centralized mega-machines can be a very cost-effective way of building complete cells. It may be unlikely for contract manufacturers to commit to dedicating equipment and floor space for cellular manufacture. Thus, this can become an effective strategy for established companies to compete with less savvy competitors who outsource batch production.

Flexible, or agile, product development is an essential enabler of Lean Production that will ensure that flexible processes are concurrently designed (Chapter 2), products are designed around standard parts (Chapter 5), products are designed for easy manufacture (Chapters 8 and 9), and quality is designed into the product (Chapter 10). Specifically, products and processes must be designed to be built in a "batch size of one" by eliminating any setup delays to kit parts, find and load parts, position workpieces, adjust machine settings, change equipment programs, and find and understand instructions.

4.1.2 Prerequisites

In general, Lean Production implementations will go faster and have greater success if companies perform two prerequisites first. For companies seeking to implement cellular manufacture to flexibly build any product in the family on demand (called *build-to-order*), these prerequisites may be required for success:

1. Rationalize products first to eliminate the most unusual products, which have the most unusual parts and procedures, as discussed in Appendix A. This will provide immediate benefits.
2. Standardize parts and materials so new products will be designed with a fraction of the part and material types. Chapter 5 shows effective ways to standardize parts and materials.

4.2 BUILD-TO-ORDER

Spontaneous build-to-order is the capability to quickly build standard products upon receipt of spontaneous orders without forecasts, inventory, or purchasing delays.[3] These products may be shipped directly to individual customers, to specific stores, or as a response to industrial customers' "pull signals" (signals that specific parts are needed right away for assembly).

Similarly, your suppliers may need to use spontaneous BTO to respond to *your* pull signals. And yet, if suppliers cannot actually build parts on demand, then they will be tempted to meter them out from inventory, in essence, transferring your parts inventory to their finished goods inventory.

The basic strategies for implementing spontaneous BTO are supply chain simplification, concurrent design of versatile products and flexible processes, the mass customization of variety, and the development of a spontaneous supply chain.[4] Spontaneous BTO can actually build products on demand *at less cost* than mass-produced batches, if "cost" is computed as *total cost*. Therefore, total cost measurements should also be part of this process (see Chapter 7).

4.2.1 Supply Chain Simplification

Although supply chain management is a much discussed topic today, most implementations fail to apply the basic lessons of Industrial Engineering 101: *"Simplify before automating or computerizing."* The simplification steps for supply chain management are standardization, automatic resupply techniques like kanban, and rationalization of the product line to eliminate or outsource the unusual, low-volume products that contribute to part variety way out of proportion to their profit generation ability (see Appendix A). The goal of supply chain simplification is to drastically reduce the variety of parts and raw materials to the point where these materials can be procured spontaneously by automatic and pull-based resupply techniques. Reducing the part and material variety will also shrink the vendor base, further simplifying the supply chain.

4.2.2 Kanban Automatic Part Resupply

Figure 4.1 shows two rows of part bins that are set up for resupply by the two-bin kanban system. Initial assembly starts with all bins full of parts. When any part bin nearest the worker is depleted, the full bin behind moves forward, as shown by the empty space in the illustration. The empty part bin then is returned to its "source," which could be the machine that made the part, a subassembly workstation that assembled the part, or a

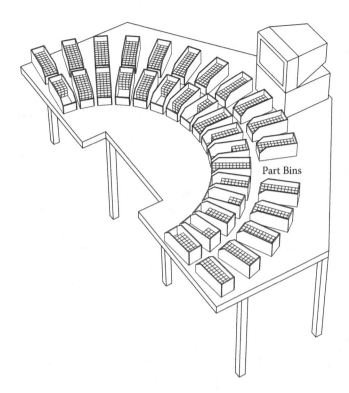

Part Bins

FIGURE 4.1
Kanban part resupply.

supplier. The source fills the bin and returns it to this assembly workstation behind its counterpart, which is now dispensing parts.

The beauty of kanban part resupply is that the system automatically ensures an uninterrupted supply of parts, *without forecasts or complicated ordering procedures.* The number of parts in a bin is based on the highest expected usage rate and the longest resupply time. The sizes of the bins are determined by the bin quantity and size of the parts. For large parts, some companies use two-truck kanbans, in which parts are drawn from one truck trailer while the other trailer goes back to the supplier for more parts. Alternative systems include kanban squares marked on the floor for larger parts and a two-card system where the cards travel (or are faxed or e-mailed) back to the source instead of the bins.

In order for kanban systems to work, there must be enough room to dispense all parts at the points of use. This, again, emphasizes the importance of part standardization, as discussed in Chapter 5.

4.3 MASS CUSTOMIZATION[5]

Mass production thrived in the bygone era of stable demand and little product variety. However, today, once-stable markets are now very dynamic and product variety is increasing to the point where products may need to be *mass customized* for niche markets or individual customers. Trying to satisfy this volatility and variety in the mass production mode will be a slow and costly ordeal. Because mass customization handles variety *proactively,* it is fast and cost-effective, with no extra cost or delays to handle various models, options, configurations, and customizations.

Mass customizers can use their versatile design and manufacturing capabilities to offer new value-added products or services that expand the scope of their markets. These offerings may cost little extra to add, especially if they can be done in existing CNC operations. And yet, they may save customers so much time and money that they would gladly pay for the options, even at premium prices, especially if these features are difficult for customers to obtain. For instance, one of the author's clients, Hoffman Engineering, built a plant in Lexington, Kentucky, that can spontaneously build-to-order a wide variety of standard or mass-customized electrical enclosures. Its high-profit, value-added opportunity is to use the very same CNC laser cutters that make the boxes to also make the holes and cutouts that would have cost their customers much time and money to do on their own.

There is a whole spectrum of ways that mass customization methodologies can benefit companies. At the most visible end of the spectrum, companies can mass customize products for individual customers.

Further along the spectrum is niche market customization. For instance, a company that makes telephones has only a few customers (telephone companies) who want several dozen models in many colors all with specific phone company logos. Exporters have to deal with many niche market products—usually a different set of products for each country exported. Even if the differences are minor, the sheer variety of SKUs (stock keeping units) can have significant cost and flexibility implications. Most companies could benefit from expansion into niche markets *if* they could do it efficiently.

At the other end of the spectrum are companies that have tremendous varieties of "standard" products; for instance, industrial suppliers of valves, switches, instruments, or enclosures, or any company with a catalog over a few dozen pages. As with product customization, there is a great contrast between how mass producers and mass customizers manufacture a variety of standard products. The mass producer has the dilemma of trying to keep a large enough inventory to sell a wide variety of products from stock or alternatively using the slow, reactive, costly process of ordering parts and building products in small batches after receipt of orders. The mass customizer can use flow manufacturing and CNC-programmable machine tools to quickly and efficiently make different products in a "batch size of one"—either mass-customized products or any standard product from a large catalog.

As with Lean Production and build-to-order, mass customization implementation also benefits from, and may even require, the prerequisites discussed in Section 4.1.2: rationalization of products to eliminate the most unusual products and their unusual parts (Appendix A) and standardization of parts and materials (Chapter 5).

4.4 DEVELOPING PRODUCTS FOR LEAN, BUILD-TO-ORDER, AND MASS CUSTOMIZATION

To be successful at designing products for Lean Production, build-to-order, or mass customization, companies must proactively plan product portfolios. Product development teams must design products in synergistic product platforms, design around aggressively standardized parts and raw materials, make sure specified parts are quickly available, consolidate inflexible parts into very versatile standardized parts, ensure quality by design with concurrently designed process controls, and concurrently engineer product platforms and flexible flow-based processes.

Further, product development teams need to eliminate setup by design by specifying readily available standard parts and tools (cutting tools, bending mandrels, punches, etc.), designing versatile fixtures at each workstation that eliminate setup to locate parts or change fixtures, and making sure part count does not exceed available part bins or space at each workstation.

Finally, products must be designed to maximize the use of available programmable CNC fabrication and assembly tools, *without expensive and time-consuming setup delays.*

4.5 PORTFOLIO PLANNING FOR LEAN, BUILD-TO-ORDER, AND MASS CUSTOMIZATION

For Lean Production, build-to-order, and mass customization, product portfolio planning (Section 2.3) must expand its focus to ensure that products are developed in synergistic families that can be produced on demand on flexible lines.

To accomplish this, product families need to be structured so that all products use standard parts and materials on the same flexible equipment without setup delays.

All the products within a flexible line—and possibly all the products within a flexible plant—must be compatible with respect to part standardization, raw material standardization, part/material availability, spontaneous supply chains, modularity, setup elimination, and flexible processing, including CNC machine tool operation.

Older generation products may not be compatible with flexible production lines and thus may have to be dropped, outsourced, redesigned, or, at a minimum, have material substitutions. Old products worth saving may have to be redesigned or upgraded and be added to the list of potential product development projects.

The evolution of product portfolios needs to be coordinated with implementation efforts for Lean Production, build-to-order, and mass customization. Decision making should be based on total cost and contributions to the overall business model.

4.6 DESIGNING PRODUCTS FOR LEAN, BUILD-TO-ORDER, AND MASS CUSTOMIZATION

For Lean Production, build-to-order, and mass customization, products and processes need to be designed so that:

- All the parts can be distributed at all points of use, which is accomplished by designing around standard parts and materials from aggressively standardized lists (Chapter 5). Too many parts would either clutter and confuse the work areas or force the parts to be kitted for a batch of products to be made, which is contrary to the one-piece flow of all the flexible paradigms. An important aspect of part standardization is fastener standardization, which is usually the easiest to do and can provide significant benefits. If screws can be standardized to one type at each workstation, then *auto-feed screwdrivers* can be utilized, which automatically orients screws and feeds them through a flexible hose to a powered screwdriver (Guideline F4 in Section 8.3). Of course, in order to designate a single type of fastener at each station, the design team will have to be concurrently designing the entire production flow.

- There will be no significant setup for any part or product in the family.[6] This includes any setup to (1) find, kit, load, or replace parts or materials; (2) change and position dies, molds, or fixtures; (3) change tools; (4) load, position, clamp, or calibrate workpieces; (5) adjust machine settings or calibrate machinery; or, (6) change equipment programs.
- Equipment programs can be located and downloaded instantaneously and manual instructions can be found, displayed, and understood quickly.
- Parts and materials can be resupplied spontaneously.[7] This includes specifying readily available parts and materials, designing around aggressively standardized parts, and specifying raw materials to be cut on demand from the same standard input stock sizes.[8]
- All parts can be pulled quickly into assembly on demand. This includes specifying suppliers who can build parts on demand or working with in-house manufacturing to establish in-house capabilities.

4.6.1 Designing around Standard Parts

Standardization of parts (Chapter 5) is the most important design contribution to the feasibility of spontaneous supply chains. *Aggressive standardization* can enable the easiest technique of the spontaneous supply chain: steady flows of very standardized parts and materials.[9] If there are too many different parts and material types, steady flows cannot be arranged because of the variety and unpredictability of demand. The total cost value of standardization and its contribution to the business model should motivate engineers and procurement organizations to implement such aggressive standardization.

Most part proliferation happens because engineers do not understand the importance to supply chains and operations and, even if they did appreciate this, do not know which parts are best to choose. The effective procedures presented in Chapter 5 show how to generate lists of standard parts and materials.

The usual operational mode in a part-proliferated environment is to order the parts ahead based on forecasts, either for scheduled batch production or to have forecasted parts available for on-demand assembly (the Dell model). If part variety is truly unavoidable or parts are to be mass customized, then they can be built on demand using the principles presented in the book, *Build-to-Order and Mass Customization*.[10]

4.6.2 Designing to Reduce Raw Material Variety

An important task in concurrent engineering is to specify not only the functional specifications of materials but also the order size and how they get cut into various parts. A company that made sheet metal products previously ordered 600 different shapes of sheet metal, which was a logistical nightmare. The author advised them to convert to just a handful of standard types, which were then cut on demand as they were needed. Even better would be to cut sheet metal on demand from a coil.

Another way engineers can reduce material variety is to specify a single tolerance or grade, instead of multiple grades. As pointed out in Chapter 5, any perceived cost increase of shifting all materials to the higher grade will be more than compensated by the total cost savings and value of the business model.

Families of parts should be designed so that every CNC machine tool uses the absolute minimum of raw material types; hopefully, just one, so there are no setup delays to change materials.

Multiple lengths of linear materials can be obtained by ordering reels or long lengths, which are then cut to length on demand.[11]

Inflexible raw parts, such as castings, extrusions, custom silicon, and bare printed circuit boards, should be consolidated into versatile parts that can be used for many functions on a wide range of products (Section 5.12). Programmable chips should all be programmed using the same "blank."

4.6.3 Designing around Readily Available Parts and Materials

A spontaneous supply chain depends on parts and materials that are always available. Therefore, it is an important aspect of engineers' jobs to specify parts and materials that are readily available. Usually, design engineers choose parts based on functionality and hopefully quality. But for flexible operations, availability is equally important.

Parts and materials that can be obtained from multiple suppliers tend to have better availability, in general, and are also more likely to be standard. On the other end of the spectrum, parts and materials available from only a single source may someday be hard to procure at all. On the lecture circuit, the author found several companies whose parts had become obsolete before their products were even released!

Rapid delivery is important for flexible operations, so design teams should specify local suppliers who will be able to supply materials on demand. Many companies preclude the ability to obtain parts and materials spontaneously by buying from supposedly low-cost suppliers on another continent, which take too long to arrive[12] for spontaneous supply chains.[13]

Similarly, even local low-bidder suppliers probably are not going to be able to deliver on demand and will most likely not have adequate quality either (see Section 6.11).

The guiding principle is to select parts and material to be readily available, be delivered on demand, and have the lowest total cost, which includes material overhead, ordering, expediting, routing, shipping, expedited shipping, incoming inspections, kitting, internal distribution, and all the costs of locating alternative sources of supplies to counter availability problems.

4.6.4 Designing for No Setup

Products can be designed to eliminate the following setups in manufacturing:

- *Part setup.* Product design has a profound effect on part setup. Excess proliferation of parts complicates internal part distribution and may make it impossible to have all parts available at all points of use. Even a moderate excess of part types will cause setup delays to distribute, find, and load parts into manual or machine bins. A greater excess of part types may make it necessary to kit parts for every batch, which is a significant setup. Part prep (such as cutting or bending leads on electronic components) is another setup that can be avoided by using versatile equipment or specifying parts that are delivered properly prepared for the product and the processing equipment.

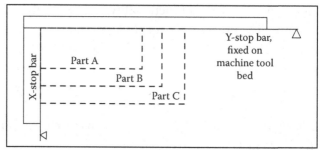

FIGURE 4.2
Flexible fixture.

- *Fixturing setup.* Designers can eliminate fixturing setup by concurrently engineering families of parts and flexible fixturing, an example of which is shown in Figure 4.2 for parts A, B, and C. All dimensions in the part family are dimensioned from the X datum and Y datum, which are shown in Figure 4.2. The Z datum is the machine tool bed. All parts in the family will be positioned against the X-stop bar and the Y-stop bar. The CNC machine tool then knows where each part is located and can thus perform all operations it is programmed to do, without any setup delays to reposition parts or change fixtures. For a milling machine bed, four flexible fixtures can be mounted on the four corners of the bed, leaving the center of the bed for general-purpose machining. Alternatively, quick-change fixtures can be developed to minimize fixture changeover times.
- *Tool setup.* Designers can eliminate tool change setup by designing parts around common tools (cutting tools, bending mandrels, punches, etc.); ideally, one tool that never has to be changed. If multiple tools are required, designers must keep tool variety well within tool changing capacity for the whole product line.
- *Instructions.* Designers need to work with manufacturing engineers to concurrently develop simple assembly procedures, with instructions that can be located and understood in a few seconds, either on a computer screen or on paper.

4.6.5 Parametric CAD

A wide variety of machined dimensions (for mass-customized or standard parts) can be performed quickly and cost-effectively using a combination

of CNC machine tools and *parametric CAD,* which *stretches "floating"
dimensions* and then automatically creates CNC programs as they are
needed by CNC machine tools.

Universal parametric templates can be created ahead of time for families of parts and structured so that, when the customized dimensions are plugged in, the drawing transforms into a customized assembly drawing which automatically updates customized part drawings.

Dimensional customization[14] can be performed quickly and cost-effectively using a combination of CNC machine tools and parametric CAD. Another use of parametric CAD is to quickly show how changing a parameter impacts the system.

4.6.6 Designing for CNC

CNC machine tools offer vast opportunities to eliminate machining setup.[15] CNC machine tools include metal cutting equipment (mills, lathes, etc.), laser cutters, punch presses, press brakes, printed circuit board assemblers, and basically any production machine controlled by a computer. Designers need to understand enough about CNC operation to use the versatility of CNC to eliminate setup.

4.6.7 Grouping Parts

The first step in designing for CNC is to structure compatible groups of parts to be processed in each CNC machine—this was originally labeled *group technology.*[16] Of course, this must be based on the overall manufacturing strategy and flow of parts and products. This evolves from a serious concurrent engineering activity in which the grouping and flow of parts are a key element of the design team's responsibilities. The grouping may determine the type of CNC needed, or existing machine tools may specify the grouping.

4.6.8 Understanding CNC

Once they thoroughly understand the range of parts to be made by each CNC machine, the designers will need to understand the capabilities and limitations of the equipment. This can be accomplished by studying the equipment specifications and talking to CNC operators. In fact, the CNC operators should be on the design team to optimize the design and processing plans. Of course, the ultimate understanding would come from actual CNC operational experience, either through prior work or a job rotation program.

4.6.9 Eliminating CNC setup

The versatility of CNC provides unique opportunities for eliminating setup if parts are designed properly. Ideally, all operations should be able to be performed on one machine in a single fixturing, as recommended in Guideline P14 in Chapter 9. Multiple machines will require extra fixturing setups. Even if multiple specialized machines have higher speed ratings, the total flow time through all operations, *including setup*, is what counts. The value of eliminating these setups may justify a more sophisticated CNC machine, compared to more setups on many cheaper machines.

In order to process parts in a single fixturing, designers need to specify a suitable *datum*[17] in each plane, from which all dimensions are referenced and which is suitable for clamping to a milling machine table or lathe chuck. The part also must be designed so that all the operations can be done in this fixturing. If all operations cannot, it is important that the most critical dimensions are cut in the same fixturing, which will routinely achieve the best tolerance of the machine tool, usually ±.001″ or better. However, removing a part to reposition for a subsequent cutting lowers the accuracy of these critical dimensions because the tolerance will then depend on the accuracy of the second positioning, which is usually much worse than machine tool accuracy.

4.6.10 Developing Synergistic Families of Products

Product families must be based on all the following criteria:

1. *Customer/marketing feasibility*, with the focus on *profitability* over *completeness*
2. *Operational flexibility*, so any variation can be built without delays, onerous setup costs, and inventory carrying costs, ideally on demand
3. *Supply chain responsiveness*, so family variations will not have to wait too long for parts and materials to be delivered and distributed
4. *Design versatility* to enable the above

The mass customization strategy is

1. Standardize everything that is feasible at the architecture level.
2. Design the remaining variety to be easy to build quickly using CNC processes from standard parts and materials that, ideally, are always available at all points of use.

4.6.11 Strategy for Designing Product Families

Design versatile standard platform(s) around standard parts. As with all standardization, *versatile platforms* may appear to cost more when looking at just BOM lines, but this strategy will result in less cost on a total cost basis. Similarly, standard parts, modules, and materials may appear to cost more when looking at just BOM lines, but this strategy will result in less total cost.

For installed equipment or machinery, determine the optimal capacity/performance steps—somewhere between (1) one-size-fits-all and (2) a unique product for every order.

The overhead savings from a platform strategy will allow the entire range of most variations to sell for no more than the lowest capacity/performance version (probably less), thus giving customers better value and possibly increasing sales at the high end of the range. Alternatively, each variation could be priced at what the market will bear, thus bringing in more profit.

> Do not "value engineer" individual *versatile products, parts, modules, or materials* without taking into account all the overhead savings, or the versatility may be eliminated "to save cost."

Such a strategy should be based on satisfying the *most profitable* customer segments, in preference to hard-to-build orders (usually the *least profitable*).

4.6.12 Designing Products in Synergistic Product Families

Design around aggressively standardized parts and raw materials. As with other standardization, some of these strategies may appear to raise a BOM line, but the total cost will be less for the whole product family.

- Standardize on the same raw materials for baseline products and anticipated customized options to avoid supply chain delays, inventory costs, and engineering changes to convert to more available materials. Select standard parts and materials that are readily available; ideally, always available at all points of use. Make sure specified parts are readily available *throughout the life of the product*, even if more available parts appear to cost more.
- Select off-the-shelf parts to be versatile enough for all anticipated applications for the current and anticipated family; for instance, with enough capacity, power, ratings, margins, and tight enough tolerances for all variations. For high-variety parts, specify *flexible CNC processes*, such as any computer-controlled process or machine tool or PC board assembler, in preference to *inflexible processes*, such as castings, forgings, extrusions, PC bare boards, etc.
- *Concurrently engineer* product families that can be built on a programmable process and flexible fixtures. Establish flexible *cells* that can build on demand any standard part or anticipated custom part. This means that any part in the family can be built on demand without setup changes. Maximize the number of custom features that can be fabricated during the initial CNC operations. Design versatile bare circuit boards that have all the traces, pads, holes, vias, and connectors for all PCB assemblies in the family. A variety of finished PCBs could be assembled on demand by CNC pick-and-place machine tools.
- Consider *additive manufacturing*[18] for low-volume variations when it avoids tooling costs. Add extra mounting holes, knock-outs, and access openings in initial fabrication, when it is almost free, to avoid trying to add these later, when it is slow, costly, and may look crude. Ensure mounting versatility for modular parts that may come in various sizes with enough space, connections, power, and mounting holes.
- Specify the original part selections to higher ratings, margins, capacities, etc. For instance, if the original requirements specify 150-volt rating, specify a higher voltage in the part selection.
- Make custom or high-variety parts easy to customize with mass customization[19] techniques. These techniques include parametric CAD, quick machine tool program generation and loading, flexible fixtures, and versatile CNC machine tools building custom and high-variety parts from standard stock on flexible fixtures.
- Avoid expensive tooling and setup charges for inflexible processes such as molding, casting, forging, stamping, extruding, or bare PC

boards. Consolidate inflexible parts into very versatile standardized parts that can be used in many products.

- Utilize a *postponement strategy*[20] and develop a versatile "vanilla" platform that can be configured by adding preplanned "flavor" modules. Another version of postponement is ordering versatile semifinished parts in quantity and then doing specific operations on demand, such as hole drilling or machining specific optional features.

4.7 MODULAR DESIGN

Modular design is a design technique in which functions are designed in modules that can be combined into subsequent designs. A related concept is "reusable engineering," where portions of previous designs become the basis of new designs. It is easy to do in CAD by copying previous design details and transferring them to other drawings.

The benefits of reuse and modular design are (1) less engineering effort; (2) more commonality among models for simpler supply chains and more flexible operations; and, (3) better reliability from using proven parts and designs (Chapter 10).

When this concept is extended to manufacturing, products may actually be assembled from "building block" modules. Care should be taken in engineering to design standard interfaces for optimal flexibility. Catalog hardware often has standard interfaces; for example, for motor mounts, shaft couplings, fluid power and electrical connections, and so forth.

An effective modular design strategy would be to intentionally create portions of designs that have general usefulness for whole families of current and future products. For instance, machinery could have modular bases, frames, drives, gearboxes, controls, and cabinetry that could be combined with specific functional modules.

4.7.1 Pros and Cons of Modular Design

Time-to-market for new products can be quicker if existing modules have already been designed, documented, debugged, and certified. Off-the-shelf modules can be utilized, like off-the-shelf printed circuit boards. Delivery may be quicker if products can be assembled from standard modules. Modularity can lead to broader product lines and be the foundation of a proactive upgrade strategy. Further, modularity can simplify maintenance and field service when defective modules can be replaced and then repaired off-line.

Modularity can lower inventory levels and overhead cost compared to several different versions of the integrated part or product. Widely used modules have less vulnerability to lead-time delays. Plug-together modularity may simplify some assembly and accept third-party plug-in modules, such as in personal computers.

Modularity may be a key element in a *postponement strategy*.[21] Postponement is a mass customization technique that is applicable for certain products that can have some variety postponed until just before shipping.

The cost of modularity can be determined only on a total cost basis. Engineering costs will be lower if existing modules can be utilized, but they may be higher if new modules and interfaces have to be designed. The cost of manufacturing the interfaces may be more than an integrated product, but using versatile modules on many products may lower costs through economies of scale. Selecting existing modules would save the costs and delays of testing and debugging, compared to an entirely new integrated design.

Similarly, product testing may be easier if existing modules have already been tested, debugged, and certified. Diagnosis efforts may be easier with less functionality on each module. However, if all the modules in the product are new, then tests will have to be devised and all the modules will have to be tested and, if necessary, repaired. After the modules are tested, the entire assembly may have to be tested again.

It may be possible that, for modular products, new modules could be developed (for lower cost or better performance) that could be introduced on *current* products and maybe even retrofitted to products in the field. Then the new modules could be the foundation for *new* products. This is one of the benefits of modular product architecture.

Modularity has some drawbacks too. Product development expenses may be increased by the necessity to design modular interfaces. Modularity may not be feasible for inherently integrated products. Modular interfaces may compromise functionality by adding weight, weakening structures, slowing electrical signals, or even causing signal interruptions at low voltage. In electronics, the extra connectors needed for modularity may cause reliability problems or degrade performance. Module interfaces may result in undesirable visual aesthetics (such as seams, joints, etc.) or undesirable acoustics (such as squeaks and rattles).

4.7.2 Modular Design Principles

Companies should invest in the design of modules that are versatile enough for many products and product families. Individual projects may not have the budget or resources to develop modules for many products. Modules should be versatile enough to have general usefulness for many current and future products. Modular principles include:

- *Total cost basis.* Do not look at module cost at the level of one project; look only at the total cost of all the products that will utilize the modules.
- *Reusable engineering.* Modular design is not limited to physical modules, but also could be reusable engineering or software code.
- *Interfaces and protocols* must be optimized at the system architecture level.
- *Standardization.* For maximum usefulness, modules must have standard interfaces. Use industry-standard interfaces, if available; if not, develop very versatile interfaces. Toyota engineers have a strong sense of the vehicle as a system and consequently focus a great deal of skill and energy at the design interfaces.[22]
- *Clean interfaces.* All modular interfaces should be clean and consistently easy to integrate together.
- *Documentation.* Document for modularity, with optimal CAD layer segregation, bills-of-material, software object identification, and so forth.
- *Debugging.* Minimize debugging costs by using existing modules that have already been debugged. Some leading companies, like Hewlett-Packard, feel that this is the best way to produce bug-free software. With a high percentage of reuse, debugging efforts can focus on new aspects.
- *Consistency.* Resist the temptation to "improve" modules, unless the improvement is substantial and justifies the cost and time of design changes, production changes, evaluation, debugging, and recertification or requalification. Do not attempt cost reduction on modules unless the total cost savings *for all applications* justifies the development of a new-generation module.

4.8 OFFSHORING AND MANUFACTURABILITY[23]

In a vain attempt to save cost, many companies are offshoring—designing products in the United States for sale in the United States, but manufacturing them overseas—because their primitive cost systems make it appear that this will save money. If all they quantify are parts and labor, like the pie chart in Figure 6.2, then moving production to a low labor rate country will appear to lower labor cost. If no other costs are quantified, like the pie chart in Figure 6.3, outsourcing would be pursued based on a back-of-the-envelope calculation.

However, manufacturing offshore for sale in the United States rarely results in a net cost savings[24] when measured on a total cost basis (Chapter 7), considering differences in labor efficiency and all the costs of shipping, quality, inventory, communications, travel, training, and transferring products, support, and complete sets of equipment needed for any manufacture.

Further, offshore manufacturing compromises six of the eight cost reduction strategies that are presented on the home page at www.HalfCostProduct. com, the most important being product development.

4.8.1 Offshoring's Effect on Product Development

As pointed out in Figure 1.1, 80% of a product's lifetime cumulative cost is determined by product design. Unfortunately, offshore production compromises all future design opportunities because it prevents concurrent engineering teamwork, which is the most promising opportunity for achieving truly low-cost products.

Offshoring prevents this teamwork because design engineers and manufacturing people are not in the same country and not even working at the same time. Without manufacturing involvement, engineers will design products alone, throw them over the ocean, and get back parts that will only be as manufacturable as the individual engineer's DFM expertise, and whose knowledge of the design guidelines may be limited. Further, dealing with all the problems of offshoring (see below) will be a resource drain that, when combined with other counterproductive practices discussed in Section 11.5, can consume two-thirds of product development resources!

4.8.2 Offshoring's Effect on Lean Production and Quality

Further, offshore plants, especially contract manufacturers, amortize their setup charges by building in batches (mass production) and then shipping across the ocean in batches. It is difficult to respond to volatile market conditions when so much forecasted inventory is at the plant and in the long pipeline in ships traveling across the ocean.

Womack and Jones, writing in *Lean Thinking*,[25] summarize it succinctly: *"Oceans and Lean Production are not compatible."* They go on to say that smaller and less-automated plants close to assembly plants and markets will yield lower total costs, considering the cost of shipping, the inventory carrying costs, and the cost of obsolescence when products built weeks ago no longer satisfy customers.

So the enormous potential cost savings from Lean Production (summarized in Section 4.1) will not be realized because products cannot be pulled[26] by customer demand across oceans, nor could they be built to-order (Section 4.2). The distance and remoteness will prevent the company from setting up flexible plants that could mass customize products for niche markets or individual customers (Section 4.3), and thus cause it to miss out on those paradigm-shifting opportunities (Section 4.9).

Without Lean Production and the design-for-quality contribution of good concurrent engineering teamwork, as discussed in Chapter 10, quality and reliability will not be designed in or built in, so the home office will have to rely on strict, and expensive, testing of all products followed by repair costs or scrap costs followed by extra setup cost to build replacement products and expedited shipping costs. Worse, if contract manufacturers are selected by low bidding, quality will suffer even more, for reasons cited in Section 6.11.

4.8.3 Offshoring Decisions

Labor cost is actually a small percentage of the selling price, and yet off-shoring decisions are usually based entirely on labor cost because it is the only processing cost that is quantified. But, any perceived savings for labor are usually exceeded by the following hidden costs. These costs would not be "hidden," and decisions would be more relevant, if all overhead costs were quantified with *total cost accounting* (Chapter 7).

Some of the total cost considerations that should determine offshoring decisions include: *labor efficiency* (which could cancel out labor cost savings); *difficulty controlling operations* that are not working at the same time; *product introduction delays* (usually months later because of the additional ramp); *supply chain vulnerabilities*; and *shipping costs and delays*, especially when expediting is used to compensate for the above problems.

Further, labor rates often rise and tax benefits only last so long, until the company ups the ante and increases the commitment, for instance, by moving engineering overseas as well.

The most ironic reason to move production offshore is that the designs are *labor intensive*. However, implementing the principles of this book can reduce labor input to the point where moving to low-labor-rate areas is no longer needed and can no longer be justified.

"Hidden" costs, all of which should be quantified by total cost (Chapter 7), would include: *quality costs* and the delays to fix them; *training costs*, which are higher than expected for hard-to-build products or where turnover is high; offshore *setup and administration costs* and resource drains; *transfer efforts* and ongoing *indirect support*, which take resources away from product development; *travel costs*, which are always more than estimated; and *"other" local costs* that pose ethical and legal dilemmas.

Unfortunately for product development, the cost of setting up offshore manufacturing and dealing with these hidden costs is usually paid for by support people at the company headquarters—otherwise, the business case for offshoring would collapse! Ironically, these support costs make headquarters look even less efficient, which, perversely, encourages yet more offshoring, thus resulting in a downward spiral.

Sadly, the final hidden cost of offshoring is the cost to move operations back to where the products are designed after realizing all the above costs or realizing how offshoring distracts from real cost opportunities in product development, operations, and quality programs.

4.8.4 Bottom Line on Offshoring

Setting up and operating offshore manufacturing doesn't save money on a total cost basis, but *just trying* compromises quality, delivery, and product development, which could otherwise provide real cost reduction and new high-profit opportunities. Rather than weakening local operations with the burdens of offshoring, or eliminating local operations entirely, local operations could then pursue more effective cost reduction by helping to design low-cost products, eliminating waste through Lean Production, lowering the cost of quality, and setting up a flexible factory that could build standard products and mass-customized versions on demand without the costs and risks of inventory.

If all of this does not make good financial sense, you need a better cost system, as presented in Chapter 7.

4.9 THE VALUE OF LEAN BUILD-TO-ORDER AND MASS CUSTOMIZATION[27]

Extending Lean Production to build-to-order and mass customization (BTO&MC) represents a business model that offers an unbeatable combination of superior responsiveness, cost, and what customers want when they want it. It enables companies to build any product on demand without forecasts, batches, inventory, or working capital.

BTO&MC companies can grow sales and profits by expanding sales of standard, customized, derivative, and niche market products while avoiding the commodity trap. BTO&MC companies are the first to market with new technologies because distribution pipelines do not have to be emptied first.

BTO&MC substantially simplifies supply chains—not just "managing" them—to the point where parts and materials can be spontaneously pulled into production without forecasts, MRP, purchasing, waiting, or warehousing.

Build-to-order is the best way to resupply stores who demand rapid replenishment, low cost, and high order fulfillment rates without the classic inventory dilemma: too little inventory saves cost but creates out-of-stocks, missed sales, expediting, and disappointed customers; too much inventory adds cost and costly obsolescence risks. When inventory carrying costs are taken into account, profits are completely wiped out after only five months of products sitting in finished goods inventory.[28]

Mass customization can efficiently customize products for niche markets, countries, regions, industries, and individual customers. There is a natural synergy between build-to-order and mass customization. They share the same batch-size-of-one operations and spontaneous supply chain. Build-to-order and mass customization operations are equally efficient and very compatible, unlike situations where a mass customization experiment must be run separately from the batch-and-queue operations of mass production. Manufacturing build-to-order and mass-customized products on the same lines will often push the combined volume over the critical mass threshold necessary to justify these implementations.

The following are several elements of the value of build-to-order and mass customization.

4.9.1 Cost Advantages of BTO&MC

BTO&MC companies enjoy substantial cost advantages, which they can use for competitive pricing, reinvestment, or enhancing profits. BTO&MC principles attack several categories of total cost to achieve the absolute lowest prices—or competitive prices with higher profits.

One of the greatest cost advantages of build-to-order is the elimination of all the costs caused by inventory: inventory carrying costs (usually 25% of value per year[29]), warehousing costs, administrative expenses, obsolescence write-offs, and discounting to sell unsold or obsolete inventory. Incoming just-in-time part deliveries minimize inventory costs for parts and materials. Flow manufacturing and setup/batch reduction efforts can virtually eliminate work-in-process (WIP) inventory costs. Building to order and shipping direct can virtually eliminate finished goods inventory costs.

Build-to-order minimizes or eliminates many other overhead costs for forecasting, MRP, purchasing, expediting, scheduling, planning, setting up production, and the extra sequence of activities required when forecasted inventory cannot satisfy demand: re-forecasting, re-purchasing, re-expediting, re-scheduling, re-planning, and re-setting-up production.

Building without batches eliminates the costs of setup changes, kitting, and the loss of expensive machine time and valuable resources. The rapid feedback aspect of flow production eliminates the cost of recurring defects, which are more likely to happen with large batches.[30]

Lean Production eliminates many categories of waste and inefficiencies, as discussed in Section 4.1. All the inefficiencies of mass production—lower productivity and the lack of the ability to shift production between lines—raise costs and cause the mass producer to need more overtime than the more efficient and flexible BTO&MC company. Eliminating inventory, incoming inspection, and kitting saves significant floor space cost, which can delay or eliminate the need to build more factory or warehouse space for growth.

Customization and configuration costs are less because they are proactively planned and executed efficiently, instead of the very inefficient craft production and fire-drill activities used in most companies. *Learning relationships* make repeat orders more efficient.[31]

Concurrently engineering products for manufacturability generally results in significant cost savings, but these are more profound for BTO&MC companies because of the greater opportunities to design out several categories of overhead cost.

Optimal utilization of flexible CNC automation saves labor cost. Flexible fixturing and setup reduction makes CNC even more efficient.

Distribution costs are less for BTO&MC goods because shipping is more direct, finished goods inventories are eliminated, and expediting is not needed to rectify shortages.

Finally, using product line rationalization (Appendix A) and total cost measurements (Chapter 7) to eliminate high-overhead/low-profit products eliminates the loser tax on cash cow products, thus letting them sell for less or make more profit. This advanced business model eliminates the need to discount or take low-margin sales.

4.9.2 Responsive Advantages of BTO&MC

BTO&MC companies build products on demand, instead of having to forecast, order, wait, build, and stock. For phone and web sales, 100% of orders can promptly be shipped directly from the BTO&MC factory. One reason the "e-commerce revolution" didn't live up to its expectations is because of poor availability and delays. Poor availability is intrinsic in any system that sells from forecasted inventory, which depends on inherently unreliable forecasts. Delays will be common when shipping from inventory or trying to build to-order in a mass production environment.

If customers want to buy something right now off the retail shelf, BTO&MC suppliers can resupply those shelves better than anyone can from inventory, so a complete selection will always be available to customers. Stores and dealers that order frequent shelf replacements from BTO&MC suppliers will develop a good reputation for availability and eventually generate a loyal customer base that keeps coming back because they know that they will always find what they want; for example, for clothing, customers will always find the style they want in their size. Learning relationships make repeat order fulfillment quicker with each order. BTO&MC companies are the fastest to adjust to changing market conditions.

New product development and introduction can be faster when new products are just "variations on a theme" that are easier to develop because of modularity, parametric CAD, and flexible processing. Production ramps can be faster on flexible lines that don't have to be "tooled up" for new products. A *Wall Street Journal* article reported that flexible automobile makers are able to release different versions of a car throughout the year instead of the traditional single release in the fall.[32]

BTO&MC companies are the first to introduce new technologies into the marketplace because they don't have to first empty the pipeline of obsolete products, which usually must be discounted to clear the pipeline. Even if that was not an issue, shorter direct distribution channels can speed new products faster to customers who can't wait for the latest and greatest.

If new product introductions result in greater than expected growth, the BTO&MC company can meet upsurge demands by transferring production to other flexible lines, as in Lean Production plants. BTO&MC implementations free up floor space, which can then be used for growth. Thus, growth will be less likely to be hampered by shortages of floor space.

Standard parts are available from more sources with more total capacity, so standard parts will be more readily available in times of rapid growth. Optimal responsiveness is also assured by supply chains that allow assemblers to pull parts quickly without lengthy hand-offs or dependence on forecasted parts inventory.

Distribution is more direct, and therefore faster, for built-to-order goods, without delays caused by shuffling inventory from factories to warehouses to distribution centers and then to customers or stores. Eliminating inventory and many nodes in the distribution chain is not only quicker but also eliminates order aberrations and demand swings caused by the order/response lag time inherent in slow, multi-node distribution chains. This is the lesson taught in the supply chain simulation, developed at MIT, called *the beer game*,[33] which can be "played" by executives in a conference room[34] or as a web-based game.[35]

For capital equipment companies, quicker product delivery itself may be a competitive advantage. In addition, responding to RFQs (requests for quotations) will be quicker with configuration software and less susceptible to subsequent delays due to order-entry errors and customer-induced changes.

4.9.3 Customer Satisfaction from BTO&MC

BTO&MC can provide unmatched customer satisfaction for industrial clients or the ultimate consumers. Consumers will be satisfied by products always available at the best prices. OEM and industrial clients will be satisfied by receiving parts on demand to support *their* build-to-order efforts.

Mass customization will enable even higher levels of customer satisfaction for customers who can quickly receive high-quality, low-cost products specifically customized to their individual needs. For customized products, customers can make better choices and consider more "what-if scenarios" with configuration software. Even if individual customization is not appropriate, customer satisfaction will be enhanced when products are customized for their culture, group, country, or region.

Certain aspects of BTO&MC enhance quality from supplier relationships emphasizing quality, continuous improvement (*kaizen*), and rapid feedback that prevents recurring defects. A higher proportion of flexible CNC operations improves quality with more consistent tolerances.

Learning relationships enable the BTO&MC company to learn and adapt from each order, thus satisfying customers better on the next order and progressively developing more committed customers.

4.9.4 Competitive Advantages of BTO&MC

Competition now is *between business models*—the company with the best business model will be the best competitor. Ironically, the subjects of several best-selling business books, *leadership* and *execution,* will only drive a company faster down the wrong path if they have the wrong business model. Dartmouth Business School Professor Sidney Finkelstein, writing in *Why Smart Executives Fail: And What You Can Learn from Their Mistakes,* concluded that, "The real causes of nearly every major business breakdown are the things that put a company on the wrong course and keep it there."[36]

As a business model, build-to-order and mass customization will compete well against competitors both large and small because of a superior combination of speed, cost, and customization. Without a superior business model, companies might have to compromise profits to enhance market share. BTO&MC avoids the worst-case competitive position, where products revert to commodity status, with purchasing decisions made solely on price.

BTO&MC company products can avoid commodity status with the differentiation aspects of build-to-order (delivery, cost, quality, etc.) and mass customizing products to better satisfy customers. Learning relationships result in greater customer loyalty with each order.

BTO&MC companies have the agility to expand business into adjacencies, niche markets, derivatives, and so forth. Finally, all the above advantages can create a reputation as a leader, which further improves sales, impresses investors, and attracts the best talent.

4.9.5 Bottom Line Advantages of BTO&MC

Lower total cost results in increased profits or a lower selling price, or both. Faster delivery, better quality, and lower cost can grow revenue and market share. Agility allows expansions into new markets. Additional value-added work offers high profit opportunities. With enough competitive advantage, premium prices may be possible.

Chapter 14 of the author's *Build-to-Order & Mass Customization* presents "The Business Case for BTO&MC."[37] Chapter 13 of the same book is on implementation. The author's customized in-house seminars (Appendix D.5) and implementation workshops (Appendix D.6) show businesses how to implement all aspects of on-demand Lean Production, build-to-order, and mass customization.[38]

NOTES

1. James P. Womack and Daniel T. Jones, *Lean Thinking: Banish Waste and Create Wealth in Your Corporation* (1996, Simon & Schuster), p. 27.
2. "Dock-to-line" may sometimes be referred to by the more ambiguous title of "dock-to-stock," which technically means parts go from the dock to some form of incoming inventory "stock" area without inspection.
3. David M. Anderson, *Build-to-Order & Mass Customization: The Ultimate Supply Chain Management and Lean Manufacturing Strategy for Low-Cost On-Demand Production without Forecasts or Inventory* (2008, CIM Press). See description in Appendix D.
4. Ibid., Chapters 3–9.
5. Ibid., Chapter 9, "Mass Customization."
6. Ibid., Chapter 8, "On-Demand Lean Production"; see section on "Setup and Batch Elimination."
7. Ibid., Chapter 7, "Spontaneous Supply Chain."
8. Ibid., Chapter 7, section on "Material Cut to Length."
9. Ibid., Chapter 7, "Spontaneous Supply Chain."
10. Ibid.
11. Ibid., see Figure 8.1, "Build-to-Order & Mass Customization for Fabricated Products."
12. Ibid., Chapter 6, "Outsourcing vs. Integration." Also see the article on outsourcing at www.HalfCostProducts.com/outsourcing.htm and a discussion of offshoring at www.HalfCostProducts.com/offshore_manufacturing.htm.

13. Ibid., Chapter 7, "Spontaneous Supply Chain."

14. Ibid., Chapter 9, "Mass Customization."

15. Ibid., Chapter 8, see the section, "CNC to Eliminate Machining Setup."

16. Charles S. Snead, *Group Technology: Foundation for Competitive Manufacturing* (1989, Van Nostrand Reinhold).

17. The datum concept is a key element of geometric dimensioning and tolerancing (GD&T), from the ANSI Y14.5 standard.

18. Terry Walters and Tim Caffrey, "Additive Manufacturing Going Mainstream," *Manufacturing Engineering*, June 2013.

19. Anderson, *Build-to-Order & Mass Customization*, Chapter 9, "Mass Customization."

20. Ibid., Chapter 9, "Mass Customization," p. 293.

21. Ibid., Chapter 9, "Mass Customization."

22. James Morgan and Jeffrey K. Liker, *The Toyota Product Development System* (2006, Productivity Press), Chapter 4, "Front-Load the PD Process to Explore Alternatives Thoroughly."

23. Anderson, *Build-to-Order & Mass Customization*, Chapter 6, "Outsourcing vs. Integration." Also see the articles on outsourcing and offshoring at www.HalfCostProducts.com/outsourcing.htm and also www.HalfCostProducts.com/offshore_manufacturing.htm.

24. Anderson, *Build-to-Order & Mass Customization*, Chapter 6, "Outsourcing vs. Integration."

25. James P. Womack and Daniel T. Jones, *Lean Thinking: Banish Waste and Create Wealth in Your Corporation* (1996, Simon & Schuster), p. 224.

26. Ibid., Chapter 3, "Pull."

27. Anderson, *Build-to-Order & Mass Customization*, Chapter 14, "The Business Case for BTO&MC."

28. Ibid., Figure 2.2, "How Inventory Erodes Profits over Time When Selling from Finished Goods Inventory," which assumes 10% profit margin and 25% inventory carrying cost per year.

29. Ibid., Figure 2.1, "Inventory Carrying Cost since 1961" which shows the average carrying cost at 25% of value per year.

30. Ibid., See the section in Chapter 2, "Defects by the Batch."

31. B. Joseph Pine, II, Don Peppers, and Martha Rogers, "Do You Want to Keep Your Customers Forever?" *Harvard Business Review*, March–April 1995, p. 103.

32. Jonathan Welsh, "A New Status Symbol: Overpaying for Your Minivan; Despite Discounts, More Cars Sell Above the Sticker Price," *Wall Street Journal*, July 23, 2003, p. B1.

33. The beer game website is http://supplychain.mit.edu/games/beer-game.

34. The on-site beer game is usually led by an experienced facilitator who provides the game boards and order tokens. This is a fun exercise that can be played in a few hours with teams that represent stores, sales, distribution, and manufacturing, who quickly experience the effects of unstable demand swings propagating throughout the distribution network.

35. The web-based beer game site is http://supplychain.mit.edu/games/beer-game.

36. Sydney Finkelstein, *Why Smart Executives Fail: And What You Can Learn from Their Mistakes* (2003, Portfolio/Penguin), p. 138.

37. Anderson, *Build-to-Order & Mass Customization*, Chapter 14, "The Business Case for BTO&MC."

38. The author offers in-house customized seminars on build-to-order and mass customization that summarize the shortcomings of mass production and how costs rise exponentially with increasing variety and market volatility in a batch-and-queue environment. The interactive seminar also shows how to establish spontaneous supply chains, build parts and products on-demand, eliminate setup, develop flexible cells, mass customize products, develop products for BTO&MC, standardize parts, and rationalize products, and then how to implement all of the above, with optional workshops focused on implementation. Also see the "seminars" page at www.build-to-order-consulting.com/seminars.htm or inquire by email at anderson@build-to-order-consulting.com.

5

Standardization

Standardization of parts and materials is a fundamental aspect of DFM that can simplify product development efforts, lower the cost of parts and materials, drastically reduce material overhead costs, simplify supply chain management, improve availability and deliveries, raise quality, improve serviceability, and support Lean Production, build-to-order, and mass customization. The standardization procedure presented herein is a powerful tool that can achieve many positive results for new designs.

Standardization *enables Lean Production*, with deliveries that are faster, more frequent, more dependable, and less affected by shortages. Standard parts can be selected to eliminate the problems of long-lead-time parts and materials by choice. Standardization also leads to more efficient internal distribution—for fewer parts—with dock-to-line deliveries possible. Automatic resupply and steady flows will be possible (Section 4.2), thus eliminating the dependence on forecasts, MRP, and purchase orders.

Standardization can *reduce cost* with better purchasing leverage and economy-of-scale savings. Standardization will reduce inventory carrying costs for parts and materials, saving 25% of the value of eliminated inventory *per year.* Widespread use of standard parts will result in less expediting and fewer change orders to solve availability problems. Further, standardization can reduce material overhead to 1/10 of nonstandard parts, which reflects savings in purchasing efforts and encourages engineers to specify standard parts. Finally, aggressive standardization enables build-to-order and mass customization, which reduce costs in many ways, as discussed in Chapter 4.

Steady flows of standard parts can be ordered and projects will be able to borrow from each other in emergencies. Minimum part variety will allow compact workstation size to encourage flow, U-shaped lines, and kanban-supplied workstations (Figure 4.1). Standardization will minimize work stoppages, shipping delays, fire drills, and penalties from parts being out of stock. Finally, standardization will improve productivity.

Standardization *improves product development,* with less purchasing effort for *purchasing,* thus allowing more purchasing participation in product development teams to establish vendor partnerships. This also ensures that there will be more thorough searches for off-the-shelf parts early before arbitrary decisions preclude their use. Standardization also results in more thorough qualifications to ensure the highest quality, lowers quality costs, and eliminates change orders to replace inadequate parts. More focused purchasing can also ensure availability *throughout the life of the product,* thus avoiding expediting, out-of-stocks, missed sales, end-of-life buys, and change orders to substitute more available parts.

With standardization, design teams can concurrently engineer Lean workstations, with each designed around one procedure, one versatile fixture utilizing one standard fastener with one torque setting, one wrench, one gauge, and so forth.

Standardization *improves quality* because procurement will have the *time* and *focus* to (1) search for the best quality parts and (2) qualify them before they can be specified. Keep in mind that high quality *parts* will raise *product* quality, especially in complex products, where product quality degrades exponentially as part count rises (see Section 10.3). These high-quality parts remove that variable from quality assurance and product development efforts.

Standardizing critical parts will focus more thorough efforts to select those parts, evaluate them, and ensure processing will be consistent with process capabilities. This avoids the design practice of "rescuing" designs with unusual parts, which may be hard to get and may complicate supply chains, put quality at risk, and thwart standardization, which is especially important for critical parts.

Finally, standardization, employing poka-yoke (mistake-proofing) ensures less chance of assembling the wrong part, as presented in Section 10.7.

This chapter presents a powerful, yet easy-to-implement standardization procedure, which can reap enough benefits to be well worth the effort as a stand-alone program. A more expanded treatment of standardization can be found in *Build-to-Order & Mass Customization,* which has one chapter on part standardization and another on material variety reduction."[1]

Hereafter, the word "part" also includes *material* and *subassembly*. The phrase "material overhead" includes all these categories.

5.1 PART PROLIFERATION

Competitive pressures are forcing manufacturers to look at all the ways to lower total cost and improve flexibility. One of these thrusts is to investigate and reduce the cost of part and material variety.

An electronics company had 1,500 different types of resistors, including 120 different kinds, sizes, and tolerances of 1,000 ohm resistors.[2] That company was able to reduce the 1,500 resistor types to less than 200 actively used for manufacturing.

One of the author's clients that made consumer products discovered that it was using the following numbers of different part types: 1,248 wire assemblies, 152 motors, 151 screws, 74 switches, 67 relays, 65 capacitors, 37 valves, 16 transformers, 62 types of tape, and 1,399 different "standard" labels.

Every company has similar "horror stories" with regard to part proliferation. One might ask, *why?*—especially considering the fact that part proliferation raises part and overhead cost, impedes flexibility, and, as will be shown later, is really unnecessary.

5.2 THE COST OF PART PROLIFERATION

Part proliferation is expensive. A Tektronix study determined that half of all overhead costs related in some way to the number of different part numbers handled.[3] Most companies do not even know how much that cost is in dollars. A survey of several Fortune 500 manufacturing companies revealed that *not a single company or division had an accurate estimate of the cost of a part over its lifetime!*[4] According to Venkat Mohan of CADIS, Inc., who markets parts management software, "intuitive estimates range from $5,000 for a standard part to as high as $60,000 or even $100,000 per part for custom parts."[5] James Shepherd, director of research for Advanced Manufacturing Research (AMR), Boston, says that in electronics, the cost of just *entering* new purchased components is between $5,000 and $10,000 per component.[6] The *Ernst & Young Guide to Total Cost Management* states, "It is not surprising that manufacturers have estimated the annual administrative cost of each part number to be $10,000 or more."[7]

In addition to these official materials costs, excessive part proliferation adds cost to field service and manufacturing in important, but rarely measured, ways related to setup, inventory, floor space, lower machinery utilization, and other flexibility issues.

Part proliferation also lowers assembly productivity. In a Wharton Business School report about the automobile industry, Fisher, Jain, and MacDuffie wrote that, "Part variety also appears to have the greatest negative impact on assembly plant productivity."[8]

5.3 WHY PART PROLIFERATION HAPPENS

Part proliferation happens for the following reasons, which are all easily avoidable:

- *Engineers don't understand.* Most product designers do not understand the importance of part standardization and, therefore, do not attempt to design around standard parts. An example of this attitude was discovered when the author was soliciting feedback from engineers on a proposed standardization list (which was generated by techniques described below) for resistors. One electrical engineer commented, *"Why are we standardizing on resistors? Aren't they cheap and aren't they in the computer?"* What this engineer did not realize is that regardless of a part's cost and the company's ordering and tracking sophistication, *every* part must be physically delivered to the plant, possibly inspected and warehoused, and then distributed to each point of use. One solution to this perception problem would be training and education that stresses the importance of part standardization and connects it to corporate goals.
- *"Not invented here."* Sometimes standardization is resisted because of the not-invented-here syndrome. That mind-set can be countered, however, by teamwork, training, and encouraging engineers to make the best choices for the product, for the customer, and for the company.
- *Arbitrary decisions.* Product designers make many arbitrary decisions when specifying parts. They may arbitrarily specify a fine-pitch

$^{5}\!/_{16}$ bolt with a button head that is $^{7}\!/_{16}''$ long when a more common course-pitch $^{3}\!/_{8}''$ hex bolt that is $^{1}\!/_{2}''$ long could have done the job just as well. Section 1.8 discusses the general problem of arbitrary decisions in product development.

Electronic engineers at Intel's Systems Group said that for digital circuitry, they did not really need any resistor values between 1,000 and 2,000 ohms. From this feedback, those values were immediately deleted from the approved parts list for new designs.

- *Many versions of the same part to "save cost."* When product designers are pressured to lower cost, and all that is measured is part cost, they sometimes specify the cheapest version of a part for each application. This may appear to lower the part cost by a few pennies, but such a proliferation will generate much greater overhead costs. For example, one manufacturer of medical equipment had 11 different versions of 1K resistors on a single circuit board! The engineers thought they were saving cost by specifying the cheapest tolerance and wattage for each application. However, not only did this practice cause rampant part proliferation, it also forced the board to be run through the assembly machine *twice* because there were more parts than bins available on the assembly machine. The standardization principles presented herein encourage using the best single version for all applications.

- *The minimum weight fallacy.* A phenomenon that may be contributing to part proliferation is the fallacy that all parts have to be sized "just right"—to have the minimum weight and be made of the minimum amount of materials. Engineers may resist standardization because the standard part may be the next larger size to ensure adequate strength and functionality. The following rules of thumb may help guide engineers get past this obstacle:

  If it doesn't fly or accelerate quickly, use the next larger size standard part.
  If it isn't made of precious metals, use the next larger size standard part.

- *Qualifying part families.* A related cause of part proliferation is the practice of *qualifying* (for entry on approved parts lists) entire families of parts, such as fasteners, resistors, and capacitors. Intel's Systems Group discovered that out of 20,000 approved parts for printed circuit boards and computer systems, 7,000 had never been used! In other words, more than one-third of the approved parts were not used on any product. And, yet, any engineer could have arbitrarily chosen one of those unused parts and entered a new part into the system without any approval or authorization. In this case, those unused parts were immediately deleted from the approved parts list for new designs.

- *Contract manufacturing.* Sometimes, a shortsighted business strategy undermines standardization efforts. Some companies, in response to downturns, respond to the pressure to "fill the factory" by, in essence, becoming a contract manufacturer. This use-it-or-lose-it approach may bring in some additional revenue, but it often loses money in the long run. This loss would be understood if computations included total cost issues such as the overhead expense to support such part diversity and, of course, the substantial learning curve expense needed to gear up to build many new products.

- *Mergers and acquisitions.* Another cause of excessive internal variety is the merger of dissimilar products through corporate mergers and the acquisition of companies, products, patents, and so forth. Products that originated in different companies are likely to have very different parts and processes. Thus, manufacturing flexibility should be added to the list of primary factors that determine such decisions.

- *Duplicate parts.* When product designers do not know what parts exist, they will often add a "new" part to the database, *even when the identical part already exists.* Even if they suspect that the needed part exists, they will probably specify a new purchased part if it takes less time than finding an existing one. Similarly, they will probably design a new part if it takes less time than finding an existing one. Typically, engineers find it much harder to search through awkward databases than to design new parts or purchase them.

Many companies have hundreds of incidents of duplicates, triplicates, or even several versions of the exact same part, existing under different company part numbers, plus even more situations where a close existing part could have been used instead of introducing a new part.

Upon the author's suggestion, one aerospace company investigated this and found that it had 900 different types of spacers! Apparently, it was easier to design number 901 than to search through all 900 existing spacers. So, every time engineers needed a spacer, *they just designed a new one!* The fallacy in their thinking was that spacer design appeared to be "easy," but, in reality, the documentation, procurement, manufacture, storage, and distribution of 900 spacers were not "easy."

When a large machine tool company investigated this phenomenon, it discovered that it had 521 very similar gears. They eventually reclassified all those gears into 30 standard gears.[9] There had been 17 times more gear types than necessary! Every gear had an average of 17 duplicates.

5.4 RESULTS OF PART PROLIFERATION

The net result of part proliferation is that most companies have thousands or even hundreds of thousands of different part types (unique part numbers). Such internal variety is rarely necessary and is usually the result of careless—often rampant—proliferation of parts. The absence of any standardization goals or awareness allows designers to simply choose new parts for new designs, without any consideration of prior usage of similar parts.

Every company can investigate the extent of part proliferation by simply looking up the total number of active part numbers for all part categories. In many cases, the proliferation will seem obvious, even to the most casual observer. Another revealing investigation would be to summarize the materials budget for all the overhead expenses related to parts. Hopefully, these investigations will provide the motivation to eliminate existing duplicate parts and to substantially reduce part types for new designs using the very effective procedures presented here.

5.5 PART STANDARDIZATION STRATEGY

5.5.1 New Products

Part standardization presents the greatest opportunities for new designs or redesigns. Usually it would cost too much to convert existing designs, part by part, and there may be issues with replacement part backwards compatibility. In addition, changing existing designs may be precluded by current qualifications of existing products.

5.5.2 Existing Products

The main opportunities for existing products are for "better-than" substitutions. If this is an opportunity, then the standard parts and materials lists would focus on the better grades. In most cases, the overhead savings and supply chain benefits resulting from standardization would far exceed any perceived cost increase for better materials.

As new products with standard parts and materials are phased in, manufacturing operations and procurement will realize more benefits. When the products with the standard parts and materials reach a certain critical-mass threshold, the factory will benefit from eliminating the remaining unusual parts and materials. At this point, the old products can be redesigned around standard parts, outsourced out of the plant, or rationalized away, using the procedures of Appendix A.

5.6 EARLY STANDARDIZATION STEPS

5.6.1 List Existing Parts

Many parts lend themselves to listing in a logical order, as examples shown in Figure 5.1. For these parts, simply list all existing parts in order, circulate the lists to the design community, and encourage engineers to use existing parts whenever possible. This step should be done immediately to at least stop designers from adding new parts when they could

Part Type	Listing Order
Threaded fasteners	Diameter, pitch, length, head type, grade
Washers/spacers	O.D., I.D., thickness, material, finish
Gears	Pitch, number of teeth, face width, material
Gearboxes	Ratio, horsepower, shaft orientations, shaft diameters
Motors	Horsepower, voltage, phase, shaft diameter, mount
Pumps	Pressure, flow rate
Power supplies	Output voltage, wattage
Resistors	Ohms
Capacitors	Microfarads

FIGURE 5.1
Examples of part type listing orders.

use an existing one. A procedure will be presented below for determining preferred part values from these lists.

5.6.2 Clean Up Database Nomenclature

It may be necessary to clean up part and material databases before proceeding further. If every part in a category does not have consistent nomenclature, which is often the case for threaded fasteners, it may be hard to sort and identify duplicates. So it may be necessary to convert multiple labels to the most common or most logical label. If it is too hard to convert the official database, then extract the information and change it in a database used for the standardization effort.

5.6.3 Eliminate Approved but Unused Parts

Many parts are approved in families and entered into the approved parts list *en masse*. Many of these parts have never been used. If they are left on the list, any engineer could add a new part into manufacturing without approval or authorization. So the first step would be to identify approved parts that have never been used and immediately remove them from the list for new designs.

5.6.4 Eliminate Parts Not Used Recently

It would be logical to assume that parts that have not been used recently would not be good candidates for any standardization list for new designs. For the standardization effort, eliminate for consideration any part or material not ordered in the last few years—this could be two to five years, depending on the product life cycles. This would not eliminate these parts from the approved parts lists, which may be needed for spare parts or infrequently built products; however, operational flexibility can be improved if these unusual products are *rationalized* to eliminate or outsource infrequently built parts and spare part production (see Appendix A).

5.6.5 Eliminate Duplicate Parts

The problem with multiple part numbers for the same part goes beyond the obvious extra material overhead cost of carrying extra parts. Most likely, the similar parts would be ordered separately for each product that needed them. This would prevent purchasing from obtaining quantity discounts and just-in-time deliveries that would be possible with a consolidated order. Further, the smaller order quantities increase the chances of shortages for a given part. Ironically, the missing part that delays production might be sitting in another bin under a different part number.

The problem becomes more severe for Lean Production, build-to-order, or mass customization if different products are using the same part under different part numbers. Automated assembly equipment, such as for printed circuit board assembly, may have to load the same part in multiple bins. Even if the machine operator notices that some of the parts seem similar, the operator does not have the time nor the authorization to consolidate parts on the spot. This duplication alone may prevent flexible operations if there are more parts (including duplicates) in the product family than there are bins in the equipment. If this is the case, then parts would have to be reloaded twice for each product, which is the type of setup that must be eliminated for flexible operations.

Using part management software, Tektronix was able to deactivate 32,000 part numbers from an active base of 150,000. Bob Vance, Tektronix VP and chief information officer, summarized the return on eliminating excess parts: "There are few areas where a manufacturing company can make such a significant impact to its bottom line with so little effort. We want to invest our resources in product innovation and customer services, not carrying an overburdened parts inventory."[10]

An easy way to stop the introduction of duplicate parts is to make it easier to find existing parts than to release new ones, as shown in Figure 5.1. Then, the duplicate parts should be eliminated and consolidated into a shorter list of unique parts.

5.6.6 Prioritize Opportunities

Because the standardization procedure presented next must be performed for every category of parts and materials, it is important to prioritize opportunities and start with categories with the following characteristics:

- Unnecessary proliferation, which is typical for fasteners, resistors, certain raw materials, and so forth
- Excessive parts/materials inventories
- Excessive material overhead costs to procure a proliferated variety

- Missed opportunities for automatic resupply, such as kanban and breadtruck, because of the excessive variety
- Kitting parts required because there are too many parts to distribute at all points of use
- Excessive setup changes if there are more part types than bins on automatic assembly machines; for instance, for printed circuit boards
- Production delays caused by shortages
- Excessive expediting

In addition to these criteria, valuable insight can be gained simply by asking employees to rank the categories. Employees can be given a list of all major categories and be asked to vote.

5.7 ZERO-BASED APPROACH

There is an easy-to-apply approach that is more effective than *part type reduction* measures, which require tremendous efforts for their return. Reducing active part numbers, say from 20,000 to 15,000 will, in fact, lower material overhead somewhat, but it may not reach the threshold (eliminating part-related setup) that would enable the plant to build products flexibly without delays and setups to get the parts, kit the parts, or change the part bins.

Instead, the most effective technique to reduce the number of different parts (part types) would be to standardize on certain *preferred parts*. This usually applies to purchased parts but it could also apply to manufactured parts and raw materials. The methodology is based on a zero-based principle that asks the simple question: *"What is the minimum list of part types we need to design new products?"*

Answering this question can be made easier by assuming that the company, or a new competitor, has just entered this product line and is deciding which parts will be needed for a whole new product line. One of the advantages of new competitors is the ability to start fresh, without the old "baggage": too many parts. Just imagine that a competitor simultaneously designed the entire product line around standard parts. Now, imagine doing the same thing internally. This is called the *zero-based approach*.

The zero-based approach, literally, starts at zero and adds only what is needed, as opposed to reducing parts from an overwhelming list. An analogous situation would be cleaning out the most cluttered drawer in a desk, a purse, or a glove compartment: removing unwanted pieces would take great·effort and still not be very effective. The more effective zero-based approach would be to empty everything from the cluttered drawer into a box labeled "garbage" and add back to the drawer only the items that are essential. The difference in these approaches is where the clutter ends up: still in the drawer, purse, or glove compartment or in the garbage can. Similarly, parts reduction efforts have to work hard to remove the clutter (excess part variety) in the system, whereas the zero-based approach excludes the clutter from the beginning. The clutter is the unnecessary part proliferation that would have not been needed if products were designed around standard parts. Not only do these excess parts incur overhead costs to administer them, they also lower plant efficiency and machine utilization because of the setup caused by too many parts to distribute at every point of use.

This approach determines the minimum list of parts needed for new designs and is not intended to eliminate parts used on existing products, except when the standard parts are functionally equivalent in all respects. In this case, the new standard part may be substituted as an equivalent part or a better-than substitution, where a standard part with a better tolerance can replace its lesser counterpart in existing products.

Even if part standardization efforts apply only to new products, remember that in these days of rapid product obsolescence and short product life cycles, all older products may be phased out in only a few years, especially when companies rationalize away their fading products before they become a drain on resources (see Appendix A).

5.8 STANDARD PART LIST GENERATION

To determine a standard parts list, the company must achieve a consensus on the minimum number of parts necessary to design new products. It is important that all engineering groups, in particular, agree on the list because they will be the ones that will be designing products using the standard parts. Consensus can be reached by forming a team of key representatives from engineering, purchasing, quality, and manufacturing departments.

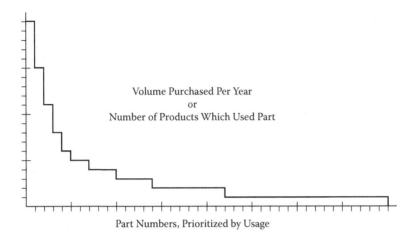

Part Numbers, Prioritized by Usage

FIGURE 5.2
Pareto chart of existing part usage.

The steps are listed as follows for each category of parts:

1. *Determine usage histories.* For each category of parts, start with a baseline list based on usage history of existing products. It would be counterproductive to create a standard parts list by adding even more new parts. Usage history can be based on total quantity of all parts purchased per year or the number of products that use the parts (which can found by generating a "where used" report from MRP systems). Exclude parts used in products that are near the end of their lives and should be phased out soon anyway. The Pareto plot would have a shape like Figure 5.2, but real-life plots usually have a much steeper shape near the vertical axis.

2. *Establish baseline list.* The next step is to establish a baseline list that will be based on the inherent commonality of the existing parts plotted above. Obviously, the "high runners" on the left of the curves should be part of the baseline list and the low-usage parts on the right should not. Where to draw the line (that separates common parts from the rest) will require some judgment. This judgment should come from a consensus of all engineering groups, manufacturing, purchasing, and those who qualify parts and suppliers (quality or materials engineering departments). Quality people should be interested because minimizing the number of part types has a positive impact on procurement and quality activities, because fewer part

types will mean that more attention can be focused on procurement, supplier qualifications, and quality issues for the standard parts.

If the usage profile looked like Figure 5.2, the left third or forth of the parts might qualify for the baseline list. Note that, when parts proliferation has been more rampant, the low-usage and single-use parts would extend much further to the right. In such a case, the standard parts may be only the top 5% or 10% of the parts.

In some cases, there may be anomalies, such as parts used in low volume but in many products. The widespread use, in itself, may qualify that part for the list, despite the low volume. In other cases, a part may be used on only one product but in high volume. This may require some investigation to determine if the design team has discovered a clever use for the part that may apply to future product designs.

These parts become the baseline list, which would then be arranged in some appropriate order, as discussed in Section 5.6.

3. *Develop procedures to add new parts.* Proactively develop procedures to add new parts (for instance, new-generation parts) to the baseline list. Ideally, this should be done by a material qualifying group, sometimes called "materials engineering," who should also qualify the new parts and their suppliers. If this is not possible, new part additions should be done by consensus of researchers and engineers from advanced design and R&D groups, manufacturing, purchasing, and quality.

4. *Consolidate duplicates.* As discussed before, many duplicates get into the system, and good parts management can eliminate the exact duplicates. When companies start investigating duplications, they often discover slight differences between the duplicate parts. This presents an opportunity to select the part that could most likely replace the others. This may involve choosing the "better" part, using the procedure discussed next.

5. *Consolidate parallel lines of parts.* If whole families of parts are available in multiple tolerances, quality levels, thread pitches, material finishes, or purity levels, the team should consolidate them into one set of parts, even if it has to standardize on the more expensive parts. Usually, any increase in part cost, due to using the "better" parts, will be dwarfed by the overhead savings from having fewer parts overall. If the company cost accounting system cannot quantify the value of this, then the company must recognize the

qualitative value to manufacturing flexibility and lowering material overhead costs (see Section 5.11, "Standardization of Expensive Parts"). Various grades of bolt strength or resistor tolerance could be consolidated, as indicated by the following example.

When the author was initiating a parts standardization program at Intel's Systems Group, there were two completely different families of resistors: 5% (tolerance) carbon resistors and 1% metal film resistors. Using this approach, they were consolidated into one line of exclusively 1% resistors, thus eliminating hundreds of part numbers for new designs. These parts were also substituted in existing products because it was considered a "better-than" substitution. A few years later in a public DFM seminar, one company reported that the purchasing leverage resulting from making all resistor purchases the same tolerance (1%) resulted in a net cost savings and each application in the factory got a better resistor.

Note that consolidation based on higher quality parts may, in reality, raise product quality, even if the lower quality parts were theoretically adequate.

6. *Optimize availability.* The availability and sourcing of all parts and materials on the standard parts list should be investigated and optimized. Standard parts and materials should be selected to be readily available from multiple sources *and be available over the expected life of the products.* Investigations should include the number of sources, the technical and business strengths of the sources, and the average amount available at any time.

7. *Structure lists.* Structure the list into appropriate order, as shown in Figure 5.1, by values of diameter, pitch, power, flow rate, voltage, ohms, microfarads, and so forth.

8. *Review the lists.* Review the tentative baseline lists for each type of part and feature by involving representatives of all relevant engineering departments for feedback and approval. This could be a formal process that is part of the procedures to generate standard parts lists. Or, it could be an informal process that would solicit informal feedback from some experienced engineers who could be assumed to be representative of their departments. Earlier participation of representatives of these departments on the task force should minimize surprises at this stage.

9. *Circulate the lists.* Circulate the tentative lists to all engineers with an explanation of why the standardization is important to company

goals to simplify product development efforts, lower part cost, reduce material overhead costs, simplify supply chain management, improve availability and deliveries, raise quality, improve serviceability, and, if applicable, support Lean Production, build-to-order, or mass customization. Solicit feedback about whether the tentative baseline list has the right parts for new designs. Query reviewers if any part on the list is superfluous or if any important part was wrongly omitted.

10. *Finalize the lists.* Review feedback from all those reviewing the lists. Investigate promising suggestions and add or subtract appropriate items to or from the lists. Finalize the lists and prepare for implementation.

11. *Determine scope of implementation.* The scope of the standardization effort should be matched to implementation resources and the general company awareness of the importance and value of designing around standard parts. Some companies may choose to start with the "low hanging fruit" first (e.g., fasteners or resistors). Success here may then be leveraged to other types of parts. As companies embrace operational flexibility, it becomes imperative to implement strong standardization efforts.

12. *Educate the design community.* Before standard parts lists are issued, the design community needs to be educated on the importance of using standard parts in new designs. Point out how important this is to manufacturing flexibility and lowering overhead costs. Another educational and motivational technique is the "embarrassing statistics" technique: reveal the scope of past part proliferations and discuss their causes, which are usually designers who arbitrarily chose a low-usage part when a high-usage part would have worked.

    Design engineers need to realize that *no matter how simple a part appears, every part number incurs a material overhead burden* to document, procure, store, distribute, resupply, and, most significantly, to manufacture in low volume.

13. *Determine the strictness of adherence.* Generally, adherence from 90% to 95% should provide significant benefits without inhibiting design freedom or generating resistance to standardization. For instance, General Electric Lighting has established a goal of 90% parts reuse in all new designs.[11]

    Stricter adherence to the standard parts list may be necessary for manufacturing flexibility, automation utilization, overhead cost reduction, and ease of service. A high-volume, flexible plant with expensive automation might require 100% adherence to the standard parts list so that the equipment would not have to stop to load nonstandard parts. Lean Production, build-to-order, and mass customization environments may require 100% adherence for manufacturing flexibility. Companies contemplating adherence less than 100% need to analyze the effect of this *"un*commonality" on the flexibility of their operations.

14. *Issue the standardization lists.* Designate the parts on the standard lists officially as *standard, common,* or *preferred* parts and give them special emphasis, with at least an asterisk or bold type on the larger approved part lists. A more effective method is to present the preferred standard parts on a separate list, perhaps in the front of the section containing that category of parts—on paper, on a company database, or on an internal web page.

Intel's Systems Group presented preferred parts on gold paper, followed by the existing approved parts list on white paper. When standard parts are to be used exclusively, the standard parts list would be the only parts list issued to engineers.

5.9 PART STANDARDIZATION RESULTS

This part standardization approach was implemented by the author at Intel's Systems Group. Starting with 20,000 parts for printed circuit boards and computers, this standardization approach generated a preferred parts list of 500 parts! For resistors, capacitors, and diodes, 2,000 values were reduced to 35 values, one set for leaded axials, and another set for the surface-mount equivalents. Fasteners for computer systems were standardized on one screw!

This is how the standardization process worked: Service wanted a Phillips head so they, and customers, could keep using the same tools. Quality wanted a captivated crest-cup washer to protect surface finishes and yet still have a locking effect. Engineering wanted the 6-32 size screw to be only ¼″ long. Manufacturing recommended that the screw be ⅜″ long so that it would not tumble as it was fed to auto-feed screwdrivers (see Guideline F4 in Section 8.3).

Previous designs had so many different screws that manufacturing could not use their auto-feed screwdriver at all. The next design used the standard screw in 40 locations. This, in addition to the correct screw geometry, made use of the more efficient auto-feed screwdriver practical. In order to feed the screw, it had to be ⅛″ longer, but this meant that the screw would protrude beyond the fastened material. This violated a workmanship standard that prohibited such protrusions; some people even thought the standardization was doomed. But the workmanship standard was modified to allow the protrusion as long as it did not pose a safety hazard or compromise product functionality in any way.

Intel's enforcement goal was not 100%, as might be required for a totally flexible operation, but they felt that even 95% usage would result in significant material overhead savings.

In general, it should be possible to generate a preferred parts list that is 2% to 3% of the proliferated list. For very standard parts, such as fasteners or passive electronic components, it should be possible for the preferred parts list to be less than 1% of the current list.

5.10 RAW MATERIALS STANDARDIZATION

If raw materials can be standardized, then processes can be flexible enough to make different products without any setup to change materials, fixturing mechanisms, or cutting tools. This is an extremely important prerequisite to build-to-order operations, which strive to make *any product on demand from standard raw materials* without having to forecast and order materials. Raw material standardization can apply to bar stock, tubing, sheet metal, molding plastics, casting metals, protective coatings, and programmable chips:

- *Bar stock and tubing.* If raw materials can be standardized on one size of bar stock or one size of tubing, then computer-controlled cut-off machines can be programmed to cut off required lengths from the same stock. This flexibility may determine the feasibility of "cut-to-fit" (dimensional) customization.[12] For manual cutoff operations, material standardization would simplify instructions to the length only. This would minimize the chances for mistakes related to picking the wrong material.
- *Sheet metal.* If sheet metal can be standardized on one shape, thickness, and alloy, then computer-controlled laser cutting machines can

cut all the sheet metal parts needed without changing sheet types. Automatic sheet feeders could reload the machine as needed. This is even more important if the parts are so small that many parts could be cut from the same sheet without having to change sheets. Standardization of sheet metal would allow heavy users to save money by ordering sheet metal in reels, maybe directly from the mill.

Some manufacturing operations can be eliminated by ordering *prefinished material* that is prepainted, preplated, embossed, expanded, anodized, or clad with a different surface alloy. Painting operations for sheet metal can be eliminated by switching to stainless sheet metal. This might be justifiable if the total cost of painting is considered. Prefinished material can be ordered with the finished side protected by adhesive-backed paper that can be peeled off after assembly.

- *Molding and casting.* Part of a flexible operation strategy may involve offering a wide variety of molded or cast parts. Molding and casting operations will be more cost-effective if they can standardize on the same raw material, so that many different parts could be made in the same machine without changing over the equipment for raw materials. It even may be possible to make many different parts in the same mold, thus sharing processing time and tooling expense. Standardizing materials avoids the setup of changing materials and cleaning the equipment. If molds are designed with *fixturing standardization*, they could be changed rapidly to minimize setup time. With casting material standardization, several molds could be filled with the same "pour" from a single "melt."
- *Protective coatings.* Standardizing on protective coatings simplifies processing and makes painting and coating operations more flexible by eliminating the setup to change coating materials and clean equipment. As with parts standardization, coatings could be standardized on the better coating. Even if that coating appears to cost more, the net result would be overall cost savings, considering the process value of this standardization. The logic of this concept is discussed further in Section 5.11, "Standardization of Expensive Parts." Coating standardization could also apply to paint if the purpose of the paint is purely functional, with little aesthetic considerations; for instance, for industrial equipment or inside major appliances. In fact, many industrial and agricultural products enjoy brand recognition because of a standard paint color. For instance, farmers immediately recognize a green tractor or combine as a John Deere product.

- *Programmable chips.* Many integrated circuits (ICs) can be programmed separately or in the product. For programmable chips, standardize on the fewest types of "blanks." This allows the flexibility to program these devices on demand by online programming stations as they are assembled into the product. Ideally, each programming station would be dedicated to one blank device to avoid setup changes and errors. Thus, programmable chip standardization minimizes the number of programming stations and may make it possible to program chips as they are inserted or placed onto circuit boards.
- *Standardization for linear materials.* Standardization can also be applied to material purchased by length: wire, rope, plastic tubing, cable, chain, and so forth. Linear material variety can be reduced in the following ways:
  - *Cut as needed:* Linear materials can be standardized by type, with the length cut as needed. Available equipment will even cut and strip the ends of wires. This approach would require a dispensing machine at each point of use.
  - *Kanban system:* Alternatively, wire and tubing can be cut ahead of time, but, to keep overhead low, not be given individual part numbers. The following system was developed by MKS Instruments, Inc., makers of vacuum and flow measurement and control instrumentation. Instead of issuing part numbers and treating many cut lengths as many different parts, a predetermined amount (for instance, one day's worth) of wire or tubing can be cut to the needed lengths and placed in each of two adjacent kanban bins at each point of use (see Figure 4.2). The kanban bins are arranged one in front of the other, as shown in Figure 4.1. When the front bin is emptied, it is sent to a central dispensing machine. The label on the bin tells the machine operator the material type, length, and quantity and the bin's return destination in the factory. After the bin is filled, it is returned to the point of use and placed behind its counterpart so it can be moved forward and used when the other bin is emptied. This simple approach can eliminate hundreds or thousands of part numbers from the plant, thus lowering costs, minimizing delays, improving flexibility, and encouraging flexible operations.
  - *Printing while dispensing:* The number of types of linear materials can be reduced further by using equipment that prints on wire and tubing as it is dispensed, so that many various colors of the

material need not be ordered and stocked. Printing on wire and tubing can minimize mistakes in assembly and service, because workers do not have to memorize or refer to color codes. Using words rather than color also solves problems associated with color blindness. Examples of printing opportunities: Ground, +12 volts, Supply, Return, 100 psi, and so forth. For international products, these codes can be printed in multiple languages, which may reduce internal variety due to labeling differences. Such printing and dispensing equipment could be utilized either at each point of use or at a central location to feed kanban bins.

5.11 STANDARDIZATION OF EXPENSIVE PARTS

Usually, standardization programs for inexpensive parts, like fasteners, do not meet serious resistance, because the standard parts are perceived to cost no more than nonstandard parts (in reality, they cost less). But, as the cost of the parts increases, standardization efforts confront more resistance because of the perception that specifying the next larger (and more expensive) standard component would cost more than one that just satisfies its requirements in the product. However, consideration of the total cost savings of standardization can encourage its use even for expensive components.

The following experience illustrates the resistance and the opportunities involved in the standardization of expensive parts. While training a company that manufactured heating, ventilating, and air conditioning (HVAC) equipment, the author discovered the company used 152 different types of motors. When he challenged the designers, they insisted they needed every size to specify "just the right" motor for every application. Then we asked their supplier, GE Consumer Motor Division, "What would be the savings if those 152 motors could be reduced to five or ten?" The one word answer was *"Massive!"* Why? Because each of those five or ten motors would be ordered in volumes that would be ten times their current order volumes, thus resulting in greater economies of scale. Further, the five or ten chosen would have been the most cost-effective in their line—the best-designed motors that they produce in high volume for other customers too.

The upper graph in Figure 5.3 shows the apparent implication that the standard parts would cost more than parts that have "just enough"

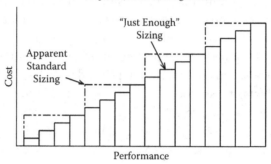

Component Cost Step Functions
With Independent Purchasing Decisions

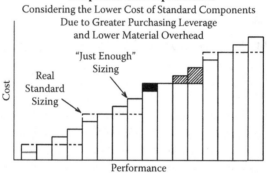

Component Cost Step Functions
Considering the Lower Cost of Standard Components
Due to Greater Purchasing Leverage
and Lower Material Overhead

FIGURE 5.3
Standardization of expensive parts.

performance for a given product. However, if all company products use company standard parts, the cost of those parts will be less, due to purchasing leverage and material overhead savings. Thus, there would be a net company savings for expensive parts, as shown in the lower graph in Figure 5.3.

Some products may be forced to use a more expensive part than is required (as shown for one size by black shading), but most products would be able to use less-expensive standard parts (as shown by the crosshatch shading). The result is a net cost savings for the company plus the flexibility that is essential for Lean Production, build-to-order, and mass customization.

Further, when a standardization task force standardizes on expensive parts, it can select the most cost-effective parts made in large enough quantities to take advantage of suppliers' economies of scale. These parts are usually the ones that have better availability and are more likely to have better quality and reliability. Design engineers, however, may not be aware of these nonlinear relationships between performance, price, availability, and quality.

5.12 CONSOLIDATION OF INFLEXIBLE PARTS

The logic of Section 5.11 can be applied to the *consolidation of inflexible parts*. Inflexible raw parts, such as castings, moldings, stampings, extrusions, wiring harnesses, bare circuit boards, and so forth, can be consolidated into versatile common parts that can be used on many products. The typical problem with these parts is that their processes are inherently inflexible, with lengthy setups, so they are normally built in batches, which complicates supply chain management, disrupts operations with frequent setup changes, and incurs inventory carrying costs.

Several different raw parts can be consolidated in a single versatile part with enough extra metal, features, functions, mounts, or circuitry to be useful in many applications. If this type of consolidation is done throughout the product line, substantial reduction in raw part variety can occur. This will result in greater purchasing leverage, less dependence on forecasting (with steady flows possible), less storage space needed, and simpler supply chains and internal logistics.

Figure 5.4 tabulates the costs added and costs saved from consolidation efforts that combine "N" parts into one part.

For example, water meter and gas regulator castings can be consolidated, with extra metal included on all castings for options such as test ports. Similar logic applies to plastic moldings: all the features for many products can be molded into a few versatile plastic parts. This will have all the above advantages *plus* saving the cost of multiple molds.

Versatile bare printed circuit boards could contain all the traces and holes or pads needed for several products, instead of the common practice of designing a different bare board for every product or variation. Then, steady flows of the versatile bare boards could be arranged, knowing that they will be used one way or another. Widespread use of expensive or long-lead-time parts can allow projects to borrow from each other in emergencies. Automatic resupply techniques can be set up for these parts, allowing them to be built in batches (Section 4.2). Material overhead and floor space requirements will be less for fewer types of parts and materials.

Cost added:
- Extra cost per part on some of the parts for extra material, wire, circuitry, etc.
- One-time cost to implement the consolidation to existing products (does not apply to new product developments)

Cost saved:
- Economy of scale savings (purchasing leverage) for ordering N times the volume at 1/Nth of the number of part types
- Tooling cost cut by a factor of N if N fewer parts result from the consolidation
- Material overhead savings on fewer parts
  - Bill-of-materials and MRP expenses
  - Ordering expenses
  - Warehousing/stocking cost for raw materials inventory
- Setup cost saved by not having to set up the eliminated parts; the cost savings include:
  - Setup labor reduction
  - Machinery utilization improvements
- WIP inventory savings from fewer types of part types
- Design cost saved by eliminating N parts in new designs
- Prototyping and debugging cost savings from fewer new parts
- Documentation and administration cost of the eliminated parts for new and existing designs
- Value of fewer work stoppages due to part shortages
- Value of the consolidation's contribution to flexible operations

FIGURE 5.4
Cost trade-offs for part consolidations.

Consolidated parts will be more likely to benefit from economies of scale and arrange deliveries that are more frequent and more reliable.

Printed board assembly machines are CNC machine tools and can be programmed to place or insert unique combinations of components onto versatile standard bare boards. Automotive wiring harnesses are evolving toward this principle, with a single wiring harness used regardless of the number of options ordered. In the old paradigm, many different wiring harnesses were manufactured so that a minimally optioned car would not have extra, unused wires. However, when one considers total costs, the excess variety costs of manufacturing and assembling multiple wiring harnesses greatly exceeds the cost of the unused wires.

It may be hard to justify part consolidations if the cost system does not quantify total cost. The "extra" features will show up immediately and clearly as an "extra cost." However, it may not be possible to quantify the substantial benefits in conventional cost systems. Until a total cost accounting can be implemented, the criteria presented in Figure 5.4 may help itemize the benefits of these consolidations and serve as a basis for quantification or, at least, for educated leap-of-faith decisions.

5.12.1 Custom Silicon Consolidation

Smaller companies may not think they are "in the league" for custom silicon. However, versatile ASICs (application-specific integrated circuits) with widespread use may be viable. Custom silicon can be designed to be versatile enough to be used in a wide range of products, *even if each product uses a small portion of the chip.* The increased volume for the versatile chip spreads out the NRE (nonrecurring engineering) and tooling costs to encourage custom silicon even on small- to medium-volume products. Figure 5.5 compares the traditional analysis for ASICs (which usually discourages their use) and the total cost analysis for widespread use.

5.12.2 VLSI/ASIC Consolidation

Hewlett-Packard offers a wide range of specialized calculators for many niche markets, such as business analysis, engineering calculations, mathematical analysis, and so forth. In the book, *Product Juggernauts*,[13] Arthur D. Little consultants Jean-Philippe Deschamps and P. Ranganath Nayak noted:

> "Viewed in terms of functions offered, these calculators appear substantially different. However, HP has developed common architecture, subsystems, and components to an extreme degree. This strategy often means that higher grade components find their way into low-end products. In these cases, HP spends more than necessary on parts for low-end products in the interest of minimizing component diversity. The payoff comes in the efficiencies gained in production creation across the whole product line."

Nokia designs it cellular phones with common custom silicon to gain economies of scale, spread out the design investment, simplify supply chain management, and eliminate setup changes in manufacturing.[14]

Traditional Analysis for Independent Application

ASICs	Status Quo
High NRE charge/chip	No NRE
High tooling charge/chip	No tooling charges
High cost per ASIC in low volume	Cost per several discrete chips
Lead time	

Analysis for Versatile ASIC with Widespread Application

Low NRE charge/chips	No NRE
Low tooling charge/chip	No tooling charges
Lower cost per ASIC in higher volume	Cost per several discrete chips
Lead time relevant only first time	

Total Cost and Other Implications

Reduces variability = quicker development, more robustness, higher design quality

Higher quality when dozens of discrete chips are replaced by one ASIC. This avoids the cumulative degradation of product quality caused by the quantity and quality of the parts (Figure 10.3).

Higher quality because fewer parts are left that could be replaced by lower quality versions

Eliminate the quality costs incurred when too many components are crowded onto PC boards

Better reliability with fewer interconnections

More compact; get these developed now to be ahead of the miniaturization curve

Higher density = fewer circuit boards, maybe one, eliminating card cages

Faster performance coming from shorter signal paths for circuits in ASICs and other chips brought closer together, especially if circuitry can be condensed onto the same circuit board

Faster factory throughput with fewer components and fewer circuit boards

Better equipment utilization and lower equipment charges

Supports a strategy to simplify assembly with a few integrated modules

Versatility can be enhanced with "extra" code embedded for a wide range of variation and customization

Simplified supply chain management

Better flexibility for quicker delivery and build-to-order

Scare off competition with your logo on custom chips

Thwart reverse engineering and better protect IP

Impress customers

FIGURE 5.5
Decisions for ASICs.

5.12.3 Consolidated Power Supply at Hewlett-Packard

HP's staff of operations research PhDs used total cost analysis to prove that money could be saved by utilizing one worldwide power supply for HP laser printers. The part cost of the universal power supply appeared to be higher and was thus resisted by engineers. But when worldwide logistic costs were figured in, they could see that the universal power supply resulted in substantial total cost savings.

5.13 TOOL STANDARDIZATION

A subject related to part standardization is *tool standardization*, which determines how many different tools are required for assembly, alignment, calibration, testing, repair, and service. Tool standardization enhances manufacturing flexibility by eliminating the setup to locate and change tools needed in the manufacturing process. If adjustments are to be provided by dealers or users, ideally no tools should be required. But if tools are required, the product should be designed around standard tools that are easy to use and would be available to dealers or users. A single tool that performs all repairs and adjustments could be supplied with the product.

Some designs may require several lengths of screws, but if they have the same head geometry, then one screwdriver could be used for all of them. Tool standardization becomes even more important if service people have to be mobile or have to perform service in awkward situations, such as clean rooms, crawl spaces, catwalks, utility poles, under water, in "space walks," and so forth. Tool standardization can also help minimize the expense of providing repair toolkits that usually come with complex products to enable the user to perform more repairs.

Tool standardization should be based on standard, readily available tools. Often, special tools are required because tool specification was not part of the design process. Sometimes special tools are required because tool access was not designed into the product.

Company-wide tool standardization can be determined as follows: First, analyze tools used for existing products. Prioritize usage histories to determine the most common of existing tools. Work with people in manufacturing and service, in addition to dealers and users, if appropriate, to determine tool preferences. Coordinate standard tool selection with standard part selection. Issue standard tool lists with standard parts lists.

5.14 FEATURE STANDARDIZATION

Features such as drilled holes, reamed holes, punched holes, and sheet metal bend radii all require special tools, such as drills, reams, hole punch dies, and bending mandrels. Unless there is a dedicated machine for each tool, the tools will have to be changed, and this will result in a setup change for every tool changed. Exceptions would be machines with automatic tool changing capabilities, but they are limited in the number of tools they can store, and that tool selection must be broad enough for all parts built in a flexible family.

Ironically, most sheet metal bends do not need to be any specific value within a reasonable range. But designers must enter a bend radius value to complete the drawing. Unfortunately, most designers specify an arbitrary bend radius, which often requires the shop to locate and change mandrels to bend the sheet metal to arbitrary radii. In the worse case, a special tool would have to be fabricated for an arbitrary decision! Using feature standardization, designers would use the shop's most common bend radii. At an in-house DFM seminar[15] at Hewlett-Packard, where key vendors had been invited, the sheet-metal vendor stood up and said he could generate only four bend radii for bending sheet metal. And then he identified the *one* mandrel that was usually on the bend brake, so if designers used that bend radius, there would be fewer setup delays and cost.

When designing parts for milling, designers should specify geometries for the whole product family so that a single cutter or "end mill" may be used. This ensures that all parts in the family can be milled without setup changes and, thus, ensure flexibility and high machine tool utilization on expensive equipment.

To implement feature standardization, standardize features around standard production tools, making sure not to exceed the tool storage capacity. Investigate the tools used by the plant and by key outside vendors (whether currently used or not). A safe approach is to choose only features that can be easily built by all (or at least most) potential production facilities and vendors. Based on the production tool availability and capabilities, compile a feature list and issue it with the standard parts lists, hand tools, and raw materials.

5.15 PROCESS STANDARDIZATION

Standardization of processes results from the concurrent engineering of products and processes to ensure that the processes are actually specified by the design team, rather than being left to chance or "to be determined later." Processes must be coordinated and common enough to ensure that all parts and products in the product families can be built without the setup changes that would undermine flexible manufacturing.

One of the processes affected by part standardization is mechanized screw fastening. The auto-feed screwdriver is a very cost-effective mechanized tool that orients and feeds a screw and then blows it down a tube to the screwdriver head, where it waits for the operator to activate the power drive by pushing down on the screwdriver handle (see Guideline F4 in Section 8.3). A preset torque-limited mechanism makes sure the screw is fastened consistently. This useful production tool can feed any style of screw (machine threads, self-tapping, and so forth), but *only one size and type at a time*. Changing screw sizes is possible but would cause too much of a setup to be used in flexible operations. Thus, auto-feed screwdrivers can be utilized effectively only if there is fastener commonality.

Another concurrent engineering issue is that the screw specified must be longer than it is wide so that it will not tumble as it is blown down the feed tube. Each manufacturer has detailed guidelines for specifying these dimensions.

Similar devices, based on the same principle, can fasten screws automatically when mounted on robots or on assembly mechanisms specifically designed to dispense screws, such as those used in Hewlett-Packard's DeskJet factory in Vancouver, Washington.

5.16 ENCOURAGING STANDARDIZATION

Given the importance of standardization for cost, supply chain simplification, and optimizing current *or future* flexible operations, it is imperative that manufacturing companies encourage standardization implementation as early as possible. This really means encouraging design engineers to design around standard parts (even ones that might appear too expensive), specify standard design features, select standard tools, base designs on standard materials, and concurrently design products to be built on standard processes. The author's experience indicates that design engineers are not naturally committed to these goals. In fact, engineers may be pushed the other way by poorly conceived metrics, such as emphasizing part cost and low bidding on individual projects instead of using standardization to minimize total costs on all product families and all part-related expenses.

The following steps can be taken to encourage standardization. They involve "discounting" material overhead rates for standard parts, prequalifying standard parts, making samples and specifications of standard parts readily available, and emphasizing total cost thinking, preferably incorporated into total cost accounting systems. In addition to these procedural steps, managers should take every opportunity to emphasize the importance of standardization in goals, policies, directives, "pep talks," and training.

1. *Material overhead rate.* The procurement of standard parts and their distribution through the plant will incur less overhead burden than the usual excessive internal variety. Therefore, the material overhead rate for standard parts should be lower, to reflect the lower actual overhead. In addition to being a more accurate reflection of overhead costs, lower material overhead rates for standard parts should motivate engineers to specify standard parts.

    This is a logical approach because standard parts really do consume less overhead expense, for the reasons pointed out throughout this chapter. In order to compensate for a lower material overhead rate for standard parts, the general material overhead rate may have to be raised from the previous single rate. It is also logical to assign a higher overhead rate to low-usage parts because of their higher overhead demands. Thus, if engineers choose standard parts, their design will be "rewarded" with the lower material overhead rate. Conversely, if they choose nonstandard parts, the overhead rate will be higher than even the previous single rate. Section 6.16 shows that this difference could often be *10:1*, so that standard parts should have one-tenth the overhead of oddball parts. Consequently, the only "cost" that designers should be shown is the part cost *plus this material overhead charge*, which would steer them to the standard part.

    Another method of establishing overhead rates for standard parts would be a variable rate that would be inversely proportional to volume, so that a very high-usage part would have a very low material overhead rate and a very low-volume part would have a much higher overhead rate. This approach was used by the Portable Instruments Division of Tektronix as a "cost driver" to discourage engineers from using low-volume parts.[16]

2. *Prequalified standard parts.* Companies using standard parts will use many fewer types of parts for new designs than without standardization. Thus, these standard parts can be more thoroughly evaluated and their suppliers can be more thoroughly scrutinized than is possible with many times the number of parts in the system. The standard parts can be prequalified for immediate use. This can accelerate product development, because design teams would not have to wait for this qualification.

3. *Floor stock.* Usually, the list of standard parts is small enough to allow a floor stock to be kept in the engineering area, so that engineers always have samples of the standard parts available for evaluation. Having floor stock samples can help the design team visualize concepts based on standard parts and, thus, encourage standard part usage. Floor stocks can make standard parts readily available for engineers to evaluate parts, conduct experiments, and build breadboards. Floor stocks can also be mounted on display boards in prominent places near the design team.

4. *Personal display boards.* For small, inexpensive parts, like fasteners, every engineer could be issued a personal display board with the standard parts mounted with labels that list the generic value plus the company part number. Hewlett-Packard, at the author's recommendation, followed the above procedure and reduced the number of fasteners for large-format plotters from dozens to only seven. Samples of these seven fasteners were mounted on an aluminum plate, with values and part numbers printed on paper that was affixed to the plate. These personal display boards were issued to all engineers to encourage them to use the standard fasteners.

5. *Spec books.* Part specifications for standard parts can be reproduced and compiled into a single "spec book" or an easily accessible electronic form. This would encourage engineers to use the standard parts because they can see all specifications in a single reference.

6. *Cost metrics.* If cost metrics are based on total cost for product families, they will encourage standardization, especially when material overhead is included in what engineers see when they look up "part cost." If accounting systems cannot quantify total cost, then engi-

neers should be encouraged to balance the costs that are reported with the qualitative benefits of standardization, as addressed earlier.

7. *Encourage designers to use standard parts and materials* through training, directives, policies, and procedures.

8. *Determine targets for compliance.* Usually, standardization percent usages in the 90s can offer significant results. Automated or flexible factories may have stricter criteria.

9. *Avoid actions that are counterproductive to standardization:*
   - *Avoid changing materials for "deals."* Purchasing departments should avoid changing to different materials because of perceived purchase cost savings. The value of flexibility—or the cost of more material variety—would far exceed any perceived purchase cost savings.
   - *Don't merge dissimilar products in the same plant from mergers or acquisitions.* If so, acquisition expenses should include a budget for the efforts to integrate the products and add or convert the additional parts, so that this does not drain resources from product development or standardization efforts.

5.17 REUSING DESIGNS, PARTS, AND MODULES

How to avoid reinventing the wheel

The most obvious way to make parts more common is to use parts that already are in production. In addition to the automatic standardization benefits, existing parts have gone through the learning curve and have been debugged and stabilized in production. Thus, products designed around existing parts will have fewer introduction problems.

A key quality principle from Chapter 10 is reusing proven designs, parts, modules, and processes to minimize risk and ensure quality, especially on critical aspects of the design (Guidelines Q20 and Q21). This was one of the *design strategies* presented in Section 3.1.

For product designers, one benefit of using existing parts is the time saved from not having to "reinvent the wheel." Before using previously designed parts, the designer should check whether they are still in production, how easy they are to manufacture, and how well they perform in use.

If the exact part or design cannot be found, existing designs can be modified to produce a new design with less effort than would be necessary starting from zero. Previous designs can be copied and modified more easily using computer-aided design, which is easier if various details are drawn on unique "layers" or "views."

In addition to checking part history, designers should perform enough analysis, as with any design, to ensure that the design will satisfy all its design objectives.

In order to use existing parts, designers will need to know what has already been designed. Comprehensive coding and classification schemes have been devised by some companies, but such schemes are not generally available. However, most parts can be listed in some logical order, as discussed in Section 5.6.

5.17.1 Obstacles to Reusable Engineering

Some engineers resist using previous engineering because of the not-invented-here (NIH) syndrome. Others want to start with a "clean sheet of paper" or a clear computer screen. Sometimes poor documentation discourages reusing previous designs. Incomplete documentation is bad enough, but incorrect documentation is even worse.

The NIH syndrome might be overcome by team training or more selective hiring. Documentation issues can be corrected by insisting on good documentation as part of every project team's responsibilities.

5.17.2 Reuse Studies

A survey of 53 companies that were using group technology indicated *the need for designing new parts was down by 50%:*[17]

- 20% of their needs could be satisfied by an existing part.
- 18% required only slight modification of an existing part.
- 12% required extensive modification.

One aerospace company discovered that a virtually identical part had been designed independently five times. The part had been purchased from five suppliers *at prices ranging from 22 cents to $7.50 each!*

5.18 OFF-THE-SHELF PARTS

For standard parts with little variety, designers can greatly reduce development time and cost, in addition to reducing product cost, by *never designing a part that is available out of a catalog.* Rarely can a company save any money or time by designing and building parts that are available off-the-shelf.

Designers are often misled by their own accounting systems into thinking that internally produced parts can be designed and built less expensively than off-the-shelf hardware, because accounting systems rarely report the total cost of parts (see Chapters 6 and 7). Usually, when designers look up part cost in an accounting database, they are provided only the material cost and labor cost, with very little "burden" (overhead). Considering the enormous overhead in some companies, part costs may be substantially understated.

Using off-the-shelf parts has an inherent standardizing effect on a company's parts list because many catalog parts are used broadly enough to become de facto standards.

Paradoxically, designers should choose the off-the-shelf parts first and design the product around them, or else they may make arbitrary decisions that may preclude their use. By contrast, incorporating off-the-shelf parts into the design early will greatly simplify the design and the design effort.

Off-the-shelf parts are less expensive to design, considering the cost of design, documentation, prototyping, testing, and debugging, in addition to the overhead cost of purchasing all the constituent parts and the cost of non-core-competency manufacturing. Off-the-shelf parts save time when one considers the time to design, document, administer, build, test, and fix prototype designs.

Suppliers of off-the-shelf parts are more efficient at their specialty, because they are more experienced with their products, continuously improve quality, have proven track records on reliability, design parts better for DFM, have dedicated production facilities, offer standardized parts, and sometimes pick up warranty or service costs.

Off-the-shelf part utilization helps internal resources focus on their real missions, which are designing and building products.

5.18.1 Optimizing the Utilization of Off-the-Shelf Parts

Make sure that off-the-shelf parts are thoroughly considered early in the concept/architecture phase before arbitrary decisions limit the opportunities.

Conduct a thorough search of all potential parts, especially commonly available parts that are made in high volume on automated machinery, compared to less-efficient in-house manufacture in small runs with high setup charges. Look for parts that are used in your industry with your demands and life spans. Strive to use standard off-the-shelf parts; specials may have availability and processing impacts, unless the supplier can mass customize them.

Be sure to base make versus buy decisions on total cost measurements, because off-the-shelf part cost reflects the total cost, but in-house production cost may not include all the overhead.

Be careful when using off-the-shelf parts in unusual ways. The parts may not be designed for that usage or the suppliers may change dimensions or features that they think will not affect their customers.

5.18.2 When to Use Off-the-Shelf Parts

Designers should specify off-the-shelf parts when the following conditions exist:

- *Parts are standard.* Off-the-shelf parts are especially applicable when they are de facto standards, always available, and can be purchased from multiple suppliers.
- *Volumes are high.* Standard parts can be manufactured in high volumes. Even companies with small order quantities can benefit from suppliers' economies of scale. High-volume parts may be mass produced on dedicated lines, which further lowers cost and improves quality.
- *Quality and reliability are important.* Good suppliers are able to focus more on their products and thus achieve higher quality levels than occasional production at an assembler's plant, especially suppliers who have been making them for a long time and are already "up the learning curve." Parts that have been produced for a while will have track records that can be used to predict and ensure reliability targets will be met.
- *Specialized skill, expertise, or costly equipment.* Suppliers may have specialized skill and expertise that may be hard to equal in house. High-volume suppliers may have installed sophisticated equipment that can make parts faster, at lower cost, and at higher quality.
- *Different processing.* If certain parts require different processing than can be efficiently done in house, it will be better to get those parts from off-the-shelf suppliers.

- *Growth or constrained capacity.* If in-house capacity is limited or growth is anticipated, off-the-shelf parts can relieve capacity constraints and make that capacity available for product growth.
- *Lowest total cost.* Off-the-shelf standard parts will usually have a lower total cost, even though in some applications standard parts may appear to cost more than the "just right" sizing in Figure 5.3. But when the effects of purchasing leverage and material overhead are included, the standard parts will be the lowest cost solution for the company.

5.18.3 Finding Off-the-Shelf Parts

The first task in finding off-the-shelf parts is to search all the potential sources for the type of parts needed. A common shortcoming here is to execute a superficial search and then conclude that no one makes the needed parts. Then the designer feels justified in designing and building the part.

Many directories and other sources can show engineers where to find off-the-shelf parts, including Internet searches, the Thomas Register, trade journal annual "directory issues," the yellow pages, and trade shows. Purchasing agents know where to find parts, which is why they should be early participants on teams.

Catalogs of off-the-shelf parts can be obtained on the Internet and from company headquarters, local representatives, reader service cards ("bingo cards") in trade journals, trade shows, company libraries, and peers' private collections. Many websites contain complete catalogs online; but if the site is too cumbersome to navigate, the complete printed catalog may be better at visually showing the range of parts available.

Company websites and catalogs contain a wealth of information, including specifications, selection guidelines, design guidelines, cross-references to other brands, listings of local reps, and policies on specials or custom parts.

Typically, catalogs do not list prices. Prices for specific parts can be obtained by asking the headquarters or local rep. Price lists for the entire line are more useful for doing "what if" analyses. They can be obtained with enough persistence. One method is to inform the supplier that in the pricing stage of the design, you must use a catalog that has a price list and that these tentatively chosen parts often end up in the final design.

Designers can get some indications of relative availability by cross-referencing other brands to reveal which sizes or models are supplied by the most suppliers. These are likely to be the most standard parts, with greater availability than parts from only one or two sources. The best indication of availability is to find "quantity in stock" data. The size or model with the highest quantity in stock is probably the most standard and has the best availability, not to mention the best pricing.

After the needed part is located, the designer will want to evaluate the actual part. The quickest way to get a part for evaluation is to ask for a *sample*. Samples are usually sent out immediately, without charge, unless the parts are very expensive or semi-custom in nature. This is quicker than purchasing and saves administrative expense at both ends. To maximize success in obtaining samples, be sure to mention yearly or lifetime projected consumption of the parts being considered.

Specify off-the-shelf parts from multiple suppliers that are interchangeable for functionality and quality. If not drop-in interchangeable, design versatile mounts that can accept all or prepare adaptors. Extra mounting holes are almost free if made on CNC machines.

If customers insist on one brand, emphasize that the optimal (or quoted) delivery time is based on your supply chain versatility and specific brand requests may result in extra costs and delivery delays.

For rare materials and potentially scarce parts, identify trends early and proactively investigate alternative materials and parts.

If these solutions don't work for critical parts, consider bringing back outsourced production, learning how to build your own parts, establishing affiliations with the best supplier, or acquiring a captive supplier.

> For projects with challenging cost and time goals, it is essential that off-the-shelf opportunities be thoroughly pursued early.

5.19 NEW ROLE OF PROCUREMENT

5.19.1 How to Search for Off-the-Shelf Parts

For important parts and materials, *don't specify one spec and ask purchasing to buy it*. Obtain data on a wide range of candidates and plot them (cost vs. performance). In many cases, higher performing parts may cost

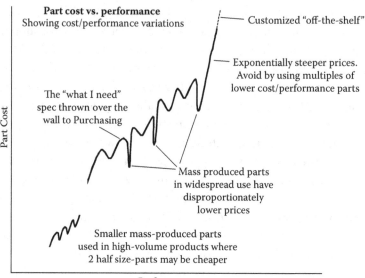

Part cost vs. performance
Showing cost/performance variations

Customized "off-the-shelf"

Exponentially steeper prices.
Avoid by using multiples of
lower cost/performance parts

The "what I need"
spec thrown over the
wall to Purchasing

Mass produced parts
in widespread use have
disproportionately
lower prices

Smaller mass-produced parts
used in high-volume products where
2 half size-parts may be cheaper

Part Cost

Performance

FIGURE 5.6
Searching for ranges of parts.

less but would not normally be considered. In some cases, two smaller high-volume parts may cost less than a single large version for applications where parts can be used in parallel. Plot all relevant parts for availability and delivery (see Figure 5.6)

The *procurement* function needs to shift from just purchasing "the" part that the engineers want to searching for ranges of parts to optimize the following:

- Maximizing availability throughout the life of the product. *Be prepared to pay for this availability*; but any increase in BOM lines will be saved many times over by avoiding change orders to solve availability problems.
- Standardizing on *versatile parts* for many applications. This may raise a BOM line, but will save much more in total cost.
- Eliminating long lead times by selecting standard parts that are quickly available.
- Finding suppliers whose other customers have similar challenges, quality demands, and life spans.

- Prequalifying parts that will be available for engineers for immediate deployment.
- Qualifying suppliers and vendors for quality, ability to deliver, and stability.

Given the importance of the above, companies need dedicated specialists to do these tasks. To justify, quantify the value of the above and compile past costs from not doing this.

5.19.2 Maximizing Availability and Minimizing Lead Times

Search thoroughly for parts with the best availability, even if their purchase prices appear to be more expensive, knowing that more money would be saved by avoiding expediting cost, customer dissatisfaction, delay penalties, lost sales, or, worst of all, compromising product development efforts.

For parts that still have long lead times, specify the most versatile version for many products, so that:

- Better delivery terms could be negotiated; for instance, a steady flow that is used one way or another. This could avoid separate orders waiting in queues.
- These standard parts could be inventoried because you know they will be used one way or another, without the excessive cost and risk of stocking many unusual versions.
- Various projects could borrow from each other in emergencies.
- Service and spare parts availability will be improved for the customer, with fewer stocked parts that could service more products, better uptime, simpler maintenance, easier training, and fewer availability problems for the customer.

Do not let these versatile, short-lead-time parts be changed for "cost reduction" or when switching to vendors who want to change your parts to their parts from their supply chains.

Minimize cumulative queues from multiple sequential steps, such as preprocessing, multiple primary processors, or postprocessing. Prevent this by eliminating the extra steps; for example, with more versatile machine shops or with through-hardened materials and harder cutters. Another version of this is designing more with the supply chain's process capabilities. If necessary, make quick turnaround a condition for all steps in the sequence.

Standardize those parts to economically enable a steady flow or stocking. If queuing is because volumes are too low, setups too large, and parameters too unusual, consider upgrading in-house capabilities or bringing in new capabilities. Make versus buy decisions must include the value of the shorter lead time and all the costs of delays.

Consider arranging for a vendor/partner to run a "captive" operation in your factory to overcome the above shortcomings and eliminate the delays. This may appear to cost more but actually provide a net cost savings while improving responsiveness and competitiveness.

5.20 STANDARDIZATION IMPLEMENTATION

Standardization can be implemented by forming a standardization task force, which would include key people from engineering, manufacturing, purchasing, quality, and finance, plus appropriate managers and implementers. A key member of this effort would be a "database wizard"— a designated person who has the skill and availability to quickly extract data and generate many Pareto plots from various IT systems. The standardization steps are as follows:

- Generate interest in standardization by creating an overall Pareto chart showing the periodic consumption of all parts (on the vertical axis) in descending order, with only the descending count on the horizontal axis, in the format of Figure 5.2 but without part numbers.
- Arrange training on standardization methodologies (or at least have all implementers read this chapter) followed by workshops[18] to standardize each category of parts and materials.
- Discuss any Pareto charts previously generated, including any anecdotal and documented consequences of standardization shortcomings.
- Start with the early steps: list existing parts (Figure 5.1) to immediately stop the proliferation, clean up database nomenclature (e.g., for the many labels for bolts and screws), eliminate approved but unused parts, eliminate parts not used recently, eliminate duplicate parts

(Section 5.6), combine parts with multiple grades and tolerances, replace lower versions with better-than substitutions, and scrutinize long-lead-time parts to see if they could be converted to standard parts with better lead times.

- Prioritize opportunities of what categories of parts to standardize first and start from the top of the list, creating standard parts lists using the procedures presented in Sections 5.7 and 5.8. This prioritization could be done with surveys or by having the workshop group vote for their preferences.

- Similarly, standardize raw materials (Section 5.10), tools (Section 5.13), features (Section 5.14), and processes (Section 5.15).

- Determine related investigative tasks to identify standardization issues, like backward compatibility, "copy-exact" policies, and the implications of Lean/cellular manufacture.

- Start the process of developing standardization implementation procedures. Discuss and assign tasks to change procedures or policies that may hinder standardization.

- Assign tasks to generate more Pareto plots.

- Designate people to do the above.

- In subsequent sessions, analyze new Pareto plots, specify additional investigations, start creating lists of standard parts/materials, obtain approvals, and issue the lists.

- For the standardization of expensive parts (Section 5.11), quantify the benefits of the standardization (in the format of Figure 5.3) for specific cases and, in general, start the process of creating a financial model to help quantify the overall cost savings and overcome resistance to some products getting what appears to be a better standard part.

- Issue the standardization lists (Section 5.8, point 14) and incorporate into subsequent DFM training.

- Encourage standardization through appropriate material overhead rates, prequalified standard parts, floor stock, personal display boards, spec books, and cost metrics (Section 5.16).

- Perform *product line rationalization,* ideally as a first step if time permits, to eliminate or outsource the most unusual products, which usually have the most unusual parts (see Appendix A).

NOTES

1. David M. Anderson, *Build-to-Order & Mass Customization: The Ultimate Supply Chain Management and Lean Manufacturing Strategy for Low-Cost On-Demand Production without Forecasts or Inventory* (2008, CIM Press). Chapter 4, "Part Standardization" and Chapter 5, "Material Variety Reduction."

2. Brian H. Maskell, *Software and the Agile Manufacturer* (1994, Productivity Press), p. 335.

3. Robin Cooper and Peter B. B. Turney, "Internally Focused Activity-Based Cost Systems," *Measures of Manufacturing Excellence*, edited by Robert S. Kaplan (1990, Harvard Business School Press), p. 293.

4. Timothy Stevens, "Prolific Parts Pilfer Profits," *Industry Week*, 244(11), June 5, 1995, pp. 59–62.

5. Venkat Mohan, President and COO, CADIS Inc., Boulder, Colorado.

6. George Taninecz, "Faster In, Faster Out," *Industry Week*, 244(10), May 15, 1995, pp. 27–30.

7. Michael R. Ostrenga, Terrence R. Ozan, Robert D. McIlhattan, and Marcus D. Harwood, *The Ernst & Young Guide to Total Cost Management* (1992, John Wiley & Sons), p. 150.

8. Marshall Fisher, Anjani Jain, and John Paul MacDuffie, "Strategies for Product Variety: Lessons From the Auto Industry," The Wharton School, University of Pennsylvania, October 9, 1992. Revised January 16, 1994.

9. Eric Teicholz and Joel N. Orr, *Computer Integrated Manufacturing Handbook* (1987, McGraw-Hill), p. 96.

10. Stevens, *Industry Week*.

11. CADIS case study, CADIS, Inc., Boulder, Colorado.

12. Anderson, *Build-to-Order & Mass Customization*. (2008, CIM Press). Chapter 9, "Mass Customization," presents three ways to customize products: (1) modules or building blocks, (2) adjustments or configurations, and (3) dimensional customization, which involves a permanent cutting-to-fit, mixing, or tailoring. See book description in Appendix D.

13. Jean-Philippe Deschamps and P. Ranganath Nayak, *Product Juggernauts: How Companies Mobilize to Generate a Stream of Market Winners* (1995, Harvard Business School Press), pp. 35–36.

14. David Pringle, "How Nokia Thrives by Breaking the Rules," *The Wall Street Journal*, January 3, 2003.

15. For more information on customized in-house DFM seminars, see Appendix D in this book or www.design4manufacturability.com/seminars.htm.

16. Cooper and Turney, *Measures for Manufacturing Excellence*.

17. Urban Wemmerlöv and Nancy L. Hyer, "Cellular Manufacturing in the US Industry, a Survey of Users," *International Journal of Product Research*, 27(9) 1511–1530, 1989.

18. For more on the author's customized in-house seminars (that all include standardization), as well as information on the author's standardization workshops, see Appendix D.6.4.

Section III

Cost Reduction

6

Minimizing Total Cost by Design

Cost must be designed out of the product and production processes, because it is very difficult to remove cost through cost reduction measures after the product has been designed. Many cost reduction efforts may not even pay off the expense of the cost reduction effort, which may even lose money on a total cost basis.

Most of this book shows how to lower parts cost (Chapters 5 and 9), labor cost (Chapters 4 and 8), and system cost (Chapters 1, 2, 3, and 10). *This chapter focuses on how not to lower these costs and how to lower overhead cost.*

The key to achieving the lowest product cost is to base all thinking and decisions on a *total cost* perspective, as will be discussed further in Chapter 7. Unfortunately, the typical company cost system reports only material and labor costs. All other costs are called *overhead,* which is spread across corporate activities according to some arbitrary allocation (averaging) algorithm; for instance, proportional to material, labor, or processing cost. However, not all products have the same overhead demands. In fact, much can be done to lower overhead costs by design, which makes a rational allocation of overhead charges even more important.

This chapter will discuss two categories of cost: *reported costs,* such as labor and materials, and *overhead costs.* It will show how the use of advanced design techniques can minimize labor and material costs and achieve significant reductions in overhead costs.

This chapter presents many ways to minimize total cost by design. Chapter 7 will show how to *think* in terms of total cost and *measure* total cost so that product development teams will make the best cost decisions to minimize total cost.

FIGURE 6.1
Common cost reduction scenario.

6.1 HOW *NOT* TO LOWER COST[1]

First, let's look at traditional cost reduction at Dysfunctional Engineering Inc., as shown in Figure 6.1. What is wrong with this very common picture? The first shortcoming of this approach is that it begins cost reduction efforts *after* the product is designed and already in production. In their haste to rush early production units to market, many companies defer cost concerns until later with "cost reduction" efforts. The first problem with this strategy is that it probably will not happen because of competing priorities, and thus, costs remain high for the life of the product. The second problem is that cost reduction simply cannot be very effective!

6.1.1 Why Cost Is Hard to Remove after Design

Cost reduction after the product is designed is an ineffective way to lower cost because cost is designed into the product and is hard to remove later: 80% of the cumulative lifetime cost is committed by design, and by the time the project gets to manufacturing, only 5% is left, as shown in Figure 1.1.

After the design phase, so much is cast in concrete and boxed into corners; thus, *systematic* cost reduction will be almost impossible because system design determines 60% of costs.

Cost reduction efforts on one product will not have the time or bandwidth to reduce overhead costs, which may be more than half the cost. Besides, overhead costs are not quantified unless the company has a total cost program (Chapter 7).

Thus, the focus usually shifts to specifying cheaper parts, cutting corners, omitting features, beating up suppliers, switching to a new low bidder, or letting labor costs dominate sourcing and plant location decisions.

The changes will cost money, which may not be paid back within the life of the product. They will also take time, especially if requalifications are required, which may delay the time-to-market.

Changes may induce more problems, thus requiring yet more changes, thus expending more hours, calendar time, and money to make the subsequent changes and possibly compromising functionality, quality, and reliability. Toyota believes that "late design changes... are expensive, suboptimal, and *always degrade both product and process performance*"[2] [emphasis added].

In addition to the above, cost reduction can't be counted on because it may just not happen due to competing priorities, such as mandatory changes and designing new products. Committing valuable resources to cost reduction after design takes them away from other more-effective efforts in product development, quality, Lean Production, and elsewhere. If too many resources are committed to cost reduction, then:

1. There will not be enough available for *real cost reduction* through new product development. If this continues over time, the result will be little, if any, real reduction in cost, while such a drain of resources will impede new product development efforts.
2. It will prevent the transition from back-loaded efforts to the more-effective front-loaded methodology that uses complete multifunctional teams to design low-cost products right the first time, as discussed in Chapters 2 and 3.
3. The company will be lured into thinking it is doing all it can to lower cost, when, in fact, costs are not really being reduced and opportunities for real cost reduction are not being pursued.

Finally, cost reduction attempts, coupled with incomplete cost data, may discourage innovative ways to lower cost, maybe even thwarting promising attempts. While teaching DFM to companies, the author often suggests innovative ways to lower cost by design, only to be countered by a chorus of "we looked into that, but it didn't work out." However, it is hard to be innovative when so much is cast in concrete that there are very few opportunities available.

6.1.2 Cost-Cutting Doesn't Work

Mercer Management Consulting analyzed 800 companies from 1987 to 1992. They identified 120 of these companies as "cost cutters." Of those cost-cutting companies, "68% did not go on to achieve profitable revenue during the next five years."[3]

There are also intangible impacts of an excessive focus on cost reduction: it absorbs effort and talent that could be applied to more productive activities, such as developing better new products and improving operations. One division of a large international company did not have time for the author's training on low-cost product development because they were too busy with 31 cost reduction efforts!

6.2 COST MEASUREMENTS

6.2.1 Usual Definition of Cost

Traditional cost systems provide the cost breakdown shown in Figure 6.2, where only parts and labor are quantified. The rest of the costs are lumped together in several categories, collectively called *overhead*, which is then averaged (allocated) over all products. This practice results in the distortion of product costing, which leads to:

- Distorted pricing, because overhead costs vary[4]
- Cross-subsidies, where good products subsidize bad ones[5] and standard products subsidize customs and specials
- Overpricing of good products = less competitive

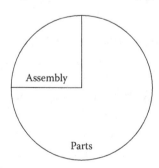

FIGURE 6.2
Typical cost breakdown.

- Underpricing of bad products = lose money[6]
- Distorted profitability = poor product planning
- Bad cost decisions when the focus is only on parts and labor[7]
- Parts and labor "cost reductions," which often raise other costs

This approach to cost encourages product development teams to focus only on material, labor, and tooling costs. Considering only these costs allows a limited perspective and might lead to shortsighted conclusions that most of the product's cost consists of parts (and tooling). Therefore, cost reduction measures often focus only on minimizing parts costs, usually by buying cheaper parts.

However, the following slogan states what is really important to customers: You don't compete on *cost*; you compete on *price*. Customers don't care about *your* cost; they care only about *their total cost*, which is *your price*. There, the only relevant pie chart is the selling price.

6.2.2 Selling Price Breakdown

Total cost measurements enable the creation of the selling price breakdown chart shown in Figure 6.3 (for an integrated company) and encourage

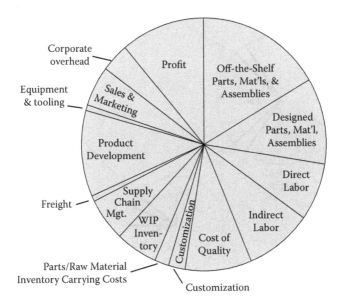

FIGURE 6.3
Selling price breakdown.

everyone to make decisions that will minimize the total cost while maximizing profits.

6.2.3 Selling Price Breakdown for an Outsourced Company

Unlike an integrated company (see Figure 6.3), heavily outsourced companies buy a lot of parts (like the left-most pie in Figure 6.4) and therefore don't think that they have to consider overhead costs. However, each OEM's part is a supplier's product, which, in turn, has parts that are its suppliers' products, so the net result is a collection of pie charts (far right), all with the same proportion of overhead costs of an integrated company.

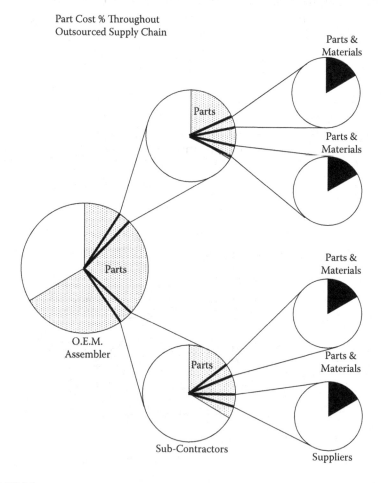

FIGURE 6.4
Part cost percentage throughout outsourced supply chain.

Many overhead costs, such as inventory, may be spread throughout the supply chain (in the white spaces of all the pie charts). For the OEM to have visibility and control of the overhead costs throughout an outsourced supply chain, it must have either very good vendor relationships or an ownership stake in its suppliers.

6.2.4 Overhead Cost Minimization Strategy

Total cost can be minimized by addressing all the costs that contribute to the selling price. Figure 6.5 shows the overall strategy, with design for manufacturability and Lean Production efficiencies minimizing direct labor, indirect labor, and quality costs. Concurrent engineering, standardization, and product family synergies minimize material overhead, product development, shipping, and sales and marketing costs. On-demand Lean Production, build-to-order, and mass customization (Chapter 4) can virtually eliminate setup changeover costs, customization costs, raw material inventory, work-in-process (WIP) inventory, and finished goods inventory, both at the factory and in the distribution channels.

The remainder of this chapter discusses specific cost minimization opportunities for the various elements of total cost.

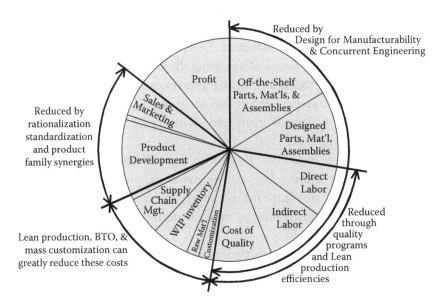

FIGURE 6.5
Programs that reduce specific costs.

6.3 STRATEGY TO CUT TOTAL COST IN HALF

The author's book-length website, www.HalfCostProducts.com, presents eight strategies for comprehensive cost reduction, each of which can offer significant returns as a stand-alone program. When combined, these coordinated cost saving strategies support each other synergistically for a whole-is-better-than-its-parts effect, instead of the more common but ineffective parts-and-labor-focused initiatives that can raise other costs, compete for resources, and compromise longer-term cost reductions. The site contains the equivalent of a 250-page book and 700 hyperlinks. The eight half-cost strategies are summarized here:

1. *Cost reduction by design.* Product development determines 80% of product cost (Figure 1.1). The concept/architecture phase alone determines 60% of cost! Half-cost products depend on *breakthrough concepts*. Sometimes these come from the brainstorming sessions in workshops (Section 3.7). More ambitious cost goals may need innovative *concept studies.*[8]

   Other strategies are enhanced by designing for quality, designing for Lean Production and build-to-order, concurrent engineering with vendors (Section 2.6), which saves more than bidding, and designing around standard parts, which simplifies supply chain management.

2. *Lean Production cost reduction.* Lean Production benefits[9] include eliminating many forms of waste and the ability to double labor productivity, cut production throughput times by 90%, reduce inventories by 90%, and cut errors and scrap in half, thus lowering other costs of quality, procurement, and inventory (see a summary of benefits in Section 4.1).

3. *Overhead cost reduction.* Build-to-order can build standard products to order without forecasts of inventory, which costs 25% of its value to carry per year[10] (see summary in Section 4.2). Specials can be quickly and easily built by mass customization[11] (see overview in Section 4.3).

4. *Standardization cost reduction.* Standard part lists can be 50 times smaller than proliferated lists, thus generating purchasing leverage, lowering material overhead and inventory costs, and improving quality (see Chapter 5).

5. *Product line rationalization.* Eliminating or outsourcing unusual, high-overhead, low-volume products and options can lower total cost immediately and free up valuable resources for other cost strategies. It eliminates the loser tax on cash cows to subsidize products that have low margins or are losing money (see Appendix A).

6. *Supply chain management cost reduction.* Supply chain resources can be most effective in reducing cost by supporting product development teams, establishing vendor partnerships, buying high-quality parts, driving and encouraging standardization, rationalizing products (because the most unusual products have the most unusual parts and materials), supporting Lean efforts and automatic resupply of parts (which eliminates forecasts, purchase orders, inventory, and expediting costs), and keeping control of local in-house manufacturing, which supports product development, Lean Production, build-to-order, inventory reduction, and quality programs[12] (some of these are summarized in Sections 4.2.1 through 4.2.2).

7. *Quality cost reduction.* The cost of quality can be a significant proportion of revenue or selling price (see Section 6.9). Improvements are supported by designing for quality, Lean Production, and rationalization because rationalizing away unusual products raises net factory quality and avoids wasting quality resources on inherently lower quality products (see Section A.9 on how rationalization improves quality).

8. *Total cost measurement* to support, justify, and quantify the savings of all cost reduction strategies. Until total cost can be quantified, everyone should make decisions based on *total cost thinking* (see Chapter 7).

6.4 MINIMIZING COST THROUGH DESIGN

As shown in Figure 1.1, 80% of the lifetime cumulative cost of a product is determined by the product's design. An even more important fact is that 60% of a product's cost is determined by product architecture. Low-cost product design is based on the premise that cost is designed into the product, especially by early concept decisions.

Tools such as design for manufacturability can help design products that are easier, and thus less costly, to build. Concurrent engineering can ensure the lowest cost processing when the processes are concurrently designed with the product. Quality can be designed into the product with robust design techniques (Taguchi Methods,™ based on design of experiments) and then built into the product with process controls instead of the more expensive inspection paradigm. Maximum utilization of off-the-shelf parts can minimize part cost (Section 5.19). Involving vendors early can result in lower cost outsourced parts. Total cost accounting data (Chapter 7) can lead to decisions that result in the lowest total cost. Applying all these DFM techniques should enable companies to develop products at half the *total cost*, with special emphasis on the key points discussed above and in Section 3.8.

The highest leverage opportunities for minimizing cost are in the architecture stage. And yet, this high-leverage opportunity is virtually ignored in many product development projects, when designers make snap decisions or just assume that the product will have the same architecture as previous or competitive products. The architecture phase of product development (Section 3.3) abounds with opportunities to greatly lower cost through creative concept simplifications using creativity (Section 3.6) and brainstorming (Section 3.7).

6.5 MINIMIZING OVERHEAD COSTS

Engineers, purchasing agents, and cost reduction managers often expend great effort trying to reduce reported costs, such as labor and materials. They precisely calculate these costs *and then multiply by three or four* to get the selling price. Yet they rarely challenge overhead charges, fittingly called "burdens."

Product development can have significant effects on overhead costs by designing to minimize them. Overhead costs can be reduced by: concurrent engineering, which minimizes product development expense; designing in quality, which minimizes the "cost of quality"; designing for Lean Production, which minimizes inventory and other factory overhead costs; and standardization, which minimizes material overhead.

Overhead costs can be significantly lowered with concurrent product and process design, parts and process standardization, reuse of engineering and software code, modular design, designing quality into the product, designing it right the first time, and designing to optimize manufacturing flexibility.

6.6 MINIMIZING PRODUCT DEVELOPMENT EXPENSES

Advanced product development can significantly reduce product development and related expenses, as detailed below.

6.6.1 Product Portfolio Planning

Total cost is a key element of product portfolio planning. Before a company can decide in which market segments to compete, it should understand the *true* profitability of its existing products. However, the *reported profitability* of existing products is only as meaningful as the *reported costs*. As will be discussed in Chapter 7, typical cost reporting systems are too aggregated to distinguish the real cost differences between products, thus distorting product costing.[13] Distorted product costing leads to distorted perceptions of profitability and, thus, to distorted decision making on which products to develop. Developing products for truly profitable market segments will result in the most efficient utilization of product development resources. Developing products for market segments that are thought to be making money, but are really marginal or losing money, is a waste of product development resources. Avoiding this waste will lower product development costs or provide more results from the same costs.

6.6.2 Multifunctional Design Teams

To minimize product development expenses, product development teams must be efficient. They must do it right the first time because engineering change orders are expensive and redesigns are even more expensive. Designing products right the first time requires good product development methodologies.

Using multifunctional teams to raise and resolve issues early will save the considerable expense of trying to do it later, after things are "cast under several layers of concrete." Multifunctional teams can concurrently design and select the optimal processes and vendors for the lowest total cost. When manufacturing and vendors are involved early, the team will be better able to design low-cost products and make rational decisions regarding tooling and automation.

Product development expenses can be minimized by utilizing the most efficient designers, which may not always be in-house engineering working alone. The multifunctional team *with active manufacturing or vendor assistance* would be more efficient than an isolated group of engineers working alone. Working with vendors who make the parts will result in the lowest part costs.

And remember, the most efficient part design effort is none at all! This is accomplished with optimal use of off-the-shelf hardware (Section 5.18).

6.6.3 Methodical Product Definition

Similarly, product development expenses can be minimized by *defining* the product right the first time, because it is very expensive to make product definition iterations at the prototype stage when the customer says, "that's not what I wanted." Designing in unwanted features wastes product development resources and causes the product to cost too much and thus be overpriced.

6.6.4 Total Cost Decision Making

All decisions must consciously be based on a *total cost focus,* even if total cost cannot be quantified (Chapter 7). The product development culture must support decisions that "just make sense" from a total cost perspective, even if they cannot be justified quantitatively by the current cost system.

Arbitrary decisions must be avoided. Arbitrary decisions are based on the assumption that all choices have the same cost impact, which is rarely the case, so everyone must select the lowest-cost choice.

On the other hand, cost concerns can have a counterproductive effect if decision makers are putting too much focus on *reported costs,* such as labor and materials, and not enough focus on *total costs.* Sometimes product development teams limit their opportunities by making major decisions based on rough estimates of reported costs, instead of using good cost models based on *total costs for multiple approaches, ideas, and scenarios.*

6.6.5 Design Efficiency

Reinventing the wheel can be avoided in product development with maximum use of off-the-shelf parts, reusable engineering, and versatile modules (Section 4.7). There is a common tendency for product development projects to ignore previous work and completely start over, when they could be more efficient by leveraging previous work (Section 5.18). This can be facilitated by proper CAD practices: a good layer convention will segregate designs into well-defined layers, so that subsequent development projects can easily find and reuse pervious engineering. Modular design can allow new products to be derived from standard modules.

6.6.6 Off-the-Shelf Parts

Basing designs on off-the-shelf parts can eliminate the cost of designing those parts and, at the same time, lower part cost and quality costs for those parts (see Section 5.18).

6.6.7 Product Life Extensions

Extending product life can be accomplished through upgrades rather than redesigns. Products can be designed to be easy to extend the product life with upgrades (see Figure 3.2). Modular design can facilitate upgrading if anticipated changes can be confined to the fewest number of modules. Many common functions of a product can be reused for many iterations of the product.

6.6.8 Debugging Costs

Minimizing debugging costs can be accomplished by using existing modules that have already been debugged. Some leading companies, such as Hewlett-Packard, feel that this is the best way to produce bug-free software. With a high percentage of reuse, development and debugging efforts can focus on new aspects.

6.6.9 Test Cost

Diagnostic test development can be avoided by designing quality into the product and building it in with process controls. In such a quality environment, failures would be so low that diagnostic test development could be avoided, because the cost of discarding failed parts would be less than the cost of test development, test equipment, and repair efforts (as described in Section 8.6). Avoiding diagnostic tests can have a significant effect on product cost because diagnostic test equipment for circuit boards can cost up to $2 million and, for some products, diagnostic test development can exceed the cost of product development!

6.6.10 Product Development Expenses

Development expenses can be paid off sooner because of quicker product development cycles. This lowers the interest or opportunity cost of the money invested in product development. Similar logic applies to tooling and research expenses for technology development.

6.6.11 More Efficient Development Costs Less

Advanced product development methodologies are more efficient because the product is well defined and the architecture is optimized around creative concepts. This thorough early development work means fewer false starts or "looping back" to do things over, fewer changes, and fewer redesigns.

6.6.12 Product Development Risk

Faster developments have less obsolescence risk. Shorter product development cycles result in less chance of market shifts and technical obsolescence by the time the product reaches the market, thus resulting in fewer changes and redesigns.

6.7 COST SAVINGS OF OFF-THE-SHELF PARTS

Specifying off-the-shelf parts and subassemblies (Section 5.19) offers several ways to minimize cost in product development, purchasing, manufacturing, quality, and reliability.

Off-the-shelf parts are less expensive to design if you consider the cost of design, documentation, prototyping, testing, debugging, and, if necessary, change orders or redesign. Purchasing one off-the-shelf subassembly is only one purchasing action, but it incurs less material overhead than the effort to purchase all the constituent parts.

The cost of off-the-shelf parts will be lower because suppliers of off-the-shelf parts are usually more efficient at their specialty because they are more experienced with their products, continuously improve quality, design parts better for DFM, and may dedicate production lines. Further, they are far up on the learning curve, so they don't have to incur costs and delays to learn how to make good parts.

Quality costs will be lower, often much lower, because good suppliers have processes under control, with the best suppliers utilizing statistical process controls and Six Sigma quality techniques. Suppliers who have dedicated product lines can ensure higher quality than lower-volume suppliers that have to keep changing over lines for different products.

Reliability costs can be lower, especially if a supplier's part has been in production long enough to get feedback from the field and corrective action has been taken. For these parts, suppliers can furnish reliability data.

6.8 MINIMIZING ENGINEERING CHANGE ORDER COSTS

Advanced product development methodologies result in a higher percentage of products designed right the first time and, thus, fewer engineering change orders. Better product definition results in fewer changes to satisfy customers. Early changes in the planning stage are less expensive than late changes at the hardware stage.

Well-designed products can minimize a substantial overhead expense that is not always included in change order cost reporting: firefighting or problem solving. This is the often-considerable effort expended to solve production problems, which is usually intense when a new product is launched into production. The cost of changes rises exponentially as the product progresses toward production, as shown in Figure 1.3.

6.9 MINIMIZING COST OF QUALITY

The *cost of quality* is really the cost of *poor* quality: the cost of finding and repairing defects. Companies without strong Total Quality Management programs can have quality costs equal to 15% to 40% of revenue.[14] Advanced product development can *design in* quality. Concurrently engineered processes that are in control can *build in* quality. This dual approach to quality can substantially reduce both the internal and external costs of quality, which are defined as follows:

> *Internal cost of quality* includes the cost of non-value-added activities such as testing, scrap, diagnostics, rework, reinspection of rework, purchasing actions to procure replacements materials/parts, analysis of quality problems, cost of corrective actions, change-induced quality costs, and change orders to correct the design.
>
> *External cost of quality* includes the cost of dealing with customer complaints, refunds, returned goods, repair of returns, warranty claims, patch costs, legal liabilities, recalls, damage control, penalties, and corrective actions on the above. External costs also include hard-to-quantify costs, such as bad publicity, damaged reputations, lost good will, and lost sales.

Companies can have poor internal quality but, with a good "test screen," can keep defects within the factory and, thus, have a high external quality. Achieving external quality totally by testing and rework is expensive, and depending on testing would result in a high cost of quality. Many expensive products, such as luxury automobiles not made in Lean environments, may enjoy a quality reputation, but they are expensive because quality is achieved at a high cost. Before TQM spread to this industry, some luxury automobiles required several times more labor effort to *fix* than a well-designed car requires to *build!*[15]

Another cost of quality is the "defects-by-the-batch" effect in which large batches of parts are made with recurring defects that are not noticed until the batches have traveled through many more workstations. By the time the defects have been spotted in final inspection, hundreds or thousands of defective parts have been made, which then have to be reworked or scrapped. In Lean Production, parts are made in a *one-piece flow*, with immediate feedback at each handoff, so there should be very little chance of recurring defects being produced.

A thorough list of cost of quality categories is presented in the American Society of Quality's book, *Principles of Quality Costs: Principles, Implementation, and Use.*[16]

Continuous improvement, or *kaizen* in Japanese, is an effective technique to keep driving costs down with incremental improvements that can have significant cumulative effects. Continuous improvement can be performed spontaneously by in-house workers,[17] as a part of quality programs, or in cooperative efforts with suppliers.[18,19]

6.10 RATIONAL SELECTION OF LOWEST COST SUPPLIER

The rational selection of the lowest cost supplier encompasses both the make versus buy decision and supplier selection. These decisions must be made rationally on a total cost basis. As discussed throughout this chapter, conventional cost systems can mislead decision makers if they report only on labor and materials. Not quantifying and including all the overhead costs for in-house manufacture creates a bias toward that option, because purchases include *all costs,* by definition, and in-house production may not.

Regarding supplier selection, total costs can actually increase as a result of choosing the supposedly low bidder if purchase cost is emphasized over such subtle, but important, characteristics as quality, delivery, flexibility, and help with product development. Jordan Lewis, in his book about customer–supplier alliances, *The Connected Corporation*, commented about the effects of General Motors' 1992 demands for double-digit price cuts from suppliers, which were instituted by the now-infamous J. Ignacio Lopez de Arriortua:

> "By emphasizing price alone with its suppliers GM won immediate savings—and ignored total cost. At GM's plant in Arlington, Texas, an ill-fitting ashtray from a new, substandard supplier caused a six-week shutdown of Buick Roadmaster production."

Another GM plant saved 5% by going with a low bidder; when the parts were delivered, one-half failed quality tests. The other supplier, that had originally lost the bid, had to gear up production in four days and fly parts to GM by chartered plane. The second supplier commented that, "My guess is that their 5 percent savings turned into a 15 percent loss."[20]

Peter Drucker, writing in *Managing in a Time of Great Change,*[21] encouraged lowering total cost by minimizing "interstitial" costs between suppliers and manufacturers or between manufacturer and distributor:

"But the costs that matter are the costs of the entire economic process in which the individual manufacturer, bank, or hospital is only a link in a chain. The costs of the entire process are what the ultimate customer (or the taxpayer) pays and what determines whether a product, a service, an industry, or an economy is competitive."

"The cost advantage of the Japanese derives in considerable measure from their control of these costs within a keiretsu, the "family" of suppliers and distributors clustered around a manufacturer. Treating the keiretsu as one cost stream led, for instance, to "just-in-time" parts delivery. It also enabled the keiretsu to shift operations to where they are most cost-effective."

Early and active participation of vendors will result in a lower net cost, because having the vendor help design the part will greatly improve the manufacturability, quality, and lead time, thus resulting in lower manufacturing, quality, and supply chain costs. Further, vendor partners who work with their customers from the beginning will be able to charge less because they: (1) understand the part requirement better, (2) are able work with their customer to minimize cost, and (3) won't have to add a "cushion" to deal with an unknown customer. The case for vendor partnerships is presented in Section 2.6.

6.11 LOW BIDDING

Going for the low bid is something management consultants have been trying to discourage for years, but now it has seen a resurgence just because it is easy to do on the Internet. This comes at a time when many purchasing functions are under heavy pressure to go for low bids,[22] and directives are coming in from as high as the chairman of the board at one company with a $9 billion/year purchasing budget.

Bidding for the cheapest parts is not only an ineffective way to achieve real cost reduction, but it can substantially raise less-obvious costs and compromise other important goals, such as quality and delivery. Quality usually takes a back seat when buying decisions are focused on purchase cost, especially in bidding situations. Some say that quality standards can be set for all bidders, but it is a dangerously naïve assumption to believe that quality can be assured simply by setting a metric. Further, focusing cost reduction efforts on part bidding distracts attention from real cost reduction opportunities, which are addressed throughout this book.

Dick Hunter, vice president of fulfillment and supply chain management for Dell Computer, says that online auctions are no "silver bullet":

> "Auctions and exchanges have fueled the thinking that price is everything. But there is more to procurement of materials than just price. Quality, service, responsiveness, and the willingness to improve common processes also are very critical to driving down the total cost of materials."[23]

6.11.1 Cost Reduction Illusion

In old-paradigm companies, cost reduction efforts are focused primarily on parts and materials (hereafter called parts), because that is all that most cost systems are able to quantify, besides labor. Many manufacturers, especially in the automobile business, beat up their part suppliers for repeated cost reductions.

However, before manufacturers fall for a magic elixir, they should consider how part costs really would be lowered under such pressure. The assumption seems to be that either purchasing agents have naïvely offered to pay too much or cavalier part makers have been gouging their customers. Although this may have been true in sleepy industries of the past, it is rarely true in today's dynamic marketplaces.

Another assumption is that inefficiencies can be corrected by pressure after a supplier wins a contract at a lower-than-usual price. However, soon after a supplier wins a bid, it is expected to deliver the goods, and there will not be time to implement a meaningful cost reduction program, as presented throughout this book. Thus, without a real means to lower costs, the supplier will have to cut its margins (which will be resisted from the corner office all the way to Wall Street), cut corners, or pressure *its* suppliers, who may have the same difficulty achieving real cost reductions. Further, if suppliers are either making disappointing profits or struggling to reduce costs, they will not be receptive nor cooperative when asked to help a customer in a bind or to implement programs to build parts on demand for assemblers.

In some cases, suppliers will temporarily lose money to "buy into the business," with the expectation of raising costs later, once they are in. And there are even suppliers out there whose strategy is to bid jobs at zero profit and make all their money on the expected change orders.

In other cases, low bidders "win" because they don't understand the problem and then are ultimately unable to deliver at all. In other cases, winning bidders are "vapor" companies, whose goal is keep bidding down until they win and then patch together a virtual network of alliances to somehow fulfill the order. This phenomenon came out at an in-house seminar[24] when the discussion topic was if anyone had noticed any problems with online competitive bidding. Within seconds, the purchasing manager was jumping up and down waving both hands in the air. He related how they had discovered that one low bidder was working out of an apartment, with the strategy to win the auction and then figure out later how to deliver the goods! This purchasing manager also said that a corporate dictate to do part bidding was alienating their valued suppliers with whom they had good relationships.

6.11.2 Cost of Bidding

Bidding keeps purchasing so busy managing the bidding process that they are not able to help their teams develop products, ensure availability, and set up vendor partnerships. The *Toyota Product Development System* book sums up the cost and wasted resources of bidding as follows:

> "Searching the globe for the lowest cost means managing very large numbers of suppliers as well as introducing a steady stream of new suppliers into your system. These suppliers are unfamiliar with your requirements and demand a great deal of attention to get up and running. While administering complex contracts, managing global bidding wars, and overseeing the constant introduction of new suppliers into the process, US automakers must maintain mammoth purchasing organizations, deal with incredibly cumbersome and slow sourcing processes, and live with constant variation of supplier performance in the development process."[25]

All of that to "save cost"!

Even after all that effort, problems often arise because low bidders may not understand the problem or may be cutting corners, which raises *other costs*, such as quality, expediting, delayed launches, warranty costs, or the costs of recalls.

The biggest cost of bidding may be the value of the resources it takes away from product development to support bidding (at both the customer and vendor), to: (1) update/change documentation, CAD files, materials, tooling, and processing; (2) complete transfers; and (3) deal with new or ongoing problems related to ramps, delivery, quality, or getting up the learning curve, sometimes through many iterations. All the above problems are much worse when offshoring to another continent, the case against which is presented in Section 4.8.

6.11.3 Pressuring Suppliers for Lower Cost

Beating up suppliers for drastic part cost reduction is what Lopez tried at GM. This tactic not only failed to generate real, lasting cost savings, but it also alienated the supplier base and drove the best suppliers to competitors. The suppliers that remained put their best people on Ford and Chrysler projects and withheld their newest developments from GM, because Lopez was using proprietary supplier information to press all suppliers for lower prices. Further, the so-called "savings" in purchasing cost caused severe costs to be incurred elsewhere. The "cost-cutting campaign of Lopez, whose heavy-handedness drove away many of the company's best suppliers ... perversely, may have helped raise GM's total costs."[26]

When Rubbermaid first encountered cost pressures from powerful retailers, such as Walmart, its first response as a leading company was to make sure customers understood the necessity of price *increases!* When they realized that they really had to reduce prices, they tried what didn't work for Lopez at GM and were similarly alienated by their supply base. According to the largest research project ever devoted to corporate failures, *Why Smart Executives Fail: And What You Can Learn from Their Mistakes:*

> "With little talent in cutting costs in-house, Rubbermaid looked to shift responsibility elsewhere. Suppliers were prodded to knock down their own prices, alienating some of the best, low-cost vendors in the process."[27]

Bidding creates a standoffish relationship between buyers and sellers that inhibits cooperative cost reduction, which is the key to real cost reduction. The book that started the Lean Production movement in America concluded this about the effects of bidding on suppler relations:

> "A key feature of market-based bidding is that suppliers share only a single piece of information with the assembler: the bid price per part. Otherwise, suppliers jealously guard information about their operations, even when they are divisions of the assembly company. By holding back information on how they plan to make the part and on their internal efficiency, they believe they are maximizing their ability to hide profits from the assembler."[28]

6.11.4 The Value of Relationships for Cost Reduction

Another common assumption is that if suppliers know they will have to bid, they will implement effective long-term cost reduction efforts. However, the most successful real progress in cost reduction has come from long-term relationships, where manufacturers work together with suppliers.[29,30]

Again, citing *The Machine That Changed the World*, which said that in Lean Production companies, suppliers *"are not selected on the basis of bids, but rather on the basis of past relationships and a proven record of performance."*[31]

Honda's criterion for selecting suppliers is the attitudes of their management.[32] As a philosophy-driven company, Honda feels it is easier to teach product and process knowledge than to find a technically capable supplier with the right attitudes, motivation, responsiveness, and overall competence.[33]

Much of the real cost reduction opportunities are not just at the assembler or at the supplier, but rather *in their relationship*. In a thorough study of Japanese Lean manufacturers, *When Lean Enterprises Collide,* Robin Cooper stated that, "it is no longer sufficient to be the most efficient firm; it is necessary to be part of the most efficient supplier chain." The key to accomplishing this is inter-company cooperation, summarized by Cooper as follows:

> "The blurring of organizational boundaries becomes critical as competition intensifies because it not only reduces the time it takes the entire supplier chain to bring out new products with increased functionality but also allows quality to be improved while reducing cost."

Cooper recommends partner companies "create relationships that share organizational resources, including information that helps improve the efficiency of the interfirm activities."[34]

Such inter-company cooperation offers significant cost reduction opportunities, especially if suppliers can build parts on demand for build-to-order assemblers.[35] In this way, both organizations avoid all the cost and risk of parts inventory in addition to minimizing many categories of overhead for procurement, material overhead, expediting, warehousing, internal distribution, and so forth.

Switching suppliers every time a competitor drops its price is incompatible with this strategy and can jeopardize ongoing relationships. A *Fortune* magazine article by Jerry Useem, Director of the Center for Leadership and Change at the Wharton Business School, summarized the failure of an online bidding site:

> "For the bulk of spending, corporations have long been moving in precisely the opposite direction, establishing deep relationships with a few favored suppliers in a "total cost" approach. Under this approach, price is but one of a host of criteria, which include quality, cycle time, service, and geography."[36]

The same article also revealed some realities about the purchasing process that question how welcome bidding would be for typical buyers:

> "In retrospect, say analysts, most B2B efforts betrayed pronounced cluelessness about how industrial buying actually works. Start with the supposition that purchasing managers would be thrilled to take bids online from dozens if not hundreds of suppliers each vying to be the lowest bidder."

Another article that proclaimed low-bid auction sites as "yesterday's darlings," said that: "Many companies just weren't willing to dump the networks of suppliers they had built up over the years and do all their buying through a new, unfamiliar medium."[37]

6.11.5 Cheap Parts: Save Now, Pay Later

Actually, this phrase should be, more precisely: *save a little now, pay a lot later*. Often, trying to save money on purchase cost has the unintended effect of driving up other costs many times the assumed savings, like the English adage: *penny wise, pound foolish*, or the more colloquial, *you get what you pay for*. The chain reaction of consequences from cheap parts is graphically shown in Figure 1.2.

Cheap parts are usually just that—*cheap parts* that usually earn the stereotypical image of poor quality, which will add significant cost in the plant and cost even more if bad products get out, not to mention potential hazards to life and limb and loss of corporate reputations.

Even though quality disasters may be thought of as infrequent anomalies, these costs must be included in the company's cost of quality metric, not just considered a one-time charge. Programs that aim to improve quality should be justified in their ability to prevent all quality costs, from an accumulation of many up to and including "the big one."

In the 1990 J.D. Power ratings of automobile reliability, Mercedes-Benz received the top rating. But by 2003, Mercedes had slipped to 26 out of the

37 cars ranked.[38] One of the reasons for the drop in quality was cited by European analysts:

> "Executives of what then was Daimler-Benz grew worried about escalating production costs in the early 90s. Executives then made a policy decision to start trimming costs by notching down specifications for many components."[39]

6.11.6 Reduce Total Cost Instead of Focusing on Cheap Parts

In addition to the cost ineffectiveness of part bidding and its detrimental effects on relationships, there is the compelling argument that other cost categories provide much greater opportunities for real cost reduction, as is emphasized throughout this book. There are many ways to minimize total cost; Chapter 7 presents easy ways to quantify total cost.

There are enormous opportunities to reduce total cost throughout the supply chain, without any negative consequences, by designing for manufacturability, specifying off-the-shelf parts, eliminating the costs of setup, inventory, and obsolescence, and substantially reducing the costs of quality, distribution, and material overhead.

6.11.7 Value of High-Quality Parts

Receiving high-quality parts is especially important to Lean and build-to-order operations because:

- Dock-to-line deliveries count on "quality assured at the source" so that incoming inspections are not necessary and parts can go straight to all the points of use. This not only saves on the cost of incoming inspections, but also enables spontaneous resupply techniques.[40]
- One-piece-flow operations are more sensitive to failed parts looping back and disrupting the flow.
- Testing large batches of identical parts is not compatible with flexible operations.
- Having part quality assured at the source plus the continuous quality feedback of one-piece flow will enable Lean plants to assure quality by process controls rather than expensive and time-consuming testing—or risky low bidding.
- Raising the quality of parts improves product quality (Section 10.3).

Fortunately, there are suppliers that practice *kaizen* continuous improvement and can provide high-quality products at a low price. Because of their cooperative nature and big picture orientation, these companies would naturally align with customers who value long-term relationships instead of participating in the bidding process.

Another related trend is becoming apparent: The best suppliers are shunning the bidding process. Philip L. Carter, professor of purchasing at Arizona State University, Tempe, and executive director of the Center for Advanced Purchasing Studies (CAPS), concluded that: "Manufacturers that are tempted to source key parts and materials through a trading exchange may find it difficult to connect with the most innovative, quality conscious vendors, since many likely will boycott the auction bidding process, viewing it 'as a margin-squeezing play'."[41]

6.12 MAXIMIZING FACTORY EFFICIENCY

Rapid product development can more quickly phase out older, more costly products with new-generation cost-effective products. The older, less-efficient products, in addition to having higher direct costs for labor and materials, have higher overhead demands for engineering change orders and firefighting. More efficient production, from better designed products, can result in more output from existing plants and equipment. For growing companies, this extra output might spare the company, or at least defer, the expense of adding new equipment or expanding facilities.

6.13 LOWERING OVERHEAD COSTS WITH FLEXIBILITY

Flexibility can reduce overhead costs significantly. Some interesting parallels exist between flexibility and quality. Decades ago, it was commonly believed that quality cost more. At that time, Philip Crosby wrote the book, *Quality Is Free*,[42] and showed that the gains from lowering the cost of quality would pay for quality improvement efforts, thus making it free.

Similarly, the financial gains derived from flexible operations can more than pay for the effort to make operations flexible. These gains will become a source of competitive advantage over companies that do not embrace Lean Production, build-to-order, and mass customization.

There is a lot of *working capital* tied up in various forms of inventory: raw materials and parts inventory; WIP inventory; and finished goods inventory in factory warehouses, at distributors, and at the dealers. *Fortune* magazine estimates that, for Fortune 500 companies, working capital averages an amount equal to 20% of sales.[43]

Ironically, inventory shows up on the balance sheet as an *asset* when, in fact, inventory is really a *liability* to the operation of any manufacturing plant, especially those needing to be flexible. This point was one of the revelations presented in Eli Goldratt's novel, *The Goal*, when managers of the fictional plant, faced with extinction, realized that they had to focus on *the goal* (making money) instead of letting their behavior be dictated by irrelevant performance and cost accounting metrics.[44]

One progressive materials manager of a processing equipment company said that after much successful work to reduce inventory, he got a call from the company controller, who was having trouble preparing the annual report because the inventory had been reduced so much that it was "lowering company assets," according to their traditional accounting rules. Companies must make sure they are pursuing the real "goal" instead of irrelevant metrics.[45]

The following discussion describes several opportunities to lower the costs of inventory and other overhead costs by designing flexibility into products and plants.

6.14 MINIMIZING CUSTOMIZATION/ CONFIGURATION COSTS

Many companies offer customized goods but do not do it cost-effectively. By *mass customizing* products, the customization process is built into the system—the product design and the manufacturing operation. For more on mass customization, see Section 4.3 or read the companion book for matching improvements in operations, *Build-to-Order & Mass Customization.*[46]

Thus, the mass-customizer has cost advantage over companies that are inefficient at customization and configuration. Two underreported costs related to customization are custom engineering and changing or modifying standard designs and processes. Both of these activities usually cost much more than is indicated by current cost systems. Many companies do not even keep track of engineering costs by project. In addition, engineers often try to get a lot of "free" help from various support people who are "just on overhead."

The extra manufacturing cost to perform ad hoc customization and configuration reactively is much more than it would be for mass-customized products. Reactive customization costs often include extra tooling, lengthy setups for small runs, inefficient production control, low equipment utilization, special programming, slow and frequent learning curves, special tests and inspections, and lots of fire drills to shove customized products through mass production factories. Further, these low-volume customized products may disrupt the manufacture of the standard products, thus increasing their cost. Thus, it is often the case, in companies with traditional accounting systems, that custom products are really being subsidized by the standard products.

6.15 MINIMIZING THE COST OF VARIETY

A large part of working capital is tied up in the cost of variety, which is discussed in depth in Chapter 3 of *Agile Product Development for Mass Customization*.[47] Eliminating setup and reducing the batch size to one eliminates most of the cost of variety.

The key element of Lean Production is setup elimination. If setup could be eliminated, then operations would be flexible, meaning that every product could be different and, yet, still reap mass production efficiencies. In low-volume operations, setup could be caused by any effort to do something "different"; for instance, to get parts, change dies and fixtures, download programs, find instructions, or any kind of manual measurement, adjustment, or positioning of parts or fixtures. Setup reduction is discussed in *Build-to-Order & Mass Customization*, in Chapter 8, "On-Demand Lean Production."[48]

6.15.1 Work-in-Process Inventory

WIP inventory can be virtually eliminated by setup reduction, JIT, and design commonality of parts and processes. WIP inventory costs rise proportional to batch size, except when the batch size is one, in which case WIP inventory can be virtually eliminated. WIP inventory carrying cost could be 25% of its value per year.[49] Thus, eliminating WIP inventory could result in substantial savings.

6.15.2 Floor Space

Floor space can be reduced because of reduced inventories, elimination of the forklift aisles necessary to move large batches of parts, elimination of kitting, and higher utilization of machinery and people. Appreciation of the cost and value of floor space varies according to the need to expand manufacturing. But floor space reduction should be constantly pursued. The cost of expanding manufacturing space is a large step function that may force a company to move away from an area that is too crowded or expensive. Reduction of the need for floor space can provide an attractive alternative to expansion or relocation. Further, floor space requirements can be reduced faster than new facilities can be built. The lead time for physical plant expansion is so large that such plans must be started well ahead of the anticipated need, often based on inaccurate long-range marketing projections. Between 1991 and 1994, Compaq Computer quintupled production without increasing factory space by implementing programs such as WIP inventory reduction.[50]

6.15.3 Internal Logistics

Internal transportation costs, such as forklift activity, can be reduced. In fact, they can be eliminated when parts and products flow individually between adjacent workstations, instead of in large, heavy bins between distant workstations.

6.15.4 Utilization

Machine tool utilization is improved with less setup, thus reducing equipment cost. This has big cost savings potential for expensive equipment, such as CNC machining centers, surface-mount printed circuit assembly equipment, or expensive testers. Machine tool utilization can be as low as 10%, which means the equipment is producing parts only 10% of the time the machine is available for work; the remainder is setup or waiting. It is important to realize that *doubling the utilization rate will double output*. If production equipment previously had utilization of 30%, the output could be doubled to a utilization of 60% and another 1.5 times to 90%. Utilization improvement is a cost-effective way to increase production output, and it is quicker, considering the lead time to procure and install new production equipment.

6.15.5 Setup Costs

Setup labor expenses can be eliminated, including the labor cost to change machine setups and to retrieve parts, tools, and drawings.

6.15.6 Flexibility

Flexibility can improve the balance of labor and machinery utilization in sequential operations such as assembly lines. Products built in flexible lines can be optimally ordered (a product with high demands on the A process and low demands on the B process followed by a product with low A and high B demands) to offset imbalances in the workloads of adjacent machinery or people, using a concept known as *product complementarity.*[51]

Production can quickly adapt to changing market conditions by building all the products on the same flexible line. Inflexible operations often face a dilemma when the demand for "model A" has exceeded capacity, while "model B" is having a sales slump. Manufacturing may have adequate overall capacity, but the model A line or plant will not be able to satisfy demand, while the model B line or plant is partly idle or laying off people. A slightly more flexible approach would be to be able to move people from the "B" line to the "A" line, but that assumes adequate equipment capacity on line "A." Fully flexible operations would simply pull more model A products, and fewer model B products, through the flexible line(s).

Operational flexibility can allow companies to transfer production from one flexible line or plant to another to respond to changing market conditions, rather than the more expensive alternatives of overtime, rapidly bringing contract labor up to speed, and spontaneous outsourcing to ease production bottlenecks for the product that is in demand. Similarly, by transferring production, companies can avoid layoffs at plants making products that are in low demand. Mazda resolved such a dilemma by moving production of the popular Miata from its Hiroshima plant to the Hofu plant, which was underutilized with Mazda 626 production. This manufacturing flexibility allowed Mazda to combine production of a niche sports car with a family sedan in the same plant to balance output at both plants.[52]

6.15.7 Kitting Costs

Kitting cost and space can be eliminated. Without flexibility, there are labor costs and space requirements to gather all the parts for a batch, kit them together, and deliver them to manufacturing.

6.16 MINIMIZING MATERIALS MANAGEMENT COSTS

Purchasing costs can be reduced if there are fewer purchasing actions for fewer part types. Standardized parts will cost less because of the greater purchasing leverage of higher volume parts. Automatic resupply techniques, such as kanban, min/max, and breadtruck, can resupply parts with virtually no material overhead.[53]

Vendor fabrication and assembly are more feasible, and thus quicker and less costly, if parts are well designed and documented, especially if the vendor was part of the design team. Sometimes it may be more cost-effective to have the vendor design the parts or subsystems, as is common nowadays in the automobile industry.

Fewer part numbers means less material overhead for raw materials and parts inventories, documentation, controls, etc. There will also be less expediting cost for seldom-used parts that are difficult to obtain. In every seminar, the author holds up a piece of paper, representing a standard parts list, and asks the most senior purchasing manager in the room the question: "How much of your department's activities are devoted to ordering these standard parts?" The typical answer is 10%, which means that material overhead for standard parts should be one-tenth that of the remaining oddball parts.

6.17 MINIMIZING MARKETING COSTS

When manufacturers keep listening to customers' wants and needs and keep designing and manufacturing products to satisfy these evolving needs, this results in *learning relationships*, which result in the ability to *keep customers forever.*[54] Not only is this good for generating revenue, but it also saves the considerable cost of acquiring new customers to meet growth objectives. Studies such as the one done by the Technical Assistance Resource Project for the US Office of Consumer Affairs have shown that the price of acquiring new customers is five times greater than the cost of keeping old ones.[55]

6.18 MINIMIZING SALES/DISTRIBUTION COSTS

There are considerable cost reduction opportunities in the warehousing and distribution of products. The physical distribution system accounts for about 10% of the gross national product.[56]

Designing modular products and concurrently engineering products and production systems to build products on demand can save a lot of money with respect to the way products are configured, packaged, shipped, distributed, and sold. In fact, build-to-order can eliminate most of the distribution chain costs. Being able to build-to-order and ship from the plant eliminates warehousing and associated costs of distribution from the plant to the customer. In industries where product variety is considerable, such as clothing, this cost can be enormous, considering the number of sizes and styles.

6.19 MINIMIZING SUPPLY CHAIN COSTS

Supply chain management has become a strategic competitive advantage, especially for companies like Hewlett-Packard.[57] Peter Drucker points out the opportunities to minimize cost in the supply chain: "Process-costing from the machine in the supplier's plant to the checkout counter in the store also underlies the phenomenal rise of Wal-Mart. It resulted in the elimination of a whole slew of warehouses and reams of paperwork, which slashed costs by a third."[58] Sections 4.2.1 and 4.2.2 show how to simplify supply chains.

6.20 MINIMIZING LIFE CYCLE COSTS

An often neglected part of total costs is *life cycle costs*, which are those costs that are incurred over time, such as service, repair, maintenance, field failures, warrantee claims, legal liabilities, changes over the life of the product, and subsequent product developments. Products can be designed to minimize life cycle costs. The cost of changes can be minimized by a methodical product definition and thorough product development. Change costs and subsequent product development costs can be minimized with modular product architecture, where many modules can remain unchanged as other modules are updated or redesigned.

6.20.1 Reliability Costs

Designing for reliability can minimize many costs related to product reliability. Several techniques to maximize reliability are presented in Chapter 10.

6.20.2 Field Logistics Costs

Spare parts logistics and field service can be greatly simplified, and thus cost less, with part standardization and modular design. Products designed around common parts have smaller spare parts kits. This could lower the effective product price for savvy customers who add the cost of spare parts kits to the product's list price. Part standardization can also result in less downtime due to part shortages. Service costs can be reduced if failed modules can be quickly replaced and repaired in more efficient facilities.

6.21 SAVING COST WITH BUILD-TO-ORDER

If products and production processes can be designed to build products to order, then many costs can be saved.

6.21.1 Factory Finished Goods Inventory

Factory finished goods inventory can be eliminated by building products to order, instead of building to forecast and then holding products in a warehouse until ordered by distributors or customers. Like WIP inventory, finished goods inventory may cost the same to carry, except that finished goods are completed and therefore are more valuable. Using 25% of value per year, $10 million worth of finished goods in inventory would cost $2.5 million per year to carry. Build-to-order can eliminate factory finished goods inventory and, thus, save its yearly inventory carrying cost.

6.21.2 Dealer Finished Goods Inventory

Dealer finished goods inventory can be almost eliminated if resupply orders can be quickly filled and delivered to the customer. As with factory inventory, dealer inventory has a carrying cost. Even if the dealer or distributor is separate from the manufacturer, the carrying cost will have to be paid, ultimately by the customer.

For example, an automobile dealer with 200 vehicles in stock, with an average value of $30,000, would have $6 million worth of inventory. Using a yearly carrying cost of 25%, the carrying cost would be $1.5 million per year. Put another way, if a car sits there for a year, its inventory carrying cost would be a quarter of $30,000 or $7,500, so it would have to sell for that much more to pay for its inventory carrying costs! This cost could be eliminated by build-to-order. The European automotive industry sponsored a research project to investigate build-to-order for cars, which resulted in the book, *Build to Order: The Road to the 5-Day Car.*[59]

When new car prices exceed customers' ability to afford them, customers buy more used cars as a cost-effective alternative.[60] Built-to-order new automobiles could compete well against used cars, with lower prices, because the new cars could avoid dealer inventory expenses, whereas used cars must be stocked in inventory by definition.

6.21.3 Supply Chain Inventory

Supply chain inventory can be minimized because build-to-order products do not need to be stocked at various warehouses along the supply chain: at distributors, consolidators, forwarders, and so forth. Similarly, a build-to-order system pulls parts from suppliers on a just-in-time basis, thus eliminating parts inventory along the supply chain. Regardless of who pays for this inventory, the customer ultimately must pay a higher price. Eliminating excessive supply chain inventory costs will allow customers to pay less for equivalent products.

Companies known for rapid deliveries, such as FedEx, are providing companies with inventory-less direct deliveries of parts and products to and from factories. Adding the value of increased sales from customer satisfaction would make inventory-less distribution even more attractive.

6.21.4 Interest Expense

Less interest expense will be incurred for expensive components in products that sell sooner because of the quicker throughput of flexible plants and the elimination of finished goods inventory.

6.21.5 Write-Offs

Inventory write-offs can be eliminated because there would be no products in inventory that could deteriorate, incur damage, or become obsolete. If there is a substantial finished goods inventory of expensive products, at the end of those products' lives the obsolescence write-offs could be enormous.

6.21.6 New Technology Introduction

Quicker transitions to new technology are possible if there are no older technology products waiting in inventory that must be sold first. Dell Computer is often the first to introduce new technology because they don't have large inventories that have to be sold first.

6.21.7 MRP Expenses

Bill-of-material and MRP (materials requirement planning) expenses could be minimized. Build-to-order could minimize overhead expenses to translate forecasts into purchase orders using materials ordering requirements with MRP systems. Mass customization can generate bills-of-material on demand for products in families instead of writing and storing an individual BOM for each product variation.

6.22 EFFECT OF COUNTERPRODUCTIVE COST REDUCTION

Companies will have a hard time achieving real cost reduction if they are trying cost reduction attempts that are, in fact, counterproductive:

- Manufacturing companies that offshore their manufacturing will have a hard time implementing concurrent engineering when there are no manufacturing people there with whom to be "concurrent." In many offshoring situations, people in engineering and manufacturing are not even working at the same time. For more, see Section 2.8 (co-location), Section 4.8 (offshoring), and the articles on outsourcing[61] and offshoring[62] at www.HalfCostProducts.com.

- Manufacturing companies that try to take cost out after the product is designed will find it difficult and a waste of resources, for reasons discussed in Section 6.1.

- Manufacturing companies who insist on bidding custom parts are, in effect, precluding vendor partnerships and, thus, preventing those vendors from helping the company design the parts. The benefits of vendor partnerships are discussed in Section 2.6. These companies also encounter all the problems of low bidding discussed in Section 6.11.

In companies that practice all three of the above, a high percentage of product development resources will spend most of their time: making change orders to try to implement DFM (because it couldn't be done with concurrent engineering); trying to take cost out using change orders after the product is designed, converting documentation for outsourcing; getting outsourcers up to speed; dealing with quality and delivery problems, and so forth. In his travels, the author has encountered several companies that spend *two-thirds of product development resources* on the above three activities (which are cited at the bottom of the home page of www.HalfCostProducts.com under the heading "How Not to Lower Cost"), which really puts their future in doubt if that future depends on new product development. More counterproductive policies are discussed in Section 11.5.

NOTES

1. See "How Not to Lower Cost" (and the linked articles) on the bottom of the home page at www.HalfCostProducts.com.
2. James Morgan and Jeffrey K. Liker, *The Toyota Product Development System* (2006, Productivity Press), Chapter 4, "Front-Load the PD Process to Explore Alternatives Thoroughly."

3. Robert G. Atkins and Adrian J. Slywotzky, "You Can Profit from a Recession," *Wall Street Journal*, February 5, 2001, p. A22.

4. Robin Cooper and Robert S. Kaplan, "How Cost Accounting Distorts Product Costs," *Management Accounting*, April 1988.

5. H. Thomas Johnson and Robert Kaplan, *Relevance Lost: The Rise and Fall of Management Accounting* (1991, Harvard Business School Press).

6. Douglas T. Hicks, *Activity-Based Costing for Small and Mid-Sized Businesses: An Implementation Guide* (1992, John Wiley), p. 20.

7. Michael R. Ostren, Terrence R. Ozan, Robert D. McIlhattan, and Marcus D. Harwood, *The Ernst & Young Guide to Total Cost Management* (1992, John Wiley & Sons), p. 146.

8. Dr. Anderson facilitates product-specific workshops (Appendix D) after customized in-house seminars. These workshops consist of many brainstorming sessions on concept simplification and architecture optimization. More challenging endeavors, such as developing half-cost products, may need his concept studies (also in Appendix D), in which he generates breakthrough ideas that concurrent engineering teams can develop into manufacturable products.

9. James P. Womack and Daniel T. Jones, *Lean Thinking: Banish Waste and Create Wealth in Your Corporation,* (1996, Simon & Schuster), p. 27.

10. David M. Anderson, *Build-to-Order & Mass Customization: The Ultimate Supply Chain Management and Lean Manufacturing Strategy for Low-Cost On-Demand Production Without Forecasts or Inventory* (2008, CIM Press), Figure 2.1, "Inventory Carrying Cost Since 1961." See book description in Appendix D.

11. Ibid., Chapter 9, "Mass Customization."

12. Ibid., Chapter 7, "Spontaneous Supply Chains."

13. Robin Cooper and Robert S. Kaplan, "How Cost Accounting Distorts Product Costs," *Management Accounting*, April 1988.

14. Philip B. Crosby, *Quality Is Free* (1979, Mentor Books).

15. James P. Womack, Daniel T. Jones, and Daniel Roos, *The Machine That Changed the World: The Story of Lean Production* (1991, Harper Perennial).

16. Jack Campanella, editor, *Principles of Quality Costs: Principles, Implementation, and Use* (1999, Quality Press, American Society for Quality).

17. Kiyoshi Suzaki, *The New Manufacturing Challenge: Techniques for Continuous Improvement* (1987, Free Press).

18. Jordan D. Lewis, *The Connected Corporation: How Leading Companies Win Through Customer–Supplier Alliances* (1995, Free Press), Chapter 8.

19. Womack, Jones, and Roos, *The Machine That Changed the World,* Chapter 6, "Coordinating the Supply Chain."

20. Lewis, *The Connected Corporation,* p. 38.

21. Peter F. Drucker, *Managing in a Time of Great Change* (1995, Truman Talley Books/ Dutton), p. 117.

22. Philip L. Carter, professor of purchasing at Arizona State University, Tempe, and executive director of the Center for Advanced Purchasing Studies, was cited in *Industry Week* (February 12, 2001, p. 43) as observing that "Many purchasing organizations are under heavy pressure form the corporate brass to implement some form of e-commerce."

23. John H. Sheridan, "Proceed with Caution," *Industry Week,* February 12, 2001, pp. 38–44. One of the cover stories on manufacturing exchanges.

24. For more information on customized in-house DFM seminars, see Appendix D or www.design4manufacturability.com/seminars.htm.

25. Morgan and Liker, *The Toyota Product Development System*, Chapter 10, "Fully Integrate Suppliers into the Product Development System, p. 200.

26. "Smart Partners," *Business Week*; review of *The Connected Corporation* by Jordan D. Lewis, December 10, 1995.

27. Sydney Finkelstein, *Why Smart Executives Fail: And What You Can Learn from Their Mistakes* (2003, Portfolio/Penguin), p. 62.

28. Womack, Jones, and Roos, *The Machine That Changed the World*, Chapter 6, "Coordinating the Supply Chain," p. 142.

29. Yasuhiro Monden, *The Toyota Production System,* Second edition (1993, Institute of Industrial Engineers).

30. Womack, Jones, and Roos, *The Machine That Changed the World*, Chapter 6, "Coordinating the Supply Chain."

31. Ibid., p. 146.

32. Jeffrey Pfeffer and Robert I. Sutton, *The Knowing–Doing Gap: How Smart Companies Turn Knowledge into Action* (2000, Harvard Business School Press), p. 23.

33. John Paul MacDuffie and Susan Helper, "Creating Lean Suppliers: Diffusing Lean Production through the Supply Chain," *California Management Review,* Summer 1997, pp. 118–150.

34. Robin Cooper, *When Lean Enterprises Collide* (1995, Harvard Business Press), Chapter 9, "Interorganizational Cost Management Systems."

35. Anderson, *"Build-to-Order & Mass Customization."*

36. Jerry Useem, "Dot-Coms: What Have We Learned?" *Fortune,* October 30, 2000, p. 92.

37. "Lessons from the Dot-Com Crash," *Fortune,* October 30, 2000, p. R8.

38. Lee Hawkins, Jr., "Finding a Car That's Built to Last," *Wall Street Journal,* July 9, 2003, p. D1.

39. John O'Dell, "Even Mercedes Hits a Few Speed Bumps," *Los Angeles Times,* July 13, 2003, pp. C1 and C4.

40. Anderson, *Build-to-Order & Mass Customization,* Chapter 7, "Spontaneous Supply Chain."

41. Carter., p. 43.

42. Crosby, *Quality Is Free.*

43. Shawn Tully, "Raiding a Company's Hidden Cash," *Fortune,* August 22, 1994, p. 82.

44. Eliyahu M. Goldratt, *The Goal,* Second revised edition (1992, North River Press), Chapter 33, p. 268.

45. Goldratt's *The Goal* makes this point often about the need to do "what makes sense" to achieve "the goal," rather than basing decisions and actions on attempts to satisfy arbitrary performance metrics and cost measurements that only make a small part of the system *appear* to look productive or cost-effective.

46. Anderson, *Build-to-Order & Mass Customization,* Chapter 9, "Mass Customization."

47. David M. Anderson, with an introduction by B. Joseph Pine, II, *Agile Product Development for Mass Customization* (1997, McGraw-Hill), Chapter 3, "Cost of Variety."

48. Anderson, *Build-to-Order & Mass Customization,* Chapter 8, "On-Demand Lean Production." See book description in Appendix D.

49. Ibid, Chapter 2, Figure 2.1.

50. Ronald Henkoff, "Delivering the Goods," *Fortune,* November 28, 1994, p. 62.

51. Marshall Fisher, Anjani Jain, and John Paul MacDuffie, "Strategies for Product Variety: Lessons from the Automobile Industry," Working paper from the Wharton School, University of Pennsylvania, January 16, 1994, p. 26.
52. Ibid., p. 31.
53. Anderson, *Build-to-Order & Mass Customization*, Chapter 7, "Spontaneous Supply Chain," or see summary of spontaneous resupply techniques in Section 4.2.
54. B. Joseph Pine, II, Don Peppers, and Martha Rogers, "Do You Want to Keep Your Customers Forever," *Harvard Business Review*, March–April 1995, p. 103.
55. Wilton Woods, "After All You've Done for Your Customers, Why Are They Still Not Happy," *Fortune*, December 11, 1995, p. 180.
56. Robert V. Delaney, *Sixth Annual State of Logistics Report*, Cass Information Systems, St. Louis, MO, June 5, 1995, Figure #8.
57. Dr. Corey Billington, "Strategic Supply Chain Management," *OR/MS Today*, April 1994, pp. 20–27.
58. Drucker, *Managing in a Time of Great Change*, p. 117.
59. Glenn Parry and Andrew Graves, Editors, *Build to Order: The Road to the 5-Day Car* (2008, Springer-Verlag, London).
60. Douglas Lavin, "Stiff Showroom Prices Drive More Americans to Purchase Used Cars," *Wall Street Journal*, November 1, 1994, p. 1.
61. See the outsourcing article at the author's website: www.HalfCostProducts.com/outsourcing.htm.
62. See the offshoring article at the author's website: www.HalfCostProducts.com/offshore_manufacturing.htm.

7

Total Cost

In order to appreciate all of the cost savings and revenue enhancements cited in Chapter 6, it would be highly advantageous to be able to *quantify all costs*. If costs were tracked on a *total cost* basis, then the cost-saving potential of well-designed products could be known and products could be assigned appropriate overhead charges and thus competitive prices. However, if total cost is not tracked, then well-designed products may be assigned the same overhead as loser products, which is an unfair burden and may ultimately compromise the well-designed product's success.

Most companies have such inadequate cost systems that it actually hinders good product development and distorts product development decisions. Merely reporting labor and material costs encourages (sometimes forces) engineers to specify cheap parts and low bidders to achieve "cost targets" and move manufacturing offshore, away from engineering, "to save cost," which, in reality, thwarts concurrent engineering. Many references cited in this chapter have been pointing out the deficiencies, and the consequences, of traditional cost measurements for the last two decades. The quotes from these authoritative sources may be helpful to build interest in your own total cost program.

Products with too much setup, inventory, firefighting, engineering change orders, excessive parts variety, low equipment utilization, and high quality costs should have a higher overhead rate. Products that are designed, using the methodologies presented herein, for quick and easy manufacture should have a much lower overhead rate. *Overhead rates should be proportional to overhead demands*, which vary by product.

The ability to quantify total cost is critical to designing for manufacturability. Let us consider the proverbial "Model-T plant"—a plant with only one product and no variations. *The Ernst & Young Guide to Total*

Cost Management discussed the cost management implications of a single-product plant:

> "If product variety were absent, the business environment would be simple. ... In a world like this, you could do product costing literally on the back of an envelope. You would simply divide the total production costs by the total production volume to calculate a unit cost."[1]

However, as Cooper and Kaplan[2] pointed out in an article with the profound title, "How Cost Accounting Distorts Product Costs," overhead costs "vary with the diversity and complexity of the product line." And, product diversity and complexity have gotten out of hand as companies keep *adding* products but *don't rationalize any away*, as discussed in Appendix A. Thus, it is important to quantify overhead cost to assign a product's overhead *charges* proportional to all its overhead *costs*.

7.1 VALUE OF TOTAL COST

Total cost systems quantify *all* costs and, therefore, should be the basis of all cost decisions. All costs should be attributed to some product or service. No varying costs should be averaged over all products, but in most companies, overhead is "spread like peanut butter." Total cost numbers should be the foundation of relevant strategies that lower cost and encourage activities that actually reduce *total cost*.

7.1.1 Value of Prioritization and Portfolio Planning

Quantifying all costs is necessary to compute the real profitability of products. Knowing the relative profitability of all product variations will help you prioritize:

- The highest return efforts in product development and improvement initiatives in quality and operations
- The highest return orders to accept and which low-return orders to turn down

For both the new product development portfolio and sales strategies, quantifying all costs will help you:

- Make the best decisions on which products to redesign to lower cost
- Decide which legacy products to drop, outsource, or improve
- Decide on investments to develop modules that can be used on many current and future products
- Optimize the product portfolio of products, product lines, customers, and market segments
- Generate a profile to predict profitability at the estimating stage
- Determine sales incentives based on profitability instead of sales dollars or units sold

7.1.2 Value of Product Development

Better order prioritization will prevent draining of product development resources on low-potential products. Knowing the real cost of cost reduction efforts can redirect resources so they can make more money through product development, Lean Production, quality improvement programs, and so forth. Investing now in the development of hardware and software modules will benefit future product developments. Quantifying the *cost of quality* will discourage selecting or substituting cheap parts to "save cost."

7.1.3 Value of Resource Availability and Efficiency

Better prioritization and portfolio planning will prevent the waste of valuable resources for hard-to-build orders and hard-to-customize variations. Total cost numbers will help rationalize them away or justify profit-and-loss centers for those items, as well as low-volume sales, legacy parts, and spare parts.

Quantifying benefits will also improve *resource efficiency,* by helping to justify design tools, training, and other overhead reduction programs such as Lean Production and Six Sigma quality improvements. Similarly, it will help justify automation, production tools, versatile CNC machine tools, and setup reduction efforts.

7.1.4 Value of Knowing the Real Profitability

Quantifying all costs will enable generation of the *real profitability* numbers for all products and product variations, which will identify low-profit and money-losing products, so that you can:

- Implement more realistic pricing.
- Turn down orders in favor of more profitable endeavors.

- Rationalize them away (Appendix A) and steer customers to newer, better (more profitable) products that can be delivered quicker. The sales force could say, *"Sorry, that old model has been discontinued, but we have a better one that we can get to you quicker at lower cost."*
- Build remaining products in self-supporting profit-and-loss centers with dedicated people, not borrowed from product development efforts.
- Outsource to contract manufacturers specializing in low-volume and legacy products.

7.1.5 Value of Quantifying All Overhead Costs

Quantifying all overhead will result in lower overhead charges allocated to products designed to lower overhead costs. Quantifying the cost of quality will encourage behavior that lowers quality costs and does not raise quality costs by trying to lower the usual quantified costs, such as parts costs.

Quantifying all costs will enable real cost reduction actions, which often appear to raise a BOM line, but actually lower total cost. For instance, it will encourage consolidation, standardization, modularity, and versatility. Quantifying all costs will enable more realistic decisions on offshoring versus integrated manufacture (discussed in Section 4.8).

7.1.6 Value of Supply Chain Management

Total cost will remove the biggest obstacles to standardization: the perception that a better part will cost more if it raises a BOM line. However, the overhead savings will be much more.

Knowing all the costs of part and material availability problems (expediting, buying lifetime inventory, change orders to design in replacement parts, and the resources to do these) will justify searching for more available parts and even paying a little more to ensure that availability, knowing that will save more on a total cost basis. Total cost numbers can lead to the best decisions on designing and/or making parts versus buying them off-the-shelf.

7.2 QUANTIFYING OVERHEAD COSTS

The first step in quantifying overhead is to acknowledge the deficiencies in the current cost system, which is the central theme of Johnson and

Kaplan's pivotal book, *Relevance Lost: The Rise and Fall of Management Accounting*.[3] Traditional cost accounting systems were designed to present operational results and the financial position of the organization *as a whole* for investors and for agencies that tax or regulate. However, managers and engineers need *relevant cost information* to make good decisions. Typical problems caused by conventional cost systems are discussed in this section.

7.2.1 Distortions in Product Costing

This costing distortion is discussed at length in the Cooper and Kaplan reference cited above. Johnson and Kaplan concurred:

> "The management accounting system fails to provide accurate product costs. Costs get distributed to products by simplistic measures, usually direct-labor based, that do not represent the demands made by each product on the firm's resources."

The *Guide to Total Cost Management* asserts that product costs "are distorted because each product typically includes an assignment of overhead that was allocated on some arbitrary basis such as direct labor, sales dollars, machine hours, material cost, units of production, or some other volume measure."[4]

Distorted product costing results in distorted pricing that can underprice some products so low that they actually lose money and overprice other products to the point where they are uncompetitive.[5] Distorted product costing results in a distorted perception of the profitability of all the company's products. This distorted view of profitability can cause managers to "feed the problems and starve the opportunities," with detrimental effects on product development priorities. Understanding the real profitability will allow companies to drop unprofitable products and focus on profitable ones.

7.2.2 Cross-Subsidies

Averaging overhead will cause cross-subsidies, where *high-volume products* subsidize *low-volume products* and *standard products* subsidize *custom products*. Johnson and Kaplan have stated unequivocally, "The standard product cost systems, which are typical of most organizations, usually lead to enormous cross-subsidies across products."[6]

Cooper, Kaplan, et al., in their Institute of Management Accountants-sponsored study, summarized what their eight case study manufacturing organizations discovered after they implemented total cost measurements:

> "The manufacturing companies generally found, as expected, that low-volume, complex products tended to be much more expensive than had been calculated by the existing standard cost system."[7]

One of the most dangerous consequences of cross-subsidies is penalizing new-generation DFM products and programs such as build-to-order and mass customization by making them pay the same overhead charges as the products that have high overhead demands. Such unfair charges could ultimately thwart new-generation products and programs.

7.2.3 Relevant Decision Making

Good decisions are the keys to success in any business venture, and this especially applies to product development. Unfortunately, many managers and engineers try to make decisions by the numbers, when the numbers are misleading or even irrelevant. Again, quoting Johnson and Kaplan:

> "Ironically, as management accounting systems became less relevant and less representative of the organization's operations and strategy, many companies became dominated by senior executives who believed they could run the firm 'by the numbers'."[8]

And again quoting the *Guide to Total Cost Management:*

> "If the costs are wrong, then all decisions about pricing, product mix, and promotion could be undermining long-term profitability."[9]

Johnson and Kaplan stated that accurate, relevant numbers, based on total cost, can lead to much better decision making:

> "The management accounting system also needs to report accurate product costs so that pricing decisions, introductions of new products, abandonments of obsolete products, and responses to the appearance of rival products can be made with the best possible information on product resource demands."[10]

Johnson and Kaplan also stated that accurate, relevant numbers are important to support improvement programs: "... An ineffective management accounting system can undermine even the best efforts in product development, process improvement, and marketing policy."

7.2.4 Cost Management

Since one of the major challenges of design for manufacturability is to produce products at low cost despite the other challenges of speed and quality, cost management takes on a new level of importance. But conventional cost management systems are of little help here:

> "Management accounting reports are of little help to operating managers attempting to reduce cost and improve productivity."[11]

7.2.5 Downward Spirals

Cost accounting distortions can create "reinforcing" behavior (reinforcing loops) that can cause a business to "spiral down." The concept of reinforcing loops was presented by Peter Senge in his book, *The Fifth Discipline*.[12] Companies making both high-volume and low-volume products, as pointed out by the above cost management references, really do have different overhead demands. However, if overhead is *averaged*, then the spiral will occur, as shown in Figure 7.1.

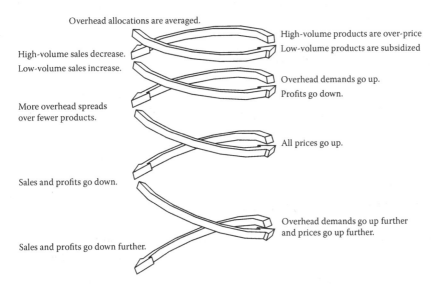

Overhead allocations are averaged.

High-volume sales decrease.
Low-volume sales increase.

More overhead spreads over fewer products.

Sales and profits go down.

Sales and profits go down further.

High-volume products are over-price
Low-volume products are subsidized

Overhead demands go up.
Profits go down.

All prices go up.

Overhead demands go up further
and prices go up further.

FIGURE 7.1
Cost distortion downward spiral.

Thus, the loop reinforces itself and the company continues to spiral down. This may drag a company down to the point of unprofitability or weaken an otherwise strong company.

7.3 RESISTANCE TO TOTAL COST ACCOUNTING

Despite the deficiencies already noted about traditional cost systems, many managers resist company-wide ABC (activity-based costing) implementation. This stems from a general resistance to change based on the following misconceptions. Many managers:

1. *Do not accept the deficiencies* in the current system. This was the topic of the first section of this discussion on total cost accounting. Many of the books referenced in this section make thorough arguments about this point, especially Johnson and Kaplan's *Relevance Lost: The Rise and Fall of Management Accounting.*[13]
2. *Underestimate the benefit.* This chapter and Chapter 6 reinforce the value of relevant cost numbers as a basis for good decision making, in general, and specifically, for decisions governing product portfolio planning, product line rationalization, standardization efforts, implementing manufacturing flexibility, and many aspects of product development.
3. *Overestimate the effort* to make any improvements. Sometimes resistance comes from horror stories that some formal activity-based costing programs have been so cumbersome that they have died under their own weight. However, the low-hanging-fruit approach, presented in Section 7.8, quickly generates benefits without consuming a lot of time and resources.

7.4 TOTAL COST THINKING

Even before any formal total cost accounting programs are implemented, companies can improve some decisions subjectively by using *total cost thinking.* The principles presented in this chapter can help individuals make better subjective decisions by correcting many misconceptions about cost and instilling the proper attitudes and beliefs.

In order for this to happen, however, the corporate culture must encourage it. Management policies can either encourage or discourage this kind of thinking. If all proposals must meet strict criteria for payback and ROI (return on investment), this will govern the decision-making process. If the criteria are based on traditional cost accounting, then the decisions will be based only on projects for part and labor savings and many truly good proposals will fail to win approval because much of their benefit comes from savings that are not quantifiable by the current system. If companies rigidly adhere to criteria based on incomplete costs, then attempts to inject *subjective* total cost decision making will fail, even if it is in the best interest of the company and its customers.

One subjective approach to this dilemma was proposed by Robert Kaplan in an article about justifying computer-integrated manufacture (CIM), "Must CIM Be Justified by Faith Alone?"[14] Kaplan's technique to work around deficiencies in accounting systems was to:

1. Compute how much the proposal fell short of the objective criteria: the *shortfall*
2. Summarize the intangible benefits (including all the benefits that could not be quantified)
3. Pose the question, "Is it worth the shortfall to gain all these intangible benefits?"

Another approach is *abductive logic*—the logic of what could be—which can be used to get around the "prove it" obstacle. A *Business Week* article recommended: "To use abduction, we need to creatively assemble the disparate experience and bits of data that seem relevant in order to make inference—a logical leap—to the best possible conclusion."[15]

One of the examples of part standardization, cited in Chapter 5, was the author's effort to standardize all resistors to 1% tolerance to replace the previous duplication of resistors in both 1% and 5% tolerance versions. There were no numbers available to justify the change. It "just made sense" to cut in half the number of resistors in all three factories. Subsequently, the author has learned of people who did the same part consolidation and concluded quantitatively that the purchasing power of the combined orders offset the "cost" of the higher tolerance.

Sometimes, subjective decisions must be made *in spite of the numbers*. One of the author's clients, who makes water meters, consolidated seven raw castings into three by adding extra brass (for test ports) to every product, whether or not they needed the optional tapped holes. This extra material appeared to add cost to some of the raw castings because the extra brass was not needed for products without test ports. In fact, the person who initiated the change felt like he would be "beat up" for raising the standard cost. Company management, however, supported the change, knowing subjectively that it would lower the cost of variety enough to be a net gain and make operations more flexible.

Management policies can encourage total cost thinking by empowering product development team leaders to make the best decisions, in their judgment, instead of trying to *limit bad decisions* by making them pass some predetermined threshold based on irrelevant numbers using incomplete criteria.

7.5 IMPLEMENTING TOTAL COST ACCOUNTING

Total cost accounting focuses on the *activities* performed to produce products; thus, the formal programs were called *activity-based costing*, although much-easier-to-implement techniques are now available, as presented in Section 7.8. Costs are assigned directly to either products or to activities, which are then assigned to products based on how much of these activity costs are incurred by each product.

Total cost measurement systems are not intended to replace the existing finance system. In most cases, these implementations create independent decision-making models. In the study that Cooper, Kaplan, et al., did for the Institute of Management Accountants, this was the case:

> "No modifications to existing financial systems were required, and companies continued to run all their existing systems in parallel with their new ABC model."

> "The activity-based model was treated as a management information system, not as part of the accounting system."[16]

The numbers from this model were more useful than those available from the existing cost system:

> "Managers found the numbers generated from the activity-based analysis more credible and relevant than the numbers generated from the official costing system."

7.6 COST DRIVERS

In any change process, there is always some "low-hanging fruit," which is always a good place to start to get some early results with little effort. Success in these high-leverage areas can then generate interest and support for more ambitious efforts. The low-hanging-fruit approach is also a good way to start the change process if there is a lack of widespread support.

In ABC implementation, the low-hanging fruit is the identification and implementation of simple *cost drivers* that make cost accounting more accurate and relevant, and encouraging behavior to lower these costs. Cost drivers are defined as the *root causes of a cost*—the things that drive cost.

Identifying cost drivers makes the root causes visible, which has two important consequences:

1. Total cost can be measured.
2. The behavior that actually lowers total cost can be encouraged.

The cost driver approach identifies key drivers of cost that should be quantified instead of lumped in with all other overhead. The cost driver approach is easy to implement and starts with the most important overhead costs that need to be quantified. New data collection efforts are focused on only a few key cost drivers. Cost drivers can be based on estimates, as long as there is universal consensus. Cost drivers can provide a more rational basis for performance measures.

For example, the activities that incur the following costs could be analyzed for significant ranges beyond the averages that usually are the basis for overhead allocation. Examples of cost drivers include:

- Relevant material overhead rates for standard parts and pulled, spontaneous resupply
- Relevant overhead charges for specials, configurations, and customizations
- Relevant overhead charges for new products that are designed to minimize overhead costs
- Setup costs and equipment utilization, which are especially important for low-volume parts
- Inventory costs and inventory-related costs
- Engineering change order costs

- Cost of quality (scrap, yield, rework, field service) and other non-value-added activities[17]
- Cost of field service, repairs, warranties, claims, litigation, etc.

The activities that cause these costs should be analyzed to find out what is causing the variation. Experienced managers will probably be able to identify the key cost drivers that are driving differences in these activities; for instance:

- *Volume*: high volume or low volume
- *Degree of customization*: standard or custom
- *Part standardization*: the current list or the standard parts list
- *Part destination*: for production products or spare parts for products that are out of production
- *Distribution costs*: direct or through channels
- *Product age*: whether the product is launching, stabilized, aging, or experiencing processing incompatibilities with newer products and/or availability challenges for parts and raw materials
- *Market niches*: commercial, OEM, military, medical, and nuclear markets have varying demands for quality, paperwork, proposals, reports, certifications, traceability, etc.

Those activities that incur difference costs from the variations in these cost drivers should be investigated. The costs of these activities should be charged accordingly. For instance, if low-volume products do incur more cost than high-volume products, then this should be reflected in the overhead allocation. If standard parts incur less material overhead, then they should be charged a lower overhead, as will be shown in the next examples. If certain operations incur more overhead than others, then the cost drivers should reflect this, as will be shown in the following examples.

7.6.1 Tektronix Portable Instruments Division

To encourage part commonality and assign accurate material overhead, Tektronix assigned a material rate that was *inversely proportional to volume*. Thus, a high-volume part had a very low overhead rate; conversely, a "low runner" was assigned a very high rate.[18]

7.6.2 HP Roseville Network Division (RND)

HP RND formerly had only two cost drivers for its printed circuit board assembly: direct labor hours and the number of insertions. A special survey showed that axial insertions were about one-third the cost of DIP (Dual In-line Package integrated circuit) insertions; manual insertion was three times as expensive as automation; and low-availability parts had an additional cost of ten times their materials cost. So they implemented the following nine unit-based cost drivers:[19]

1. Axial insertions
2. Radial insertions
3. DIP insertions
4. Manual insertions
5. Test hours

6. Solder joints
7. Board count
8. Part count
9. Number of slots

7.6.3 HP Boise Surface Mount Center

HP's printed circuit board operation in Boise, Idaho, implemented the following ten cost drivers for surface-mount printed circuit board manufacture.[20] Note driver number seven, which encourages part commonality:

Cost Pools	Drivers
1. Panel operations	Percent of a whole panel; if one panel contains four individual boards, then each board is charged 25% of the panel rate
2. Small component placement	Number of "small" components placed
3. Medium component placement	Number of "medium" components placed
4. Large component placement	Number of "large" components placed
5. Thru-hole component insertion	Number of leaded components inserted
6. Hand-load component placement	Minutes required to place all components that must be hand loaded rather than automatically placed on the board
7. Material procurement and handling	Number of unique parts in the board
8. Scheduling	Number of scheduling hours during a 6-month period
9. Assembly setup	Number of minutes of setup time during a 6-month period
10. Test and rework	Number of "yielded" minutes of test and rework time per board

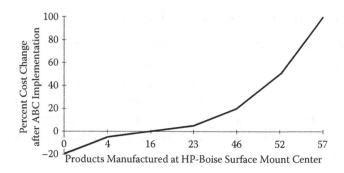

FIGURE 7.2
Changes in cost after implementing ABC.

Figure 7.2 shows the changes in product costing after implementing these cost drivers. Note that one-third of the products had their costs go down and two-thirds had their costs go up, with one product doubling in cost! The results as reported in a *Management Accounting* article:

> "Accountants now provide important inputs into product design and development decisions. Under the prior cost system, all overhead was applied as a percent of direct material cost, and it was difficult to understand how changing a board's design would change manufacturing costs. Also, designers had little motivation to optimize the board for efficient production. With ABC, however, the cost system attempts to mirror the manufacturing process, so that engineers and production managers easily can see how design changes will affect cost."[21]

7.7 TRACKING PRODUCT DEVELOPMENT EXPENSES

Some companies fail to collect important information, such as tracking product development expenses, because they feel that they cannot ask engineers to keep track of which projects they work on. Many engineers do, in fact, resist such rigor. However, such information is extremely important for making good decisions about product development and product costing. The solution to this apparent dilemma is as follows:

1. *Emphasize the importance* of the information, using all the reasoning presented herein and in the references. Some experts in the field argue that "an organization's cost accounting system can actually make or break an otherwise sound business."[22]

2. *Make it easy* to keep track of engineering time. It is better to have an approximate accountability than nothing at all. One division of Hewlett-Packard used the "bowling score" method to track engineering effort. The engineers were provided a form with lines for all the projects they may be working on. At the right of each line was a square box. An engineer who worked all day on one project would enter the bowling mark "X" for a "strike." An engineer who worked on two or more projects would enter the symbol "/" for "spare" on each of them. Here is how the system worked in practice: Engineers agreed to cooperate because the system was easy. However, because engineers are inherently precise, they eventually "corrected" the impreciseness of the system and voluntarily began to fill in more precise entries, such as hours or percent time worked.

If firefighting is common, there should be enough categories to capture these efforts by product or project, which will eventually be assigned to products.

3. *Make it required.* Total cost measurements depend on adequate data input. If senior management decides this is an important initiative, then everyone in the company will have to participate. Emphasizing the importance and making it easy will certainly make it easier to implement any mandates and ultimately make it more effective.

7.8 "abc": THE LOW-HANGING-FRUIT APPROACH

An excellent how-to book oriented toward easy total cost "abc" implementations is Douglas Hicks' *Activity-Based Costing*. It is based on the valid premise that *it is better to be approximately correct than to be precisely wrong; accuracy is preferable to precision.*[23] Knowing that a product has a negative profit margin between –55% and –65% is more valuable than thinking it has a positive profit margin of exactly 10.89% (these actual numbers are from the Harvard case study summarized in Appendix A, Section A.7).

Hicks claimed that the false pursuit of precision in product costing was unrealistic: "No cost accounting system provides an organization with precision. *All* product costing is approximate. *All* cost systems contain too many estimates and allocations to be precise."

With this focus on relevancy over precision, it is easier to implement this approach than the full-blown ABC, especially for smaller companies.

Hicks calls this "activity based costing" with the lower case acronym "abc" (to distinguish it from big ABC implementations), which he describes as follows:

> "In abc, activities are defined as groups of related processes or procedures that together meet a particular work need of the organization. Under this definition, the activities of the Accounts Payable department would most likely be Accounts Payable. Period."

Hicks makes similar arguments for treating the entire purchasing function as an activity, instead of identifying all the activities in purchasing. Many purchasing costs are avoided by kanban and breadtruck deliveries. In addition, fewer types of parts ordered in larger quantities reduces purchasing costs and increases purchasing leverage. The study published in *Just-in-Time Purchasing* reported that JIT users expected to cut expediting effort by a factor of three.[24]

7.8.1 Estimates

If good quantitative data is lacking, it would be preferable to implement some of the following shortcuts than to continue with grossly inaccurate allocations. One of these shortcuts is to *estimate* the percentage of an activity's cost caused by a particular cost driver; for instance, high-volume products compared to low-volume products. Thus, instead of averaging the cost allocation, where all products get the same charge, the low-volume products would be charged, say, 80% and the high-volume products 20% (for the typical Pareto effect).

An excellent example of estimates was mentioned in Chapter 6. When procurement managers are asked what proportion of their staff's activities are devoted to buying standard parts, the typical *estimate* is 10%, which means the material overhead rate for standard parts could be set at one-tenth that of the oddball parts and the material overhead rate for oddball parts could be set at ten times that of the standard parts—all based on that estimate. To encourage standardization, the only "cost" designers should be shown is the part cost *plus this material overhead charge*, which would thus steer them to the standard part.

7.8.2 Implementing "abc"

Understand the importance of total cost measurements for relevant costing, pricing, and decision making. For any cost reduction program, the measurement of cost is just as important as the steps to reduce cost, because total cost measurements:

- Help the company make the right strategic decisions that will be most effective
- Keep directing behavior that will continually reduce cost
- Quantify the real cost savings (or losses), which then affect subsequent decisions

The drive for implementing "abc" should come from the most *motivated* group, whereas the implementation could be done by the most *willing and able* group, which may or may not be the finance department.

Approach and label the program in a way that mitigates resistance and generates support. If "activity-based costing" is not an appropriate label, you could call it a costing or decision-making model.

Identify the cost drivers of activities that should be quantified instead of lumped in with other overhead. Roll the quantified cost driver data into the total cost model. Keep it up to date. Make sure the total cost information is readily accessible, easily understood, used for all cost-based reporting, and as the basis for decision making.

Hicks presents a simplified approach to implementing "abc", with the emphasis on accuracy and relevance rather than precision. He proposes an "abc" cost model in a format suitable for spreadsheets, such as Excel.

An alternative to creating a model on a spreadsheet would be software specifically developed for ABC analysis. The eight case studies cited in the Institute of Management Accountants' ABC study all used such software packages on PCs.[25]

7.9 IMPLEMENTATION EFFORTS

Most companies do not need a system as complex as would be needed for a multinational mega-corporation, despite misconceptions to the contrary. Implementing some degree of total cost measurement can be achieved with modest resources. Of the eight companies in the Institute of Management Accountants' study, the companies that used "medium involvement" of outside consultants took an average 6.5 months with 2.1 FTEs (full-time equivalent workers) to implement the ABC model. Companies that used "active involvement" of consultants took an average of 3 months with 1.6 FTEs.[26]

One practitioner reported that efforts to implement "abc" have "ranged from 80 hours for a small commercial printer to 500 hours for a large automotive supplier with very poor historical financial and operating records."[27]

The implementation chapter of this book includes a section summarizing total cost measurement implementation (Section 11.9).

7.10 TYPICAL RESULTS OF TOTAL COST IMPLEMENTATIONS

When ABC is implemented, companies start to see the real picture about product cost, which is often surprising. Cooper, Kaplan, et al., refer to the "typical ABC pattern," where several offerings are shown to be highly profitable, most at or near breakeven profitability, and a few highly unprofitable.[28] A Schrader-Bellows case study[29] showed that, out of seven products originally thought to be profitable, three actually were, one was barely breaking even, and three were unprofitable, with one highly unprofitable. And, the plant could not eliminate that unprofitable product until the costing changed (see Section A.7).

Total cost analyses often adjust manufacturing costs up for most products, while lowering them only for a few "deserving" products. After HP implemented the nine cost drivers cited above, they found that 72% of the products were really costing more than assumed, as shown in Figure 7.2. Cost adjustments ranged from slightly lower to double![30]

When the author implemented a parts standardization effort at Intel's Systems Group using the procedure described in Chapter 5, the result was that 500 "commonality" parts were identified as being preferred for new designs. These common parts really did deserve lower material overhead than the 13,000 remaining approved parts because they were purchased in higher quantities and were easy reorder actions. The standardization program wanted to encourage engineers to use these parts. To accomplish both these goals, the accounting department structured material overhead into a two-tiered system: one rate for the 13,000 approved parts and a lower rate for the 500 commonality parts. This reflected greater "material world" efficiencies and encouraged usage.

NOTES

1. Michael R. Ostrenga, Terrence R. Ozan, Robert D. McIlhattan, and Marcus D. Harwood, *The Ernst & Young Guide to Total Cost Management* (1992, John Wiley & Sons).
2. Robin Cooper and Robert S. Kaplan, "How Cost Accounting Distorts Product Costs," *Management Accounting*, April 1988.
3. H. Thomas Johnson and Robert Kaplan, *Relevance Lost: The Rise and Fall of Management Accounting* (1991, Harvard Business School Press).
4. Ostrenga et al., *Ernst & Young Guide*.
5. Doug T. Hicks, *Activity-Based Costing: Making It Work for Small and Mid-Sized Companies* (2002, Wiley).
6. Johnson and Kaplan, *Relevance Lost*.
7. Robin Cooper, Robert S. Kaplan, Lawrence S. Maisel, Eileen Morrissey, and Ronald M. Oehm, *Implementing Activity-Based Cost Management* (1992, Institute of Management Accountants, Montvale, NJ), p. 4.
8. Johnson and Kaplan, *Relevance Lost*.
9. Ostrenga et al., *Ernst & Young Guide*, p. 146.
10. Johnson and Kaplan, *Relevance Lost*.
11. Ibid.
12. Peter M. Senge, *The Fifth Discipline: The Art and Practice of the Learning Organization* (1990, Doubleday/Currency).
13. Johnson and Kaplan, *Relevance Lost*.
14. Robert S. Kaplan, "Must CIM Be Justified by Faith Alone?" *Harvard Business Review*, March–April 1986, p. 87.
15. Roger L Martin and Jennifer Reil, "Innovation's Accidental Enemies," *Business Week*, January 25, 2010, p. 72.
16. Cooper et al., *Implementing Activity-Based Cost Management*, p. 7.
17. For a complete list of quality costs, see Jack Campanella, Editor, *Principles of Quality Costs: Principles, Implementation, and Use* (1999, Quality Press, American Society for Quality).
18. Robin Cooper and Peter B. B. Turney, "Internally Focused Activity-Based Costing Systems," *Measures of Manufacturing Excellence*, edited by Robert S. Kaplan (1990, Harvard Business School Press), pp. 292–293.
19. Ibid., pp. 294–296.
20. Mike Merz and Arlene Harding, "ABC Puts Accountants on Design Team at HP," *Management Accounting*, September 1993, pp. 22–27.
21. Ibid., pp. 22–27.
22. Hicks, *Activity-Based Costing*. Also see the article about the "abc solution" at Doug Hicks' website: www.dthicksco.com.
23. Ibid.
24. A. Ansari and B. Modarress, *Just-in-Time Purchasing* (1990, Free Press), p. 44.
25. Cooper et al., *Implementing Activity-Based Cost Management*, pp. 6, 25, and 256.
26. Ibid., p. 296.
27. Hicks, *Activity-Based Costing*.

28. Cooper et al., *Implementing Activity-Based Cost Management*, p. 5.

29. The Schrader-Bellows case study is described in Harvard Business School Case Series 9-186-272; a summary of the findings appears in "How Cost Accounting Distorts Product Costs," by Robin Cooper and Robert S. Kaplan, *Management Accounting*, April 1988.

30. Merz and Hardy, *Management Accounting*.

Section IV

Design Guidelines

8

DFM Guidelines For Product Design

This chapter lists some general guidelines for *product design strategy* and presents assembly strategy, fastening strategy, assembly motions, and test strategy. These strategies are important aspects of the concept/architecture phase shown in Figure 3.1.

8.1 DESIGN FOR ASSEMBLY

All engineers should learn the lessons from equivalent assemblies from relevant products and formulate action plans and deliverables to leverage the best and avoid the worst. This also involves raising and resolving assembly issues based on lessons learned and DFM opportunities presented herein.

Simplify assembly with fewer parts, off-the-shelf parts that come assembled, and parts that are combined into monolithic circuit boards, castings, stampings, extrusions, and molded parts. Design for assembly without need for any skill or judgment and minimize manual tasks, for instance, by using connectors instead of wiring lugs and hand soldering. Design teams should strive to design products to eliminate the need to apply any liquids for fastening, bonding, or sealing. Eliminate the need for calibration or any kind of tweaking.

Design easy assembly features with self-jigging parts or parts that are aligned with pins, slots, or other features. Design symmetrical parts that don't have to be oriented. At each workstation, minimize part variety and standardize on one fastener per workstation. At each workstation, minimize tool variety and standardize on only one torque setting, one sealant, one type of glue, and a single procedure for each type of operation, to avoid mistakes. Concurrently engineer parts and fixturing to ease assembly and ensure correct assembly.

Eliminate the need to manually position families of parts in machine tools by:

- Locating features on the part and adjacent parts of fixtures
- Tooling pins and bushings or a reamed hole on the part and fixture (Guideline A3)
- Concurrently engineered fixtures that are versatile enough to quickly and precisely load all parts in a family, using optimal datum dimensioning, as shown in Figure 4.2

8.1.1 Combining Parts

Combining parts is a technique that can be used to reduce the part count and simplify assembly, provided the combined parts don't get so big and complex that they require expensive tooling or large mega-machine tools. The combined parts could provide the following benefits: eliminate the need to manufacture the interface features, hold their tolerances, and save the time and cost of assembly. It may be possible to fabricate *combined parts* on a single machine tool in a single setup. Examples include: many parts combined into a monolithic plastic or machined part; many integrated circuits combined into VLSI or ASICs; and multiple circuit boards combined into one, thus eliminating card cages and inter-board wiring operations.

The criteria for combining parts involves asking the following three questions:

1. When the product is in operation, do adjacent parts move with respect to each other?
2. *Must* adjacent parts be made of different materials?
3. *Must* adjacent parts be able to separate for assembly or service?

If all three answers are no, consider combining the parts into one. It is important to remember that every interface between parts requires geometrical features to be designed and manufactured plus all interface tolerances need to be held.

Eliminating interfaces eliminates the need to create interface features and hold their tolerances.

8.2 ASSEMBLY DESIGN GUIDELINES

Guidelines throughout this book will use the following numbering system for instructional clarity. Each company is encouraged to develop the numbering system optimal for its operations. The use of several categories of guidelines allows addition of new guidelines to the appropriate category rather than at the end of a single list. If new guidelines are added next to related ones, they will be considered together when the designer is dealing with that subject. In this way, newer guidelines will be less likely to be overlooked than if they were just added to the end of one long list of general guidelines.

Prefix	Category
A	Assembly strategy
F	Fastening
M	Motions of assembly
T	Test
S	Standardization
P	Part shape
H	Handling by automation
Q	Quality and reliability
R	Repair and maintenance

If guidelines are to be used as checklists, they should be worded to optimize usefulness in checklists. Then the team and management would note on the checklist whether the guideline has been obeyed or how much the product deviates from a certain goal, say, zero or 100%.

A1) Understand manufacturing problems/issues of current, past, and related products. In order to learn lessons from the past and not repeat past mistakes (Section 3.3), it is important to understand all problems and issues with current and past products with respect to manufacturability, introduction into production, quality, repairability, serviceability, regulatory test performance, and so forth. This is especially true if previous engineering is being leveraged into new designs. In a checklist, this could be checked "completed" with a lessons learned report as a deliverable.

A2) Design for efficient fabrication, processing, and assembly; identify difficult tasks and avoid them by design. Concurrently engineer the assembly sequence while designing the product. Designing for easy parts fabrication, material processing, and product assembly is a primary design consideration. Even if labor cost is reported to be a small percentage of the selling price, problems in fabrication, processing, and assembly can generate enormous overhead costs, cause production delays, and demand the time of precious resources.

A3) Eliminate overconstraints to minimize tolerance demands. An overconstraint happens whenever there are more constraints than the minimum necessary; for instance, joining two rigid frames with four bolts, guiding a rigid platform on four rigidly mounted bearings, or trying to precisely align two parts with multiple round pins inserted into round holes. (The solutions for these are shown below.)

Overconstraints are costly and can cause quality problems and compromise functionality because the design will work only if all parts are fabricated to tight, maybe unrealistic, tolerances. Fortunately, overconstraints are easy to avoid by specifying the exact number of constraints that will do the job: not enough constraints will result in an extra degree of freedom (something is loose); too many constraints will result in troublesome overconstraints. Here are some solutions:

> *Mount bearing housings or rigid members to each other on three non-collinear points, not four.* Unless the tolerances are perfect for a four-point mount, which is rare or expensive, three will determine the position and the fourth will try to warp both structures.
>
> *For critical alignment of parts, use round and diamond pins.* Use pairs of inexpensive but tight-tolerance dowel pins to locate critical parts. Matching tight-tolerance hole diameters shown in the rectangular part in Figure 8.1 can be made easily with reamers. To eliminate the tolerance match problem between holes, use one round pin to locate in *x* and *y* dimensions and a diamond pin to locate the angle from

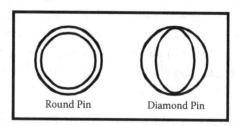

FIGURE 8.1
Alignment using round and diamond pins.

the round pin. The two pins shown in Figure 8.1 are mounted on the mating part (which is not shown). The diamond pin is precision ground to locate in the angle direction, but is relieved in the direction of the hole spacing. Although this technique was developed to locate tooling, it can also be useful for aligning parts for assembly, as shown in Figure 8.1.

This assembly technique can be used to save material, weight, and cost by pinning together machined parts to form *constant stress* assemblies in which the thickness of each part is proportional to the loads. Such parts can be made quickly and inexpensively on CNC machine tools that would automatically machine the part shape and drill and ream the pinning holes. Thus, there is no *unnecessary* material, as would be the case when constant cross-section parts are sized for the highest loads (where they are anchored or support other parts). Neither would there be *wasted* material, if the part was "hogged out" from a single piece of metal, as is common in the aerospace industry.

A4) Provide unobstructed access for parts and tools. Each part not only must be designed to fit in its destination location, but also must have an assembly path for entry into the product. This motion must not risk damage to the part or product and, of course, must not endanger workers.

Equally important is *access for tools and the tool operator,* whether that is a worker or robot arm, which usually requires more access room than a worker's hand. Access may be needed for screwdrivers, wrenches, welding torches, electronic probes, and so forth. Remember that workers may be assembling these products all day, and having to go through awkward contortions to assemble each product can lead to worker fatigue, slow throughput, poor product quality, and even worker injury. Access is also needed for field repair, where the tools may be simpler and possibly bulkier.

A5) Make parts independently replaceable. Products with independently replaceable parts are easier to repair because the parts can be replaced without having to remove other parts first. The order of assembly is more flexible because parts can be added in any order. This could be a valuable asset in times of shortages, in which case the rest of the product could be built and the hard-to-get part added when it arrives.

A6) Order assembly so the most reliable goes in first, the most likely to fail goes in last. If parts must be added sequentially, make sure the most likely to fail are the easiest to remove. This is important for both factory assembly and field repair.

A7) Make sure options can be added easily. Another advantage of independently replaceable parts is the ease of adding options later, either in the factory or in the field. Future options should be anticipated, and the product should be designed to accept these options. Considerations include allowing space for added parts, mounting holes, part access, tool access, software reconfiguration, extra utility capacity, and, of course, the safety of those performing the upgrade.

A8) Ensure the product's life can be extended with future upgrades. Early consideration of the product-upgrading strategy could be crucial to extending the life of a product. Advances in technology should be anticipated so the product can be upgraded without a complete redesign. Modular design concepts (Section 4.7) can be used to allow modules that are prone to obsolescence to be replaced with upgraded ones. Extending product life through upgrading allows products to generate even more profit after the development and introduction costs have been paid off. Figure 3.3 shows the value of upgrades.

A9) Structure the product into modules and subassemblies, as appropriate. The use of subassemblies can streamline manufacturing because subassemblies can be built and tested separately. Subassemblies could be built in specialized departments, which is especially advantageous if those processes are different from those of the product; for instance, clean room assembly (assembly in a dust-free room).

If the entire product consists of a collection of pretested subassemblies, product testing may be eliminated or reduced to only a final go/no-go test before product shipment. In designs where potential quality problems are concentrated in one subassembly, test and diagnostic attention could be focused there. The remainder of the product may then rely on process controls.

Products built from subassemblies are easier to repair in the factory and in the field by simply replacing the defective subassembly, which can then be sent back to its specialized assembly area for repairs. See the discussion on modular design in Section 4.7.

A10) Use liquid adhesives and sealants as a last resort. For *fastening*, thoroughly pursue alternatives, such as screws or nuts coated with retention compound, fasteners with deformed threads, and optimal use of lockwashers. Design to eliminate the need for *liquid sealants*; for instance, with optimal enclosures and built-in seals. Long drying times can compromise flow manufacturing.

Design to eliminate the need to use sealants for arc prevention; for instance, with optimal spacing or snap-in insulating barriers or partitions. Seal with premeasured off-the-shelf or manufactured solutions, such as rigid gaskets, compliant gaskets, custom-molded elastomeric gaskets, or O-rings, all of which should be self-jigging in the product.

If liquid adhesives and sealants are justified, make a thorough selection; be sure to optimize part alignment and repair strategy; standardize on the same application procedure (to avoid procedural errors); standardize on one adhesive per workstation (to avoid picking the wrong one); and avoid gaps, cracking, or structural weakness when glue shrinks. If justified, automate with robotic adhesive applicators or pick-and-place machines.

A11) Use press fits as a last resort. Press fits add tolerance challenges to both parts, resulting in a high potential for things to go wrong. Successful assembly and operation may be sensitive to temperature, cleanliness, procedures, etc., and parts can't be successfully taken apart for service or recycling.

The first design step is to thoroughly pursue alternatives. If justified, plan for the optimal alignment, guidance, pressing forces, etc.; ensure mating parts do not scrape, gouge, or bind; ensure engaging surfaces are free of contaminants; specify only one fixture and procedure per workstation; optimize interference/tolerance design; and ensure tolerances of mating parts.

Mating parts for press (interference) fits and shrink fits *must* have proper dimensions and tolerances.[1] Fits that are too loose will not "hold" under all service conditions; fits that are too tight may not be able to be assembled or disassembled (*see general discussion on tolerancing in Section 10.2*). Make sure press fits will not impair servicing. Press fit tolerances can be loosened by using "elastic" pins that are made by rolling spring steel sheet metal into a "C" shape or spiral wound cylinder. *More assembly guidelines (on error prevention) are presented in Chapter 10.*

8.3 FASTENING GUIDELINES

F1) Use the minimum number of total fasteners. *Fasteners may represent only about 5% of the product's direct material cost,* but all the associated labor costs can reach 75% of total assembly costs.[2] *For fastened assembly, parts must be aligned before assembly, fasteners must be found and positioned, the tool has to be positioned, torque has to be applied properly,* and the tool may need to be changed for the next job. Further, fasteners have to be ordered and delivered to the point of use. The fasteners themselves may need to be assembled (e.g., bolts to washers, but washers can be ordered captivated to bolts). In some cases, assembling with fasteners may be difficult or impossible for robots or other automation. Fasteners can be eliminated by *combining parts,* as discussed in Section 9.5.

F2) Maximize fastener standardization with respect to fastener part numbers, fastener tools, and fastener torque settings. Fastener standardization is easy to implement and has enormous benefits to manufacturing: fewer parts to order, receive, log in, stock, issue, load, assemble, and reorder. Purchasing costs will be reduced and the increased order quantities of the standard fasteners will result in quantity discounts and better deliveries, and the supplier can act like a "bread truck" and simply keep all the factory bins full.[3]

Regardless of the delivery system, standard fasteners are much less likely to run out and delay production. Further, standard fasteners are much easier to stock in the field and they require fewer tools to service—an important consideration when repairing complex products on the road.

The most effective technique for applying fastener standardization is simply discipline. The author redesigned a food processing machine that had evolved over many years with 150 different types of fasteners. The toolbox for field repair was quite cumbersome. The redesigned machine had no nuts and only two bolt types: large and small.

Careful selection of bolts encourages widespread use. Multiple grades (for strength or corrosion) could be consolidated into the better grade, which could serve well for all applications. The difference in fastener cost would be small compared to the benefits from part standardization.

F3) Optimize fastening strategy. Optimize the product fastener strategy by considering the guidelines in this section and specifying standard fasteners. Concurrently engineer workstations so that no more than a *single* fastener is used at each workstation to:

- Simplify supply chains
- Mistake-proof assembly (Section 10.7)
- Enable the use of auto-feed screwdrivers

F4) Make sure screws are standardized and have the correct geometry so that auto-feed screwdrivers can be used. A special version of the powered screwdriver feeds screws automatically from a hopper through a hose so that they are positioned under the screwdriver bit. When the screwdriver is positioned over the hole and activated, the screw is advanced into the hole and torque is applied up to a preset limit.

Auto-feed screwdrivers are inexpensive and greatly improve productivity. For automatic fastening, they can be mounted on a robot or special automation machinery. However, auto-feed screwdrivers are somewhat bulky, so designers must plan for tool access, as in Guideline A4. The geometry of the screws must meet the specifications of the equipment. Usually the screw length below the head must be greater than the diameter of the head by a margin specified by the equipment manufacturer.

Auto-feed screwdrivers can feed only one type of screw. But they are too bulky to allow more than one per workstation. Therefore, screws should be standardized on one size for each workstation. The development team will have to practice concurrent engineering and specify standard screws as they are structuring the flow of the work and laying out each workstation.

F5) Design screw assembly for downward motion. Screws are easier to apply from above, especially if downward force is needed to keep the tool bit engaged with the screw. For manual operations, applying this force from above is less fatiguing. Many robots can traverse only in a horizontal plane and apply force vertically.

F6) Minimize use of separate nuts. Separate nuts usually require a worker to position the nut while engaging the bolt. This will slow down manual assembly, especially if the nut location is hard to see or find. Semiautomatic operation of auto-feed screwdrivers will be far from optimal if the worker has to position a nut while activating the screwdriver. Robots are not advised for positioning both bolt and nut because of the expense of installing two robots. Separate nuts can be eliminated by using threaded holes, self-tapping screws, or captive nuts that are retained on the part to be fastened.

F7) Consider captive fasteners when applicable. Captive fasteners are retained in some way on the part by pressing into the part (for captive nuts or studs), by forming around the part (for threaded rivets), or by welding to the part (for weld nuts or studs). They are available to function as threaded holes (nuts) or as male threads (studs). Captive nuts or weld nuts function like thread holes in the part, but they must be applied on the *opposite* side from the bolt. There are hundreds of types of standard captive, riveted, and welded fasteners available off-the-shelf.

F8) Avoid separate washers. Separate washers increase the number of parts to order, deliver, and assemble. If forgotten, they can cause quantity problems. They are often difficult for workers to apply with the nut and are virtually impossible for automation to install. The washer can be captivated on the bolt or nut so it can still spin with respect to the fastener. Or the washer surface can be an integral part of a one-piece bolt or nut.

F9) Avoid separate lockwashers. The same arguments against separate washers apply to lockwashers. There are many solutions to fastener retention that do not rely on separate lockwashers. Captivated lockwashers are available with the lockwasher attached to the nut but free to spin. Locking ribs on the surface of the nut or bolt arc also available. Fastener suppliers use many thread-locking techniques, including deformed threads, plastic plugs or rings that bear on the threads, chemical locking agents, and part of the nut that pinches against the thread while seating. Hundreds of different *self-locking fasteners* are available off-the-shelf. Be sure to coordinate fastener selection and repair strategy with respect to the number of times self-locking fasteners can be reused safely.

8.4 ASSEMBLY MOTION GUIDELINES

M1) Design for easy, foolproof, and reliable alignment of parts to be assembled, in order of most desirable first:

- No alignment needed, using symmetry, poka-yoke (mistake-proofing) as discussed in Section 10.7, and other means
- Self-jigging parts, using clever geometries to align parts and hold them in place for fastening, pressing, gluing, soldering, brazing, or welding

- Part features allow alignment by simple fixtures or automatic equipment
- Easy manual alignment with hands-free assembly allows air presses to be used

Avoid scenarios in which parts must be positioned and held by hand during press operations.

M2) Products should not need any tweaking or any mechanical or electrical adjustments unless required for customer use. Design products so that no tweaking or adjustments are required in assembly. Adjustments slow down the assembly process and can cause quality problems if not performed correctly. Zero adjustments should be a goal for the product design, and the design team should use all the creativity at its disposal to achieve that goal. If adjustments are required for customer use, there should be a default setting that is easy to set during manufacture; for instance, at a detent, clear mark, or optimal software setting.

M3) If adjustments are really necessary, make sure they are independent and easy to make. Make sure necessary adjustments are independent of other adjustments and are easy to make consistently.

M4) Eliminate the need for calibration in manufacture; if not possible, design for easy calibration. Calibration is a form of adjustment that is time consuming and usually requires special equipment and trained personnel. Often, calibration can take place only after the product is fully assembled, making correction more difficult. If calibration is necessary, make sure it is easy to perform consistently.

M5) Design for easy independent test/certification. Design modules/subassemblies and their processing for independent test and certification to isolate corrective procedures at the lowest level. Final product certification may be avoided if modules can be designed so that if they pass certification, then the assembled product will be deemed to be certified.

M6) Minimize electrical cables; plug electrical subassemblies directly together. Electrical cable assemblies are time consuming for workers to build and install and almost impossible for automation to deal with. A better alternative for assembly is to plug electrical subassemblies directly together with the appropriate connectors designed into each part.

M7) Minimize the number of types of cables and wire harnesses. If cables must be used, minimize the number of pin types, lengths, and connector body styles. Standardize on wire harnesses with enough wires for many products, even if some applications have unused wires. Standardize on a few common lengths even if some applications have more length than needed. Standardizing on connector body types will also minimize the number of tools used for cable assembly.

8.5 TEST STRAGEDY AND GUIDELINES

Develop test strategies proportional to the need. Prioritize lessons learned by plotting severity versus frequency, as in Figure 10.1. Determine the coverage and failure modes that need to be tested based on how much quality can be designed in and built in. Decide if diagnostic testing is needed or if processing quality is high enough to discard failures. Ensure testability at the architecture level, which requires early involvement of test engineers. This can minimize the time and cost of test development and test equipment through:

- Adequate test access, including test pads and room for test probe access; this will have to be designed into circuit boards
- Standard or flexible test fixtures to minimize fixture cost and changeover times
- Compatibility with standard test equipment, programs, connectors, etc.
- Test development that is concurrent with test equipment selection, with special attention to test infrastructure needs for multiple production sites, where each site needs a complete test infrastructure
- Versatile test programming to minimize test development and changeover times
- Test ports and connectors that may have to be provided on circuit boards and systems
- Built-in tests, with remote monitoring and diagnostics if necessary and feasible
- Optimal datum dimensioning to facilitate dimensional inspections, using geometric dimensioning and tolerancing (GD&T)
- Statistical significance ensured by design of experiments

Quantify the total cost of past or anticipated testing to help justify efforts to design for quality and build in quality and possibly avoid the equipment and development costs of diagnostic tests.

T1)Product can be tested to ensure desired quality. If confidence in process control and go/no-go functional test is not high enough to ship products without complex testing, the product will need to be tested. The product should be designed in such a way to allow efficient testing. Tests may have to be developed to include diagnostics for complex products. This could be avoided with high enough process quality (Section 8.6).

T2) Subassemblies and modules are structured to allow independent testing. Guideline A9 encourages the use of modules and subassemblies to streamline manufacturing. They should be structured to allow them to be tested separately prior to assembly. The interaction between subassemblies should be predictable enough to count on the product working properly if all the subassemblies work separately. It may also be useful to be able to test subassemblies separately after assembly into the product.

T3) Testing can be performed by standard test instruments. Tests should be designed to be accomplished quickly by standard test instruments, which are easier to obtain and do not need to be designed, modified, or debugged, as may be necessary with custom test instruments. Further, field repairs will be easier for the customer to perform with standard test instruments, which the customer may own and know how to use. Consider *built-in self-test* in which the product can test itself without external test instruments.

T4) Test instruments have adequate access. Just as parts and tools need adequate access, as specified in Guideline A4, test instruments need to have adequate access. On electronic products, test points will be needed on printed circuit boards, and special test ports may be incorporated that are accessible even when the product is assembled. Test instrument access needs to be planned ahead if modules and subassemblies are to be tested separately in the product.

T5) Minimize the test effort spent on product testing consistent with quality goals. Because testing itself is not a value-added activity, product quality goals should be achieved with the minimum test effort. Process controls may dispense with much testing. If the quality "fallout" is low enough, simple go/no-go tests may suffice without the need for testing with diagnostics (Section 8.6). Subassembly testing may reduce the testing requirements of the assembled product.

T6) Tests should give adequate diagnostics to minimize repair time. If test fallout is high, the test should aid in diagnostics to minimize repair time. If the product is complex, test diagnostics may be necessary to make any repairs at all. However, manufacturing companies should strive to have their processes so well controlled that diagnostics for repair are not needed. In fact, if the fallout is low enough, it may be feasible to *discard* products that do not pass the final go/no-go test and spare the expense of diagnostic test development and the testing and repair itself.

8.6 TESTING IN QUALITY VERSUS BUILDING IN QUALITY

8.6.1 Testing in Quality with Diagnostic Tests

Diagnostic testing (e.g., using automatic test equipment for printed circuit boards) is necessary if test fallout is high and many boards need to be repaired. Diagnostic tests can pinpoint the problem component and instruct rework people to replace it. However:

- Automatic test equipment is expensive, costing millions of dollars plus significant costs in training, tooling (test fixtures), and infrastructure.
- Multiple plants require the same equipment and infrastructure, just to be complete, even if they are not needed for capacity. Thus, million dollar testers would be needed in every plant building that type of product.
- Test development, in many cases, can equal or exceed the cost and calendar time of circuit board development. Those engineers could benefit the company more by *designing* new products.

8.6.2 Building in Quality to Eliminate Diagnostic Tests

Quality should be assured by robust design, by process controls, and by ensuring part quality at the source (Chapter 10), in which case functional test yields can reach a breakeven point at which the total cost of diagnostic testing exceeds that of discarding failed products.

> *For printed circuit boards, IBM figured that if first-pass yields were above 98.5%, it could dispense with diagnostic tests and discard failed circuit boards.*

If product testing is still required and diagnostic tests are not used, then go/no-go functional tests and a built-in self-test would be needed to test all functions used in service.

8.7 DESIGN FOR REPAIR AND MAINTENANCE

The more the product will need to be repaired in the factory *or* in the field, the more important it will be to design for repair. Part of a repair strategy may be to simply replace parts or modules that are, themselves, either repaired, discarded, or recycled.

The need for ease of maintenance depends on the reliability of the product and demands for "uptime" (how much time the product needs to be available for use).

8.8 REPAIR DESIGN GUIDELINES

R1) Provide ability for tests to diagnose problems. The need for a consistent method of providing diagnostic information is proportional to product complexity and the probability of product failure in the factory or in the field. Products with a high fallout after testing can bog down a manufacturing plant if they are difficult to repair. Diagnostic information can specify where the problem is and recommend repair actions.

The need for building diagnostic capability into the tests is proportional to the inherent difficultly of diagnosing problems independently. Some complex products may take hours to diagnose with normal diagnostic tools and may need the advanced diagnostic capability available from advanced testing technology.

R2) Make sure the most likely repair tasks are easy to perform. *Anticipate* the most likely repair tasks and plan for ease of repair. This applies to part removal, part reinstallation, tools needed, and skill required. Ease of repair is especially important if customers perform repairs.

R3) Ensure repair tasks use the *fewest* tools. If fastener commonality has been designed into the product, this should have provided an inherent tool commonality also. When fasteners are selected, make sure the *minimum* number of *common* tools is specified. Repairability may be important to customers, and common tools may be part of their purchase criteria.

Avoid the need for special tools, unless the user needs to be precluded from repair for skill requirement or safety reasons. Special tools increase the number of tools that have to be supplied to repair facilities. In addition, users may be frustrated if repairs cannot be made with common tools. Users and even factory workers might not have the special tools and be tempted to use the closest common tool even if it damages a part or results in incomplete reassembly. A small repair tool set may be important if field repairs need to be made in remote sites where it would be difficult to bring a large number of tools.

R4) Use quick disconnect features. If part replacement is likely and must be done quickly, provide quick disconnect features to facilitate quick removal; for example, quarter-turn fasteners. Electrical connectors and fluid power quick disconnect fittings can be provided where quick separation is likely to be needed.

R5) Ensure that failure- or wear-prone parts are easy to replace with disposable replacements. If some parts are likely to fail or wear out during the useful life of the product, they should be easy to replace with disposable (or rebuildable) replacements. If an area is subjected to wear, cover it with a replaceable wear strip or sheet. Automobile brake shoes and pads are common examples of this principle, although ease of replacement varies from car to car.

R6) Provide inexpensive spare parts in the product. Spare parts that are expected to be needed, lost, or wear out can actually be included in the product. This practice may not cost much for inexpensive parts, but it may provide a major benefit to users. Examples of the practice are extra fuses in electronic products, extra buttons sewn on clothing, extra nozzles on spray paint cans, and extra light bulbs in flash lights, automobile tail light assemblies, and overhead projectors. In fact, some overhead projectors even allow lamp assemblies to be changed by moving an exterior lever. High-wear parts are candidates for inclusion in products; for instance, extra knife blades in retractable knives. The same principle can be applied to nonwearing parts, like extra tool bits in screwdriver handles. If the spare parts themselves cannot be included, at least provide a place to hold spare parts supplied by the user.

R7) Ensure availability of spare parts. Make sure spare parts are readily available. It is a risky business strategy to try to monopolize the spare parts business, or *inadvertently doing so* by designing in parts that are hard to find. Using industry standard parts greatly improves repairability. Customers will appreciate being able to get parts in a hurry from local sources of supply. For parts that are available only from the product manufacturer, recommend that the customer buy a "spare parts kit" for situations when downtime is intolerable.

R8) Use modular design to allow replacement of modules. One of the advantages of modular design is that it allows replacement of modules as a repair strategy. Modules can then be returned to a repair facility or to the factory for repair.

Modular repair is especially applicable when repair must be quick and for modules that need specialized facilities for repair. It also removes the actual repair function from the site of use, which is significant if the entire product is too large to move and is in a place that precludes easy repair.

R9) Ensure modules can be tested, diagnosed, and adjusted while in the product. Testing modules or subassemblies *while still in the product* will save time and prevent handling damage. Ideally, modules should be able to be tested from the controls or from software commands. If necessary, the module could be disconnected (if this is safe) and still be tested in the product. If modules need to be adjusted, make sure adjustments can be made while the module is in the product, preferably from the product's controls. Make sure adjustments are independent and do not affect other functions.

R10) Sensitive adjustments should be protected from accidental change. All adjustments should be protected from accidental change during servicing, repair, or maintenance. Adjustments and settings should be locked in position. Dial adjustments could stop at detents or be covered to prevent accidental change.

R11) The product should be protected from repair damage. The product should be protected from repair damage from workers, their tools, and the removal of other parts. Partitions and barriers may help protect parts. Subassemblies may need feet or guards to prevent handling damage after removal. Removal aids may also help.

R12) Provide part removal aids for speed and damage prevention. If it is likely that parts, modules, or subassemblies will be removed, make it easy by providing removal aids such as tracks, slides, guides, hooks, handles, and so forth. Many automobile engines have hooks installed over the center of gravity to aid in factory assembly *and* removal for repair. Inexpensive handles can be added where they would be most useful for removal. Many standard slide assemblies are available for mounting subassemblies so that they may slide out for easy servicing. This is common for many electronic systems. All of these measures not only make it easier to remove parts but also make the removal process quicker and safer for the repairer and the equipment.

R13) Protect parts with fuses and overloads. Some repair can be eliminated by protecting parts with fuses and overloads. Electrical fuses are common devices for protecting electrical equipment. Mechanical overload devices are available to protect mechanical machinery. In many applications, these devices may be necessary for safety.

R14) Ensure any module or subassembly can be accessed through one door or panel. For larger systems, this will make repairs easier. Subassembly removal should be possible through the single door.

R15) Access covers that are not removable should be self-supporting in the open position. If the system has access covers or doors, they should be self-supporting in the open position, like most car doors, hoods, and trunk lids.

R16) Connections to modules or subassemblies should be accessible and easy to disconnect. If subassemblies need to be removed or disconnected, connections must be accessible (for tools and workers), easy to disconnect, and easy to reconnect.

R17) Make sure repair, service, or maintenance tasks pose no safety hazards. Anticipate all possible repair, servicing, and maintenance tasks and make sure workers will not be exposed to any hazards from electrical shock, heat, sharp edges, moving parts, chemical contamination, and so forth. Anticipate the possibility of untrained users attempting service. Use warning signs and interlocks that cut off power when doors are opened. If unauthorized servicing may pose a safety hazard, prevent unauthorized access with locks or special access tools.

For large products and production equipment, it is common practice for repair personnel to use their own paddle lock on special "lockout" switches (that are designed into the equipment) so no one can turn on the machinery while repairs are underway. If multiple people are working on the machinery, then each repair person will lock out the switch with his or her own paddle lock; all paddle locks will have to be cleared for the machine to be turned on.

R18) Make sure subassembly orientation is obvious or clearly marked. If modules or subassemblies need to be removed for service, make sure orientation is obvious for correct reinstallation. Markings could be molded in or signs applied that indicate which end is up or which side mates with which other parts. Use polarized electrical connectors to avoid incorrect reconnection.

R19) Provide means to locate subassemblies before fastening. As recommended in Guideline M1, reinstallation of modules and subassemblies will be easier and more precise if the subassemblies can be located before fastening with guides, pins, tracks, stops, and so forth. This is especially important where correct orientation is difficult to see.

Also relevant to repair, see assembly guidelines A4–A6, detailed in Section 8.2.

8.9 DESIGN FOR SERVICE AND REPAIR

In addition to designing for repair in the factory, design teams should proactively design products for quick and easy service and repair in the field. Designing for ease of field repair may be more challenging than for factory repair, because field repair may not have access to test and repair equipment that is as sophisticated as in the factory.

- Understand the lessons learned about serviceability from current and past products, including which design features worked well and which ones impeded service.
- Don't try to make up for design shortcomings, incompleteness, or low quality with heroic field service.
- Avoid "ignorance is bliss" (not giving enough attention to service).
- Avoid "management by folklore" (focusing on well-circulated, but not statistics-based, service tales).
- Quantify the cost of service, repairs, warranties, legal, and related costs; estimate how much of that could be prevented by good *design for service.*
- Look up the customer importance rating and competitive grade of reliability and serviceability (shown in Figure 2.2) to ascertain how much effort should be applied to various issues of design for service.
- Gather service data and plot a priority chart that shows *service frequency* versus *severity* (similar to Figure 10.1, which shows quality issues in this format).
- Optimize the product concept/architecture to minimize service needs, with high enough quality and reliability designed in.
- Make sure people knowledgeable about service (yours, users, customers, and third parties) are early and active participants on product development teams.
- Focus efforts on the highest priority zones in the priority plot.
- Identify the highest priority service tasks and make them easy to do *by design*, including: architecture designed for ease of service; helpful diagnostics that either avoid service or aid in subsequent service; modules that are easy to open or disassemble quickly; parts that are easy to repair or swap when defective; minimum or no recalibration needed after service; and closures and reassembly strategies that are quick, easy, and foolproof.
- Consider modularity as part of a service strategy to allow replacement of defective modules in the field.
- Don't try to monopolize the spare parts business. Customers like standard spare parts that are widely and locally available. Custom

spare parts manufacture and distribution can drain resources from product development and customer service.

- Consider a built-in self-test to quickly ascertain product status and the service or repair approach.
- Consider remote diagnostics to monitor or ascertain product status and make decisions about service calls.

8.10 MAINTENANCE

Maintenance can be performed either after something fails (unscheduled maintenance) or at scheduled intervals to replace parts before they are likely to fail (preventive maintenance).

- *Unscheduled maintenance.* Restoring operation after a failure. Designing for ease of repair will greatly improve the ease of maintenance in general.
- *Preventive or scheduled maintenance.* If it is important to avoid downtime, then preventive maintenance can be scheduled to replace parts *before they are expected to fail.* The scheduling of such maintenance needs to be based on projected or measured failure histories of suspected parts. Useful data may be available on many purchased parts from historical performance. Critical applications require reliable parts, and maintenance programs need good reliability data.

8.11 MAINTENANCE MEASUREMENTS

8.11.1 Mean Time to Repair

The measurement of repair time is the *mean time to repair* (MTTR). This represents the mean time it takes to repair the product. In reality, it may take some time for repair personnel to respond before repairs can begin, which is called the *mean response time* (MRT). Adding the MTTR to the MRT gives the *downtime* during which the product is not available for use.

8.11.2 Availability

Availability is the measure of time that the product is available in an operative state. This is called the *uptime*:

$$\text{Availability} = \frac{\text{Uptime}}{\text{Total time}} = \frac{\text{Uptime}}{\text{Uptime} + \text{Downtime}}$$

Uptime is measured by the *mean time between failures* (MTBF):

$$\text{Uptime} = \text{MTBF}$$

Downtime is measured by the mean time to repair plus the mean response time:

$$\text{Downtime} = \text{MTTR} + \text{MRT}$$

$$\text{Availability} = \frac{\text{MTBF}}{\text{MTBF} + (\text{MTTR} + \text{MRT})}$$

8.12 DESIGNING FOR MAINTENANCE GUIDELINES

Maintenance strategy is not something to be left until after the product has been designed. Ease of maintenance should be *designed into* the product as one of the early design goals. Reliability studies should be able to predict part failure modes and frequencies and, of course, be a criterion for part selection. The usage environment should be identified early. These are some of the inputs that help develop a product maintenance strategy.

   R20) Design products for minimum maintenance. Design the need for maintenance out of the product. Parts should be carefully selected for optimal reliability. Automobiles have made great strides lately in designing for minimum maintenance by extending the maintenance periods (e.g., for oil changes). Designs should have conservative *factors of safety* so that parts will not be overstressed, even in worst-case conditions.

R21) Design self-correction capabilities into products. Design products with capabilities to correct problems they sense. Critical applications, like in aerospace, use design features that can automatically switch to backup systems.

R22) Design products with self-test capability. The product should have the capability to run its own built-in self-test to aid in diagnostics and repair. Self-test data should be stored in some form so it is available to repair personnel and new product development teams.

R23) Design products with test ports. Design the product with a *test port* to make key electrical test points easily available. This is especially useful if access to test points is difficult.

R24) Design in counters and timers to aid preventive maintenance. Products designed for ease of preventive maintenance should have counters and timers built in to determine when maintenance should be performed.

R25) Specify key measurements for preventive maintenance programs. Designers, *working with service people*, should be in the best position to know product weaknesses and so should be able to specify key measurements to determine wear or deterioration. Key parts, such as drive belts, should be measured periodically and replaced as necessary. Mechanically complex products can be analyzed by measuring the frequencies and amplitude of the noise they emit (their "sound signatures"), which may predict when parts are approaching failure.

R26) Include warning devices to indicate failures. The MRT can be minimized by signals (e.g., red lights or buzzers) and other warning devices so that repair can begin quicker. In sophisticated factories, central control panels show machine status and can instantly alert when a machine is "down." Self-diagnostics capability also minimizes the response time.

As mentioned before, plug-in modules can greatly benefit field maintenance and allow modules to be repaired off-line, where there are better repair and diagnostic facilities. The maintenance strategy may determine the order of assembly, as specified in Guideline A6, which states that the most likely parts to fail should be the easiest to remove.

NOTES

1. For tables of press fits (also called forced or interference fits), see the section "Allowances and Tolerances for Fits" in the chapter "Dimensioning, Gaging, and Measuring" in the *Machinery's Handbook*, 28th edition (2008, Industrial Press).
2. "The Best Engineered Part Is No Part at All," *Business Week*, May 8, 1989.
3. David M. Anderson, *Build-to-Order & Mass Customization: The Ultimate Supply Chain Management and Lean Manufacturing Strategy for Low-Cost On-Demand Production without Forecasts or Inventory* (2008, CIM Press), Chapter 7, "Spontaneous Supply Chain." See book description in Appendix D.

9

DFM Guidelines for Part Design

This chapter lists some general guidelines for *part design*, including fabrication, part standardization, symmetry, tolerances, part shapes, and combining parts and functions. For critical parts, the part designer should be an early and active participant in the system engineering of the product or subassembly to help optimize the systems architecture concept that determines the part's requirements. The part designer must understand the purpose of the part, how it fits in, how it relates to the whole, and the relative importance of function and performance, cost, rigidity, weight, and tolerances, especially stacks it is a part of:

> "Toyota's process does not focus on the speedy completion of individual component designs in isolation, but instead looks at how individual designs will interact within a system before the design is complete. In other words, they focus on system compatibility before individual design completion."[1]

Investigate past or similar parts to learn from their good or bad histories with respect to function, quality, cost, manufacturability, ramps, etc. First consider off-the-shelf parts. Thoroughly search for and investigate available candidates. Explore all the ways to design and make the part. Don't just jump at the first idea that comes to mind. Choose the optimal design approach. Keep thinking about how the part is to be made throughout the design process.

If the systems engineering has not been optimized, recommend ways that better system engineering or integration could improve the part's design and the manufacturability of the product overall. This especially applies to the rational apportionment of tolerances in stacks.

Understand all the candidate processes well enough to choose the best process for the optimal cost, tolerance control, quality and consistency, ramps, delivery time, compatibility with company operations and supply chains, equipment and vendor availability, tooling cost and lead time, setup time, and appearance or finish. If you don't understand all candidate processes, find colleagues who do, bring in outside experts, or call in the appropriate vendor(s), being careful to explain the nature of the inquiry.

Work with the vendor, who should be preselected, to collaboratively design the part. Decide if the vendor should design the part under careful supervision and coordination.

Keep in mind the optimal balance of design considerations: for example, function, performance, strength, weight, cost, quality, manufacturability, etc. For large or complex structures, ensure that all dimensions of each part can be made in the same operation (Guideline P14). Optimize part combinations or partitioning into multiple parts in a way that they can be accurately aligned with the techniques of Guideline A3.

Research and understand the specific design guidelines for the part and its chosen processes to optimize the function, cost, quality, and manufacturability in general. Follow the part design guidelines that appear in the next section.

9.1 PART DESIGN GUIDELINES

P1) Adhere to specific process design guidelines. It is important to use specific design guidelines for parts to be produced by each process, such as welding, casting, forging, extruding, forming, stamping, turning, milling, grinding, powdered metallurgy (sintering), plastic molding, and so forth. A good summary of design guidelines of several processes would be well over a thousand pages and is thus beyond the scope of a general book on DFM. Some reference books are available that give a summary of design guidelines for many specific processes.[2-5] Many specialized books devoted to single processes are also available.[6] Industrial organizations and suppliers of specific processes often, at no cost, furnish designers with design guidelines for their process.[7]

P2) Avoid right- or left-hand parts; use parts in pairs. Avoid designing right- or left-hand (mirror-image) parts. Design the product so the same single part can function in both right- and left-hand modes. If parts cannot now perform both functions, add features to both right- and left-hand parts to make them the same. Another way of saying this is to use parts in pairs instead of different parts for front and back, top and bottom, and right and left.

Purchasing right/left parts or paired parts (plus all the internal material supply functions) allows a company to obtain *twice* the quantity and *half* the number of types of parts. For pairs of molded parts, or any other part that is made in custom tooling, this principle can *cut tooling costs in half.*

> *At one time or another, everyone has opened a briefcase or suitcase upside-down because the top looks like the bottom. The reason for this is that top and bottom parts are identical parts used in pairs, which is done to cut the tooling cost in half.*

Consolidating *similar* parts to be the *same* part results in a fraction of the part types and several times the purchasing leverage.

P3) Design parts with symmetry. Design each part to be symmetrical from every possible "view" (in a drafting sense) so that the part does not have to be oriented for assembly. Symmetrical parts cannot be installed backwards, eliminating a major quality problem during manual assembly. In automatic assembly, symmetrical parts do not require special sensors or mechanisms to orient them correctly. The extra cost of making the part symmetrical (the extra holes or whatever other feature is necessary) will probably be saved many times over by not having to develop complex orienting mechanisms and by avoiding quality problems.

> *It is a little known fact that in felt-tipped pens, the felt is pointed on both ends so that automatic assembly machines do not have to orient the felt.*

P4) If part symmetry is not possible, make parts very *asymmetrical*; polarize all connectors. The best part for assembly is one that is symmetrical. The *worst* part is one that is *slightly asymmetrical* and thus could be installed wrong because the worker or robot could not discern the asymmetry. Or worse, the part may be *forced* in the wrong orientation by a worker (who thinks the tolerance is off) or by a robot (that doesn't know any better).

So, if symmetry cannot be achieved, make the parts *very* asymmetrical. Then workers will be less likely to install the part wrong because it will not fit wrong. Automation machinery may be able to orient the part with less expensive sensors and intelligence. For example, very asymmetrical parts may even be oriented by simple stationary guides over conveyor belts.

P5) Design for fixturing; concurrently design fixtures. Understand the manufacturing process well enough to be able to design parts and dimension them properly for fixturing, using geometric dimensioning and tolerancing (GD&T), as discussed in Guideline Q14 in Section 10.2.

Flexible operations require that whole families of parts be positioned in a common fixture without any setup changes.[8] An example of such a fixture is illustrated in Figure 4.2.

Parts designed for automation or mechanization need registration features for fixturing. Machine tools, assembly stations, automatic transfers, and automatic assembly equipment need to be able to grip or fixture the part *in a known position* for each operation. This requires *registration locations* (e.g., tooling pins or optical targets) on which the part will be gripped or fixtured while it is being transferred, machined, processed, or assembled.

Concurrently design fixtures for welding, assembly, and other processing steps to improve cost, time, and quality of both the parts being fixtured and the subsequent assembly. Fixtures could be discrete for a mass-produced part. For families of products, fixtures should be versatile enough to accept any part or assembly in the family. Versatile fixtures also could be adjusted by detents or stops, or positioned by programmable servo mechanisms.

P6) Minimize tooling complexity by concurrently designing tooling. Use concurrent engineering of parts and tooling to minimize tooling complexity, cost, and delivery lead time and maximize throughput, quality, and flexibility. Work with *preselected* vendor/partners, who should be expected to bring their process engineer and tooling engineer to the team to work early with design engineers and whoever is doing the styling.

P7) Make part differences very obvious for different parts. Different materials or internal features may not be obvious to workers. Make sure that part differences are obvious. This is especially important in rapid assembly situations where workers handle many different parts. To distinguish different parts, use markings, labels, color, or different packaging if they come individually packaged. One company uses different (but functionally equivalent) coatings to distinguish metric from English fasteners.

P8) Specify optimal tolerances for a *robust* design. Design of experiments can be used to determine the effect of variations in all tolerances on part or system quality. The result is that all tolerances can be optimized to provide a *robust design* to provide *high quality at low cost.*[9] See Section 10.2 for more on tolerancing.

P9) Specify quality parts from reliable sources. The "rule of ten" specifies that it costs 10 times more to find and repair a defect at the next stage of assembly. Thus, it costs 10 times more to find a part defect at a subassembly; 10 times more to find a subassembly defect at final assembly; 10 times more in the distribution channel; and so forth. The point here is that all parts should have reliable sources that can deliver consistent quality over time in the volumes required.

The Rule of 10

Level of Completion	Cost to Find and Repair Defect
The part itself	X
At subassembly	10 X
At final assembly	100 X
At the dealer/distributor	1,000 X
At the customer	10,000 X

9.2 DFM FOR FABRICATED PARTS

When building high-variety parts at low volumes, maximize the amount of variety done by flexible CNC machine tools, which may override the usual economic trade-offs for mass-produced parts.

P10) Choose the optimal processing. Use concurrent engineering to proactively choose the optimal processes (machining, casting, forming, molding, and so forth) for the minimum total cost and throughput time. Versatile *primary processes* can eliminate or minimize certain *secondary operations*, thus saving cost and throughput time. Design to avoid unnecessary operations. Understand how fabrication processes work and their capabilities and limitations as you learn design guidelines.

Work with preselected vendor partners from the beginning to concurrently design parts and processes and fixturing. Print 3D models early and often (rapid prototypes) to help optimize the design and processing and fixturing.

P11) Design for quick, secure, and consistent work holding. Design parts for quick, secure, and consistent work holding for clamps, collets, arbors, vises, chucks, centers, jigs, and fixtures. Provide consistent parallel, conical, or circular clamping surfaces. Design parts to be rigid enough to withstand a cutting tool and work holding forces without distortion or damage. Do not plan to clamp on parting lines or other uneven or inconsistent surfaces. Provide access room for cutting tools, clamps, and clamping tools.

P12) Use stock dimensions whenever possible. Design parts so that noncritical dimensions can be provided by stock dimensions of standard raw material, instead of requiring machining for these noncritical dimensions. For instance, if you need a bar that must be about half an inch thick, specify "½ inch stock" instead of a decimal dimension (.500"), which may have to be machined down from a larger stock size, like ⅝ inch stock.

P13) Optimize dimensions and raw material stock choices. Specify dimensions and select the raw material for the best balance of fabricating efficiency and raw material standardization for the lowest total cost. Raw material standardization (Section 5.10) needs to be even more aggressive for build-to-order and mass customization, as discussed in Chapter 4.

P14) Design machined parts to be made in one setup (chucking). Having to reposition parts or move to another machine increases cost for setups and machine time; increases the chance of errors for extra setups and repositionings; lowers accuracy, compared to the precision of locating all cuts on the same chucking; disrupts Lean flow and complicates machine scheduling; and takes more processing hours, more labor hours, and more calendar time. To take advantage of single setups, designers must:

- Make all dimensions from the most logical datum, which corresponds to the fixture, machine bed, or clamping surface.
- Concurrently design fixtures, locating features, clamping geometry, etc.
- Design geometries to minimize the number of cutting tools—ideally, one.
- Use total cost measurements to justify one 5-axis operation instead of multiple 3-axis operations by including all costs for setup, loading, zeroing, machine time, error correction, scrap, and so forth.

Single setup machining is an effective way to achieve tight tolerances between many features at low cost.

Figure 9.1 shows how the author's consulting[10] improved the manufacturability of a robot bearing holder. In the original design (on the left half

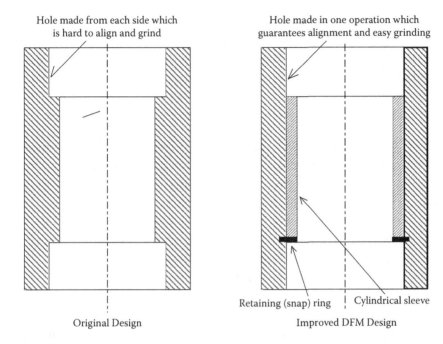

Hole made from each side which is hard to align and grind

Hole made in one operation which guarantees alignment and easy grinding

Original Design

Retaining (snap) ring | Cylindrical sleeve

Improved DFM Design

FIGURE 9.1
Improvement design for easier and better machining.

of Figure 9.1), the bearing mount holes had to be machined from each end after repositioning the part, so it was hard to align the bearing bores to the ±.001″ concentricity tolerance. Further, it was also hard to grind the bore to the exact diameter needed for the bearing mounts. The new part (on the right half of Figure 9.1) was designed so that the critical bearing bores could be made in *one* operation, which guaranteed alignment and made it easy to grind a tight-tolerance diameter. The bearing spacer function was provided by an off-the-shelf snap ring (retaining ring) and an inexpensive sleeve. Similar logic encourages doing all machining with a single cutting tool to avoid setup delays to change tools and avoid introducing inaccuracies, both of which are more likely without automatic tool changers.

P15) Minimize the number of cutting tools for machined parts. For machined parts, minimize cost and throughput by designing parts to be machined with the minimum number of standard cutting tools; for instance, end mills. Avoid tool proliferation and arbitrary decisions. Keep tool variety within the capability of the tool changer *for the entire product family*, or, ideally, all parts in a flexible plant.

P16) Avoid arbitrary decisions that require special tools and thus slow processing and add cost unnecessarily. Designers should avoid arbitrary decisions when specifying dimensions that require unique tools, such as bend mandrels, hole punches, and cutting tool bits for machine tools. Find out what are the most common tools in the shop and design around them, instead of arbitrarily requiring tools that may not be readily available.

P17) Choose materials to minimize total cost with respect to post-processing. Optimal selection of materials can minimize the cost of, and possibly eliminate altogether, post-processing steps for strengthening, hardening, deburring, painting, surface coating, and so forth. Materials that appear to cost more may actually have a lower total cost if all post-processing costs are considered. For instance, choosing stainless steel can avoid the painting costs and rusting problems of inferior metals.

As mentioned in Chapter 5, some post-processing operations can be eliminated by ordering *prefinished material* that is prepainted, preplated, embossed, expanded, anodized, or clad with a different surface alloy. Painting operations for sheet metal can be eliminated by switching to stainless sheet metal. This might be justifiable if the total cost of painting is considered. Prefinished material can be ordered with the finished side protected by adhesive-backed paper that can be peeled off after assembly.

P18) Design parts for quick, cost-effective, and quality heat treating. It is the responsibility of the designer to specify the quickest, highest quality, and most cost-effective (from a total cost perspective) post-processing for heat treating and any other post-processing step. Work with manufacturing and vendors to consider all the possible scenarios and then systematically choose the best one.

P19) Concurrently design and utilize versatile fixtures. Design families of machined parts to be processed in the same versatile fixture. If multiple fixtures are to be used in the same machine tool, design the fixtures to have standardized mounts. Flexible fixtures can speed loading and can minimize setup changes for different parts, thus improving flexibility while lowering cost. New flexible fixtures should be concurrently engineered as the product is designed (Chapter 3).

P20) Understand workholding principles. Design parts to be gripped tight enough for good machining and withstand forces from cutting tools and also from jigs, fixtures, clamps, collets, arbors, vices, and chucks.[11]

P21) Avoid interrupted cuts and complex tapers and contours.
Interrupted cuts occur when cutting tools encounter holes or other gaps in the workpiece, which results in vibrations, excessive tool wear, and inferior dimensions and surface finish. Complex tapers and contours may be difficult to manufacture and inspect.

P22) Minimize shoulders, undercuts, hard-to-machine materials, specially ground cutters, and part projections that interfere with cutter overruns. Designers should avoid features that are difficult to machine. Specially ground cutters may not yield consistent results. Although machinists are taught to grind their own cutting tools in training classes, this practice should be discouraged in production to avoid inconsistency from tool to tool. Instead, use appropriate inserts, which are standard cutting tools that have consistent cutting edges and can be mounted consistently into matching tool holders.

P23) Understand tolerance step functions. Understand tolerance step functions and specify tolerances wisely. The type of process depends on the tolerance. Each process has its practical limit of how close a tolerance could be held for a given precision level on the production line. If the tolerance is tighter than the limit, the next most precise—and expensive—process must be used. Designers must understand these *step functions* and know the tolerance limits for each process (see Figure 9.2).

P24) Specify the widest tolerances consistent with function, quality, reliability, safety, and so forth. Avoid choosing tolerances arbitrarily or from overly tight tolerance blocks

P25) Be careful about too many operations in one part, especially if the part must pass through multiple machines, to decrease the cost and delays of multiple setups and to minimize the consequences of machining mistakes.

P26) Concurrently engineer the part and processes for the best manufacturability, cost, quality, and throughput time.

P27) Avoid sharp internal corners that require sharp cutting tools, which can easily break.

P28) Proactively deal with burr removal and provide room for burrs and their removal tools. In general, ensure parts do not have sharp edges,

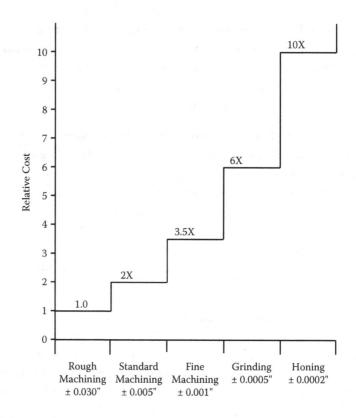

FIGURE 9.2
Cost as a function of process.

points, or burrs, which could damage other parts or injure workers and even customers.

P29) Specify 45 degree bevels instead of round external corners, to avoid special/unusual tools and tool changes.

P30) Don't overspecify surface finishes. Use "comparitors," which are gauges that show the look and feel of various surface finishes for machined, molded, and cast surfaces.

P31) Reference each dimension to the best datum for the optimal tolerance control, clear and unambiguous representation of the *design intent,* ability to make all key dimensions in the same setup on the same machine, ease of CNC programming, and ability to inspect first articles on coordinate measuring machines (see GD&T discussion in Guideline Q13 in Section 10.1).

9.3 DFM FOR CASTINGS AND MOLDED PARTS

9.3.1 DFM Strategies for Castings

Understand that casting is an *inflexible* process where every different shape requires a different die, and so *versatile* standard shapes should be encouraged. This makes castings unsuitable for customizing products, unless all versions are based on a standard raw casting. By contrast, metal fabrication is a *flexible* process where CNC machine tools can machine many different shapes flexibly.

P32) Obey all the guidelines for design of castings and molds/dies using handbook guidelines or, preferably, with the help of the casting vendor working with the team. Optimize draft angles, surface finish, wall thickness, thickness transitions, ribbing, features, holes, corners, parting planes, die filling, sprue/riser locations, ejection, cooling times, and so forth while minimizing the effect of shrinkage, warpage, and surface variations.

P33) Standardize cast parts to minimize the number of parts and the number and cost of the molds/dies.

P34) Design *versatile* raw castings that have all the shapes and features for all versions of the casting to minimize the number of raw castings, minimize the number of expensive dies, and allow these standard parts to be inventoried knowing they will be used one way or another, without the excessive cost and risk of stocking many different versions. The cost of the extra metal (to make the part versatile) will probably be saved many times over in die cost, setup costs, and inventory management for multiple raw castings. For weight-sensitive applications, it may incur less total cost to machine away any extra metal rather than specifying many different raw castings.

P35) Capitalize on opportunities to avoid machining with "as cast" shapes whenever possible. The comparitors mentioned in Guideline P30 will help you ascertain when this is feasible.

P36) Carefully plan out the sequence of machining castings starting with the machining reference points so that the raw castings are properly positioned in the machine tool.

9.3.2 DFM Strategies for Plastics

Like casting, plastic molding is an *inflexible* process where every different shape requires a different mold. Thus, *versatile* standard shapes should be encouraged.

P37) Obey all the guidelines for part design and mold design using handbook guidelines or, preferably, with the help of the molding vendor working with the team. Optimize draft angles, surface finish, wall thickness, thickness transitions, ribbing, features, holes, corners, parting planes, mold filling, sprue/riser locations, ejection, cooling times, and so forth while minimizing the effect of shrinkage, warpage, and surface variations.

P38) Standardize molded parts to minimize the number of parts and the number and cost of the molds/dies.

P39) Design *versatile* molded parts that have all the shapes and features for all versions to minimize the number of designs and tooling. The cost of any extra plastic will probably be saved many times over in mold/die cost, setup costs, and inventory management for multiple plastic parts.

P40) Standardize raw materials for all parts, or at least all parts in a product family. Even if it appears that some parts may be getting better material than needed, the total cost to the company will be less because of purchasing leverage (economies of scale), lower material overhead with fewer types of materials to procure, and fewer setup changes at the molder to change raw materials. In addition, delivery will be faster without setup changes and procurement for multiple types.

P41) Choose raw materials commonly used, especially at the chosen vendor or partner, to eliminate extra procurement cost and setup changes. Try to choose common materials used throughout the vendor base. Be sure these cost savings are factored into the vendor's charges.

P42) Consider all adjacent parts when substituting plastics. Instead of limiting thinking merely to one-for-one replacements when replacing other materials with plastics, look for nearby opportunities to combine several parts or functions into an optimal system.

P43) Optimize the number of functions in each part. Optimize decisions between part or mold complexity and the total cost savings in assembly and supply chain management:

- *Mold complexity/cost.* Compare the cost of one complex mold to several simpler molds.
- *Assembly labor.* Compare the assembly cost of multiple parts to one monolithic part.
- *Material overhead.* Compare the purchasing and logistics costs of one versus multiple parts.
- *Vendor base.* Complex molds and unusual processing may limit the vendor base.
- *Tolerance control.* Monolithic parts control tolerances between features and avoid tolerance stacks. On the other hand, precise part alignment techniques (Guideline A3) may be able to ensure alignment tolerances between multiple parts.
- *Appearance.* Monolithic parts eliminate seams. However, clever styling could mitigate this problem.

P44) Methodically *choose* tolerances for molded parts. Avoid unnecessarily tight tolerances and finishes. Understand tolerance step functions (Section 9.2). Specify optimal tolerances for a robust design using the Taguchi Method™ for Robust Design (Chapter 10, Section 10.2.5).

P45) Work with preselected vendors/partners from the beginning to concurrently design parts and tooling.

P46) Print 3D models (rapid prototypes) to help optimize the design and tooling. Order quick-turn parts before making tooling.

9.4 DFM FOR SHEET METAL

P50) Buy off-the-shelf sheet metal boxes by making this an early consideration in the architecture phase.

P51) Optimize sheet metal in the concept/architecture phase. Minimize demands on sheet metal manufacture by optimizing it as part of system architecture instead of considering it after everything else is designed. Minimize the need for skilled TIG welding; consider spot welding or roll seam welding. Consider tab-in-slot assembly, where small TIG torch actions weld the tab to the slot. Consider self-jigging geometries.

P52) Optimize sheet metal processing. Use CNC shearing and bending to automate sheet metal fabrication. Use CNC shears, laser cutters, or plasma cutters to maximize the number of operations done in one setup (Guideline P14). Such cutters may have slower cutting speeds than the equivalent operations on shears or punch presses, but doing all operations in one laser cutter setup may be quicker (including setups saved) and may have a lower total cost.

Use optimal *nesting software* to minimize sheet metal waste, with small parts nested between larger parts. This could be a kanban source for sheet metal parts that could be nested into various sheets on a "space available" basis. These could then be sent to their kanban stations, as described in Section 4.2.

Sheet metal cost can be minimized by standardizing enough usage on one thickness and grade to allow the purchase of coils, possibly directly from the mill. This reduces waste even further if the sheet metal is fed from the coil through straighteners directly into a CNC shear, laser cutter, or plasma cutter, which will have much better nesting without a length limit, as would be the case for an eight foot long sheet. As metal prices and shipping costs rise, this greater material utilization—resulting in less wasted metal—will become more attractive.

Sheet metal surfaces should be kept clean for subsequent welding, and visible surfaces could be ordered with "papered" protective film that can be removed after shipping and installation.

P52) Standardize sheet metal. All sheet metal pieces should be made from one gauge and type to minimize supply chain costs and delays, in addition to allowing the purchase in coils. Proactively, make this selection correspond to the vendor's operations. If sheet metal parts are made from discrete sheets, minimize the number of different sheet size varieties and keep sheet sizes within locally available stock.

P53) Standardize sheet metal tools. Standardize on the fewest types of holes and slots, unless done by programmable plasma or laser cutters. Standardize on one bend radius for all sheet metal to eliminate mandrel setup changes. Choose the mandrel that is usually on the machine tools, making sure the radius is not so sharp that it cracks the metal or so large that it runs into something inside.

For punched holes, slots, and louvers, avoid the cost and delays to have special tools ordered or built by *standardizing on shapes* that correspond to the tool sets already in use at both the intended local vendor and at other suppliers in other potential manufacturing locations. Preferably, select the shapes whose tooling is normally on the machine, to avoid setup changes. Limit variety to the number of tool sets in the automatic tool changer for the *whole product family.*

P54) Follow sheet metal design guidelines. Make sure sheet size and forces needed are within process capabilities. For bends, allow room for bending mandrels and avoid tight tolerances between two bends, which are hard to hold.

Tolerances should not be too tight; generally, ±.020″ and no tighter than ±.010″. A flatness of .005" per inch is the best that can be achieved without secondary straightening.

Obey spacing requirements between holes and edges or bends. Obey all spot welding design guidelines, including tool access. Avoid welding warpage to sheets by minimizing spans or using annealed sheet metal.

9.5 DFM FOR WELDING

9.5.1 Understanding Limitations and Complications

Before considering welding, understand the limitations and complications of welding with respect to: warping; dimensional uncertainties; tolerances issues for the whole weldment; appearance and finish; welder skill required; torch and welder accessibility; cost-of-quality, including straightening, touch-up, rework, scrap, etc.; extra assembly labor to position mounted parts when weldment mounting features are not precise; reduced hardness and strength to annealed levels in heat-affected zone; corrosion vulnerabilities in the heat-affected zone; outgassing issues; and total labor time, including setup, welding, touch-up, grinding, rework, scrap, and so forth. Using clamps and jacks to correct for warpage *during* welding adds more cost, takes longer, and increases *residual stresses*, unless weldments are annealed after welding.

Formulate a strategy for mounting holes. First, ascertain if mounting tolerances are wide enough to allow welding of pieces with predrilled holes. If not, then devise strategies for mounting and alignment in slots.

9.5.2 Optimize Weldment Strategy for Manufacturability

Optimize trade-offs between monolithic weldments and various modular strategies. A strategy for mounting holes should include considerations such as:

- Ensuring tolerances are wide enough to allow welding pieces with predrilled holes
- Strategies for mounting and alignment in slots
- Use of mega-machines that can machine mounting surfaces and holes after welding
- Predrilling holes in small parts that are assembled precisely by DFM techniques, as discussed in Section 9.6

9.5.3 Adhere to Design Guidelines

Adhere to welding design guidelines for: the dimension and tolerance strategy of the weldment; appearance demands; piece preparation (dimensions, tolerances, and edge preparation); concurrently designed fixtures; pre-weld fit-up and clamping; preheating of parts and fixturing; penetration specifications; the optimal welding sequence; distortion and warping control; tool and welder access; post-weld straightening; post-weld machining; post-weld grinding; and quality control procedures. Be sure to minimize the *combined cost* of welding labor and grinding labor, because less-skilled welding can increase grinding cost. Consider making welders grind their own welds, which would be a form of kaizen (continuous improvement), providing immediate feedback and correction and teaching the value of making better welds.

9.5.4 Work with Vendors/Partners

Work with preselected vendors or partners from the beginning to concurrently design the weldments and welding procedures and fixturing. This teamwork will be aided by early 3D models (rapid prototypes) to help optimize the design, procedures, and fixturing.

9.5.5 Print 3D Models

Printed 3D models (rapid prototypes), scaled if necessary, help optimize the design and welding procedures and fixturing. Discuss with welders or vendors/partners.

9.5.6 Learn How to Weld

Even one welding course from a community college or trade school will help engineers design more manufacturable weldments. If you want to make the best trade-offs between welding and alternatives, learn to operate machine tools too (as the author has done).

9.5.7 Minimize Skill Demands

Explore alternatives to manual TIG/MIG/stick welding, such as robotic or automated MIG welding. Other alternatives include spot weld or roll seam welding, both of which require less skill than hand welding. For sheet metal weldments, consider tab-in-slot assembly where a TIG torch can quickly weld the tab to the slot. To minimize setup costs, develop self-jigging geometries that can be fabricated automatically on CNC machine tools, hopefully in the same setup, to make the required features.

9.5.8 Thoroughly Explore Non-Welding Alternatives

For large or tight-tolerance assemblies, thoroughly explore non-welded alternative concepts, such as bolted assemblies of machined parts that are accurately aligned by the low-cost round and diamond pin technique presented in Guideline A3 in Section 8.2.

9.6 DFM FOR LARGE PARTS[12]

9.6.1 The Main Problem with Large Parts

For weldments, the heat of welding creates enough warpage to preclude depending on the accuracy of holes predrilled in the constituent parts. Similarly, for cast parts, precise holes cannot be cast in.

Thus, mounting holes must be machined *after* welding or casting large parts. If the weldment or casting is too large for an ordinary machine tool, then the post-processing machining must be done on large mega-machine tools, which:

- Have high hourly charges for setup, machining, repositionings, and inspections
- Usually involve labor-intensive online setups, which adds to expensive machine time, especially if the weldment is too heavy to position manually
- May involve transportation and queuing delays
- May require lengthy workplace setups and tool changes
- May require the large parts to be repositioned for subsequent machining, which can be slow if they are too heavy to move manually, consumes expensive machine time, and violates DFM guideline P14

9.6.2 Other Costs

Welding requires skilled labor to make consistently good welds plus other labor to position, fixture, clamp, straighten warpage, and grind. Casting metal is labor intensive and time consuming and involves making the molds, pouring the metal, letting it cool, and removing it from the mold.

9.6.3 Residual Stresses

Welding induces residual stresses. The choices are to:

- Live with residual stresses, which may require more metal if the residual stresses lower the payload or cause structural failure modes. Residual stresses may also cause warping after metal removal.
- Anneal the weldment after welding, which will require a large furnace that will also have expensive hourly charges and transportation and queuing delays.

9.6.4 Loss of Strength

Welding causes loss of strength in the *heat-affected zone* from the welding and annealing, thus requiring more steel compared to assembled steel, which will be used at its full cold-rolled strength. Material used for castings usually has less strength than cold-rolled bar stock.

9.6.5 Strategy

The strategy would be to *commercialize* proven parts with backward-compatible replacements with the same functionality and strength (possibly enhanced), but with much less total cost, weight, and material consumption. This would provide cost reduction now on existing products. It would also encourage a *leap-frog strategy* where these low-cost parts could then become the basis for new-generation products or the approach could be applied to new products.

The specific strategy to eliminate the abovementioned costs would be to create an optimized concept/architecture for *constant-stress trusses* and *structures* (which, by definition, use the least material) utilizing the following approach.

9.6.6 Approach

The approach would be based on the following premises:

Fabrication: All machined parts would be small enough to be set up and made quickly on ordinary CNC machine tools in a single setup (Guideline P14). Welded or cost parts would be limited to those that are small enough to be annealed and machined after welding by the typical *in-house machine tools and furnaces*. This may be appropriate for bearing blocks and other *junction parts* if it is not possible to machine them from a single block.

Assembly: Precise alignment of these assembled pieces would be assured by DFM principles, such as DFM Guideline A3 (Section 8.2), in which mating parts would be aligned to submill tolerances by inexpensive pairs of round and diamond down pins in reamed holes. Aligned parts would then be bolted or riveted together with appropriate bolts, torque settings, and retention strategies. The benefits would be multiplied several times when applied to many large parts and multi-part assemblies.

9.6.7 Procedure

The procedure to convert hard-to-build weldments and castings to manufacturable assemblies is as follows:

- Identify all the parts that are attached to the current large part. These, and the mounting feet, would represent the *loads* on the part.
- Identify the fastening interfaces for all the attached loads and the mounting feet. These would represent the *load points*.
- Ascertain the maximum *load values* on each *load point*. When these load values are drawn as 3D arrows on the CAD drawing, they represent the *load paths*. The lengths and directions of the arrows should be roughly proportional to the load values.
- Arrange a workshop[13] to do brainstorming (Section 3.7) to generate several ideas of structural concepts; for instance, various concepts for plates, bars, tubing, 2D trusses, or 3D space frames. The leading candidate(s) would be explored further until a "winner" emerges. An alternative would be to assign this to an experienced company designer or commission a design study[14] by an experienced practitioner who would then turn it over to a company engineer for completion.
- Design structural members to correspond to the load paths and be sized to carry the maximum loads along these paths. Smaller structures could be made from plates or bar stock, with all mounting and aligning features machined in the same setup (P14) on ordinary CNC machine tools. Larger structures could be trusses or 3D space frames comprised of *struts* and *nodes*, where the struts are tubes or rods and the nodes are machined blocks with all the fastening, alignment, and mounting holes premachined.

9.6.8 Results

The results would be much lower cost from:

- Quick machining on ordinary automated CNC machine tools
- Quick setup concurrently engineered to further reduce machine setup time for whole part families that could include many nodes in the same assembly
- Quick assembly with accuracy assured by the machined features
- Higher strength per weight (meaning higher strength per material cost) because (1) it is a more structurally *efficient* design (lower part stresses to support a given *applied load*) and (2) stock material (like bar stock) would be stronger than annealed weldments or cast materials. Stock material would remain at cold-rolled strength and

heat-treated strength could be preserved, which is not the case for the heat-affected zones in weldments.

- Less material cost and better material availability, especially if many pieces are made from readily available materials
- Reduced shipping costs with lighter-weight, maybe hollow, structures that, if needed for rigidity or vibrations, could be filled after shipping with water, sand, concrete, or other fillers.

NOTES

1. James Morgan and Jeffrey K. Liker, *The Toyota Product Development System* (2006, Productivity Press), Chapter 4, "Front-Load the PD Process to Explore Alternatives Thoroughly."
2. James G. Bralla, Editor, *Design for Manufacturability Handbook* (1998, McGraw-Hill).
3. R. Bakerjian, Editor, *Tool and Manufacturing Engineers Handbook, Volume 6, Design for Manufacturability,* (1992, Society of Manufacturing Engineers, Dearborn, MI). Chapter 1 is by David M. Anderson.
4. H. E. Trucks, *Designing for Economical Production,* 2nd edition (1987, Society of Manufacturing Engineers).
5. G. Pahl and W. Beitz, *Engineering Design: A Systematic Approach* (1988, Springer-Verlag), translated from German; 400 pages with 50 pages on DFM.
6. J. Hicks, *Welded Design: Theory and Practice* (2000, Woodhead Publishing), 160 pages.
7. John Campbell, *Complete Casting Handbook: Metal Casting Processes, Techniques and Design*: (2011, Butterworth-Heinemann), 1220 pages.
8. David M. Anderson, *Build-to-Order & Mass Customization: The Ultimate Supply Chain Management and Lean Manufacturing Strategy for Low-Cost On-Demand Production without Forecasts or Inventory* (2008, CIM Press), Chapter 8, "On-Demand Lean Production." See book description in Appendix D.
9. Lance Ealey, *Quality by Design* (1998, ASI Press, Dearborn, MI).
10. For more on DFM consulting and design studies, see Appendix D or visit the Consulting page at www.design4manufacturability.com. Dr. Anderson also offers concept studies (Appendix D) that solve challenging design problems and address major design opportunities.
11. "Workholding" DVD (2010; 28 minutes; DV09PUB6) by Society of Manufacturing Engineers; www.sme.org. See also a thorough, but dated, book that shows a wide range of off-the-shelf workholding products in one volume: *Workholding* (1982, SME; ISBN 0-87263-090-0).
12. For more on converting hard-to-build large weldments and castings to more manufacturable machined assemblies, see http://www.design4manufacturability.com/steel-reduction-workshop.htm.
13. Dr. Anderson has facilitated large parts conversion workshops (Appendix D.6.3) and done consulting design studies for the following types of products: frames for underground mining vehicles; frames for power plant scale current inverters; 12-ft-high

frames for large box-making machines; large filters for nuclear power plants; 400-ft-long vacuum chambers for coating window glass; and multi-story framework structures for postal sorting facilities. He has also proposed these workshops for farm machinery, large engine blocks, machine tool frames, large medical equipment framework, utility-scale transformer enclosures, oil industry equipment, and large vehicle frames. He can be reached at anderson@build-to-order-consulting.com.

14. See Appendix D for more on Dr. Anderson's workshops, consulting, and design studies.

Section V

Customer Satisfaction

10

Design for Quality

> *"One third of quality-control problems originate in the product design."*
>
> — **Dr. Joseph Juran**

Design is more responsible for quality than most people realize, as noted long ago by quality control guru Dr. Joseph Juran.[1] Designers determine the number of parts, decide which are purchased off-the-shelf, select the purchased parts, design the rest of the parts (indirectly specifying how they will be made), determine how the parts must be assembled, and specify how the parts function together. The product's design determines the factory processes, whether or not the designers realize it. If designers realized this, they might be more inclined to work with manufacturing as a team to simultaneously design the product and its processing.

Manufacturing or quality departments are often held responsible for quality. These groups may expend substantial effort to *try to* ensure high quality to compensate for insufficient attention devoted to the *inherent* quality of the design. If diligent, the factory may be successful at keeping quality problems from escaping the plant, but the company may be incurring a high *cost of quality*, as discussed in Section 6.9 and quantified in Chapter 7. Manufacturing, and often product development, can be seriously disrupted by diagnostics, rework, excess inventory, firefighting, and quality control efforts.

Merely talking up quality or having a quality program is not enough, if the product is not designed well for quality and there is a lack of understanding about how customers value quality. If the steps presented here are not followed and the product has quality problems in manufacturing or in the field, engineering resources will be depleted by excessive change orders and factory troubleshooting. In the worse cases, quality problems can force a redesign—the ultimate drain on engineering resources.

As pointed out in Chapter 1, 80% of a product's lifetime costs are determined by the first 8% of the process, which is the design. No matter how hard the factory tries, it can affect only the remaining 20%. Similarly, a large percentage of the product's quality is determined by design.

10.1 QUALITY DESIGN GUIDELINES

The following is a list of proactive guidelines, procedures, and cultural perspectives that product development teams can use to design quality into products.

Q1) Establish a quality culture. This is a culture in which "quality is everyone's responsibility," not just the quality department. Understand that quality starts with product development.

Q2) Understand past quality problems and issues. Thoroughly understand the root causes of quality problems on current and past products to prevent new product development from repeating past mistakes. This includes part selection, design aspects, processing, supplier selection, and so forth. It may be useful to have manufacturing, quality, and field service people make presentations to newly formed product development teams, showing, hopefully with some real-life examples, the past problems that can be avoided in new designs.

Usually, there is so much quality data that it is hard to comprehend all the lessons and prioritize which lessons are most important for new product development. This problem can be solved by plotting quality issue *frequency* versus *severity*, as shown in Figure 10.1.

The *frequency* would be the number of times per year each category of issue appeared. The *severity* could be based on the costs incurred, product downtime costs, resources consumed to fix the issue, resulting change-order costs, product development delays, production delays, and other quantified consequences. Severity could also include nonquantified issues such as safety risks, injuries, and corporate liabilities; these could be ranked on a subjective scale, say, from 1 to 5.

This will present a *visual prioritization* of quality and reliability issues to help product development teams prioritize their efforts to solve quality problems, with the most effort directed toward the most severe issues that

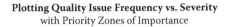

Plotting Quality Issue Frequency vs. Severity
with Priority Zones of Importance

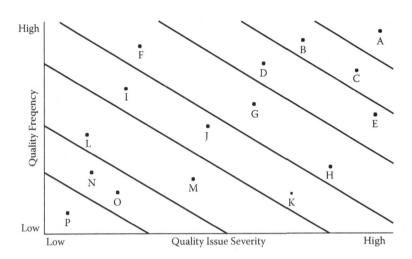

FIGURE 10.1
Quality issue frequency vs. severity.

happen most often. For the graphing format, be sure to label each *issue within its zone* for clarity.

Q3) Methodically define the product. Define the product so it will meet the customer's needs with quality function deployment (QFD), discussed in Section 2.11. This ensures that the first design will satisfy the voice of the customer without the cost and risk of changing the design.

Q4) Make quality a primary design goal. *Proactively* design in quality by making sure that quality is a primary design goal as important as functionality, manufacturability, and all the considerations discussed in Chapter 2. In older, reactive product development cultures, quality issues were dealt with only after problems surfaced in manufacturing, or worse, in the field.

Q5) Use multifunctional teamwork. Break down the walls between departments with multifunctional design teams (Deming's 9th point) to ensure that *all* quality issues are raised and resolved early and that quality is indeed treated as a primary design goal.

Q6) Simplify the design and processing. Simplify the concept/ architecture with the fewest parts, interfaces, and process steps. Elegantly simple designs[2] and uncomplicated processing result in *inherently* high-quality products.

Q7) Select parts for quality. Too often, parts are selected only for functionality and cost. However, to ensure quality by design, parts must also be selected for quality. If part cost goals seem to be at odds with quality goals, then cost is probably not being computed on a total cost basis (Chapter 7), which would include quality costs, as discussed in Section 6.9. When total cost is taken into account, product development teams would never choose the proverbial lowest-bid part. Selecting parts and suppliers on a total cost criterion that includes quality is Deming's 4th point.

Q8) Optimize processing. Select or concurrently engineer manufacturing processes to ensure the highest quality production. Be careful to make sure that new processes are robust enough to ensure high-quality products *in production quantities from production environments.* Design within process capabilities; design for processes that are in control and can reliably produce quality parts.

Q9) Minimize cumulative effects. Understand the cumulative effect of part quality on product quality, as discussed in Section 10.3. Product quality *degrades exponentially as part count increases.*

Q10) Thoroughly design the product right the first time. Use the techniques presented in Chapter 1 to ensure that the product is designed right the first time. If quality is not assured by the initial design, then expensive change orders (Section 1.10) will have to be carried out, wasting valuable engineering resources and possibly inducing further quality problems in the process. Be sure to comfortably satisfy all the design goals and constraints without having to compromise the product just to get it out the door.

Q11) Mistake-proof the design with poka-yoke. Proactively prevent defects by design and in manufacture with poka-yoke, which is Japanese for mistake-proofing or "idiot-proofing." This is discussed in Section 10.7.

Q12) Continuously improve the product. Use continuous improvement, or kaizen, to make incremental improvements to the product and processing. The old paradigm for tolerances was that quality would be assured if each dimension was made "within spec" or within the specified tolerance band. If parts are made in a low-quality environment, a tight tolerance will raise costs because parts outside the tolerance band must be scrapped. However, this can become problematic when many parts are rejected outside the acceptable tolerance range; when this happens, it means that the parts that *do* pass will have a high percentage of parts at either extreme of the tolerance range. This could result in problems when they are combined with other mating parts that are skewed toward the worst-case combinations.

Continuous improvement programs, on the other hand, strive to continuously tighten the accuracy of manufactured parts, so that the dimensions become closer and closer to the target. This results in the population of parts being closer to the center of the range, with few, if any, parts near the original limits.

Q13) Document thoroughly. In the rush to develop products, many designers fail to document every aspect of the design thoroughly. Documentation should be 100% complete, correct, and accurate, with updates made immediately as changes occur.

Drawings sent to manufacturing or to vendors need to convey the design *unambiguously* for manufacture, tooling, and inspection. Imprecise drawings invite misunderstandings and interpretation, which add cost, waste time, and may compromise quality.

Geometric dimensioning and tolerancing (GD&T) is an unambiguous methodology that can clearly convey the *design intent,* which then eliminates delays and quality problems due to documentation problems. In GD&T, each dimension is dimensioned from the most logical and precise *datum.* For instance, several holes to be machined in a piece of metal would be dimensioned from a single datum instead of from each other (which would cause a cumulative error) or from various edges (in which the edge tolerance would affect the hole spacing).

GD&T optimizes dimensioning for (1) function, ensuring the parts are made as intended; (2) manufacturing, to optimize processing and fixturing; and, (3) inspection, allowing the use of coordinate measurement machines. Proper datum referencing also allows the maximum number of operations to be done in the same setup without repositioning the part (Guideline P14).

Q14) Implement incentives that reward quality. In many organizations, individuals do what they are *rewarded* to do. If they are rewarded for releasing a design "on time," they will, effectively, throw it over the wall on time, ready or not! If they are rewarded for achieving "cost targets" without total cost accounting, they will do so by buying the cheapest parts available, probably of inferior quality. Thus, reward systems must be structured to include quality metrics.

10.2 TOLERANCES

Q15) Optimize tolerances for a robust design that is compatible with manufacturing processes. Properly specifying tolerances is one the most important steps to making designs manufacturable. Tolerances that are unnecessarily too tight often force the use of a more precise process, which results in more cost and delays. Designers must understand fabrication processes so they will know the effect of tolerancing on processing.

Do not neglect to specify tolerances, leaving it to chance or interpretation. Do not let tolerances be determined by block tolerances in CAD or drawing formats, especially for noncritical parts, like brackets. Do not repeat inappropriate tolerances from previous work. Do not make tolerances too loose because of a lack of understanding. Do not make tolerances unnecessarily tight "just to be sure."

10.2.1 Excessively Tight Tolerances

If tolerances are perceived by manufacturing as excessively tight, many undesirable things can happen. Manufacturing people who challenge the tight tolerances may question why they are needed and if they really do warrant a more expensive process. If they bring the concern to engineering, the designer who specified the tolerance should welcome the input and adjust the tolerance accordingly. The worst thing that can happen would be for the designer to act too busy or stubborn and refuse to make changes when changes are warranted.

Unfortunately, what happens too often is that manufacturing receives a verbal tolerance change without documenting the change for the next build or for similar products. This only ensures that the problem will recur and implies to manufacturing that the tight tolerance was not really necessary, thus eroding the credibility of tolerances in general.

In other cases, manufacturing people interpret tolerances because of a perceived low credibility of the tolerances. This risky procedure can backfire if some parts are made with inadequate tolerances on some really critical dimensions. The tolerance problem is intensified when parts are sent out to suppliers who do not have an understanding of which tolerances must be held precisely and which can be changed. The supplier will (and should) bid on the basis of the *stated* tolerances on the drawing. The unfortunate result of this double standard on tolerances might be improperly biased make/buy decisions in favor of internal manufacture that may be getting by with looser tolerances for the same part.

If tolerances are specified (or interpreted) too loosely, the product may fail functional tests, encounter random difficulty at assembly, have quality problems, wear out prematurely, or pose a safety hazard in use.

10.2.2 Worst-Case Tolerancing

A related problem is *tolerance stack-up* and *worst-case tolerancing*. Tolerance stack-up refers to the cumulative effect of all the tolerances in a "string" of dimensions (the combination of which affects the same overall dimension). Worst-case tolerancing refers to combining the "worst" of all the tolerances to analyze what the net effect will be. For instance, to do a worst-case analysis on the clearance between a shaft and a hole, one would consider the *largest* possible shaft in the *smallest* possible hole and the *smallest* possible shaft in the *largest* possible hole. This analysis will yield the extremes in clearance, which should conform to design requirements.

The product must be able to function reliably and safely in all worst-case tolerance situations. Tolerance stack-up analysis should be done adequately on relevant part stacks, modules, and subassemblies, as well as on the product itself.

When several dimensions combine to determine an overall dimension (which has *its* own desired tolerance), the tolerances of all the elements should be apportioned rationally based on the lowest total cost for holding all dimensions in the chain. If there is not a planned apportionment of tolerances, the result may be that the last part designed ends up with excessively tight tolerances because previously designed (and maybe built) part tolerances have already been set.

10.2.3 Tolerance Strategy

The goal should be to optimize tolerances for a *balance* of function, quality, safety, and manufacturability. Tolerances should have enough credibility to be respected *as specified* so manufacturing can then concentrate on meeting them. All tolerances specified need to be carefully thought out with respect to the processes. Working with manufacturing engineers very early in the design will help, but the best approach is for the designers to be thoroughly familiar with the processes and their limitations. Tolerances that are changed on one build should be immediately documented for future builds.

Tolerances should be methodically specified by the Taguchi Method™ for robust design (Section 10.2.5).

10.2.4 Block Tolerances

Another problem is the block (or blanket) tolerance printed on the drawing: for instance, ±.005″ for linear dimensions and 63 RMS for surface finish "on everything not otherwise specified." This often results in many tolerances that are tighter than necessary, and, if strictly enforced, would prohibit the use of standard stock; say, ½″ stock for a ½″ dimension. Block tolerances technically could even apply to chamfers and radii, which are only needed for clearance, appearance, or safety.

If a company *must* use a tolerance block, it should specify several blanket tolerances for different needs; for instance:

- .XXX dimensions to indicate a ±.005″ or a ±.001″ tolerance
- .XX to indicate a ±.015″ or a ±.020″ tolerance
- Fractions to indicate a ±1⁄32″ tolerance.

Each company should determine its most common tolerances and specify them in the tolerance block. Designers will need to realize that a two-digit dimension of ⅝ inch will have to be specified as .62 or .63, not .625.

10.2.5 Taguchi Method™ for Robust Design

Proactively specify optimal tolerances for a robust design to ensure high quality by design. The Taguchi Method™ for robust design is a systematic way to optimize tolerances to achieve *high quality at low cost.*[3] It does this by using design of experiments to analyze the effect of all tolerances on functionality, quality, and manufacturability to analyze tolerance stacks and worse-case situations. The procedure can identify critical dimensions that need tight tolerances and precision parts, which can then be toleranced methodically. The unique strength of this approach is that it can minimize cost while ensuring high quality by identifying low-demand dimensions that can have looser tolerances and cheaper parts.

Such a design would be considered *robust*, if it could be manufactured predictably with consistently high quality and perform adequately in all anticipated usage environments.

Without a methodical way to determine tolerances, the alternatives would be to:

1. Make all tolerances tight just to be sure, which is expensive. Tolerances that appear to be overly tight may have credibility problems and invite interpretation.
2. Inadvertently (or deliberately) make tolerances too loose, leading to manufacturability and quality problems. Performance, quality, and manufacturability problems may be inconsistent and thus hard to troubleshoot and rectify.

10.3 CUMULATIVE EFFECTS ON PRODUCT QUALITY

It is important to understand the *cumulative effect* of the number and quality of parts. Computations will be based on the assumption that any single part failure will cause the product to fail. This is valid unless the product has redundancy or backup features that are used in critical applications (like aerospace).

Statistically, this situation is similar to a *series reliability model*, which states that the reliability of the system is computed by *multiplying* the individual reliabilities of *all* the parts of the system, assuming any single failure causes a system failure (in series):

$$R_s = R_1 \cdot R_2 \cdot R_3 \cdot R_4 \cdots R_n = \prod_{I=1}^{n} R_i$$

where
R_s = the reliability of the system
R_n = the reliability of component n

The equivalent formula for product quality would represent the probability of the product functioning properly or having no defects, given the individual probabilities of the parts are free from defects:

$$Q_p = Q_1 \cdot Q_2 \cdot Q_3 \cdot Q_4 \cdots Q_n = \prod_{i=1}^{n} Q_i$$

where

Q_p = the quality level of the product measured as the probability of proper function, or being defect-free

Q_n = the quality level of part n

The quality level of parts is easily measured in the percentage that are defect-free. Quality levels have been rising so much that the "percent good" is becoming a cumbersome number; for instance, 99.95%. In these cases the quality level is expressed in *defects per million (DPM)*, which, for 99.95% would be 500 DPM—a more manageable number. The DPM measurement psychologically fits in better with zero-defect programs because the goal is 0 DPM. When making these calculations, be sure to convert to the decimal equivalent of percent (e.g., .9995 to represent 99.95%).

Product quality can be approximated by the *average* quality level of the parts, using the formula:

$$Q_p = (Q_a)^n$$

where

Q_p = Quality level of the product

Q_a = Average quality level of parts

n = Number of parts

10.3.1 Example

A product consists of 25 parts that are all 99% good, which some suppliers might contend is adequate:

$$Q_p = (Q_a)^n = (.99)^{25} = .78$$

This means that only 78% of products will be good because 25 parts are only 99% good! And this assumes perfect processing quality.

If, in the above example, the parts had half the chance of defects (99.5%), then the product quality level for 25 parts would be

$$Q_p = (Q_a)^n = (.995)^{25} = .88$$

Product quality jumped to 88%—a significant change in product quality *just from specifying a higher grade of part.*

10.3.2 Effect of Part Count and Quality on Product Quality

Figure 10.2 graphically shows how product quality varies with part quality and part quantities. The lines are plots of the equation,

$$Q_p = (Q_a)^n$$

for varying part quality levels. Note that each line represents *twice* as "bad" a part as the one above it.

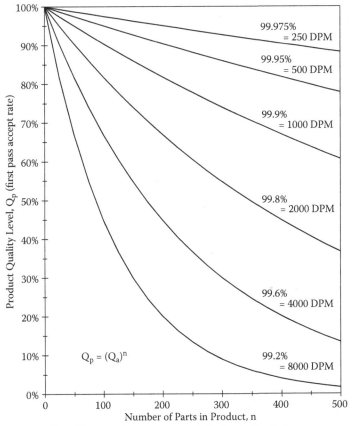

Plotted lines are for average part quality levels, Q_a cited as percent good and DPM (Defects/Million)

FIGURE 10.2
Quality as a function of part count for average part quality levels.

Printed circuit boards are built with several hundred components covering the whole range of lines on the graph. All designers should carefully use these formulas to predict product quality degradation just from the number of components.

10.3.3 Predictive Quality Model

For products with groups of parts with different quality levels, the formula would be:

$$Q_p = Q_{a1}^{n1} \cdot Q_{a2}^{n2} \cdot Q_{a3}^{n3} \cdot Q_{a4}^{n4} \ldots \text{ and so forth}$$

where

Q_p = Quality level of the product
Q_{a1} = Quality level of the first group
n1 = Number of parts in the first group

This formula is an extremely powerful tool that can be used as a *predictive model* to estimate product quality based on a parts list, which may be available early enough to compare several product architectures. This formula can easily be incorporated into a spreadsheet so that various "what if" scenarios can be calculated.

It is important to realize that product quality drops *exponentially* with increasing part count. Thus, unless part quality levels are very high, approaching six sigma (about 3 DPM), *an excessive number of components will have an exponential degradation on product quality!*

10.3.4 Quality Strategies for Products

Given the aforementioned part quality issues affecting product quality, the following strategies can be developed.

Maximize $(Q_a)^n$ by:

1. Maximizing the average part quality level, Q_a
2. Minimizing the number of parts, n
3. Optimizing both of the above

Designers should continually think in terms of maximizing $(Q_a)^n$. When doing a trade-off analysis on competing designs, they should choose the solution with the highest $(Q_a)^n$.

If the product trend is toward more complexity, *and thus higher part count,* and the part quality is the same, then the cumulative exponential effective of part quality will cause the *product* quality to suffer. Senior managers may be puzzled why quality for a more complex product drops when the incoming part quality is the same.

Part of the solution involves continuous improvement[4] of part quality and factory process quality to compensate for this increase in complexity. Another solution is to keep trying to decrease the part count, despite increasing product complexity. Designers of complex electronics circuitry have the option of combining dozens of small integrated circuits into a single standard VLSI chip or custom ASIC (application-specific integrated circuit) device with about the same quality level as any of the chips it replaces. This type of option can make a major improvement in product quality, or, in some cases, may be the only way to achieve an acceptable level of product quality. Based on the total cost factors presented in Figure 5.5, it can be seen how widespread use of versatile ASICs can be justified.

10.4 RELIABILITY DESIGN GUIDELINES

Reliability can be defined as quality in the time dimension. A product with good reliability has freedom from failure in use. The classical definition is the probability that a product will perform satisfactorily for a specified period of time under a stated set of use conditions. The elements of reliability include probability, performance, time, and usage conditions. Here are some guidelines for optimizing reliability by design.

Q16) Simplify the concept. Concept simplicity is the key to *inherent reliability,* although this rarely is mentioned in statistically oriented reliability handbooks. Significantly reducing the number of parts, interfaces, connectors, interactions, and complexity, in general, will greatly improve product reliability.

Q17) Make reliability a primary design goal. Proactively design in reliability by making sure that reliability is a primary design goal as important as functionality, cost, manufacturability, and all the considerations discussed in Chapter 3. In older, reactive product development cultures, reliability issues were dealt with only after problems surfaced in the field.

Q18) Understand past reliability problems from lessons learned. Thoroughly understand the root causes of reliability problems on current and past products from early *lessons learned* data (Section 3.3) to prevent new product development from repeating past mistakes. This includes part selection, design aspects, processing, supplier selection, usage conditions, and so forth. It may be useful to have manufacturing, reliability, and field service personnel make presentations to newly formed product development teams, showing what lasted and what caused past reliability problems that can be avoided in new designs, hopefully with some real-life examples.

Investigate the reliability of similar products, and reliability challenges in similar environments, to find out how other systems work, or don't, over time, and learn the keys to their successes or failures.

Use *failure modes and effects analysis* (FMEA) to understand reliability failure modes and their consequences. Then generate strategies and action plan deliverables to minimize failures and their consequences. FMEA can also be used as a proactive design tool to help teams identify failure modes and prevention strategies for a wide range of potential concepts. This may be the deciding criterion among various product concepts.

Q19) Simulate early. Use simulations and computer models to simulate and maximize reliability early to optimize early design decisions, which are much easier and more effective to incorporate than any changes that would be implemented later based on data from prototypes or results from the field.

Conduct FMEA early to predict most likely failure modes and develop strategies to minimize failures and their consequences. Do early selective experiments and accelerated reliability testing early on the most problematic aspects of the design.

Q20) Optimize part selection on the basis of substantiated reliability data. Select parts on the basis of *substantiated* reliability data, not just advertised claims. The lack of substantiated reliability data, which is more common for new parts, may encourage greater use of proven parts.

Q21) Use proven parts and design features. Use *proven* standard parts and design features that have been used successfully before and would be most likely to provide reliable service. Past performance data can steer designers quickly to the best parts and design features to help them maximize reliability. A key goal of design teams should be to reuse proven designs, parts, and modules. Be careful of any reuse that may be less verified and may be below current quality and reliability thresholds.

As mentioned in Section 3.1, a high percentage of complaints, field failures, recalls, and lawsuits do *not* involve new features or new technology. Rather, they involve "boilerplate" functions that should be based on proven designs, parts, and modules. For instance, in electronics, many problems arise from the mundane power supply. In the automobile industry, the most serious problems and consequences involve fuel systems, seat belts, steering, brakes, suspension, tires, and so forth. These are not the parts that companies are advertising or customers are clamoring for, which are more likely to be things like styling, cup holders, stereos, and navigation systems.

Q22) Use proven manufacturing processes that are in control and have a history of producing reliable parts, so as to avoid the added variables encountered when new processes are introduced. Ironically, some products cannot utilize existing proven processes because they are not designed well enough for manufacturability for those specific processes.

Q23) Use precertified modules. Use proven, precertified modules that can be individually certified. If the product's architecture was optimized for this, then it may be possible to consider the product certified if all the modules were precertified.

Q24) Design to minimize errors with poka-yoke. Design to proactively minimize errors in fabrication, assembly, installation, maintenance, and repair with poka-yoke, discussed in Sections 10.7 and 10.8.

Q25) Design to minimize degradation during shipping, installation, or repair. Design products and packaging so that products do not suffer any damage during shipping. Specify the installation process so that reliability is not degraded by all the steps involved in installation or repair.

Q26) Minimize mechanical electrical connections, especially for environments that are corrosive or subject to shock or vibration and for low-voltage connections. A common illustration of this problem is the automatic reflex of shaking a flashlight when it does not come on right away even though the battery is touching both contacts. Some specific solutions to avoid low-voltage mechanical connectors:
- Combine circuit boards to eliminate cables and connectors.
- Use a means of connection with the minimum mechanical connections, such as *flex cable,* which can connect different circuit boards

with flexible traces (the conductors in circuit boards) that are soldered to components on both circuit boards.

- Minimize use of sockets.

Q27) Eliminate all hand soldering. Hand soldering is the least reliable way to make electrical connections. The reliability people at Intel's Systems Group discovered disturbing data that hand-soldered joints can pass tests in the factory and fail later in the field. In contrast, automatic soldering of circuit boards (reflow or wave solder) is about the most refined industrial process, with plants routinely achieving six sigma quality levels. Hand-soldered joints can be avoided with three strategies:

1. Design out of the products the need for hand-soldered joints by combining circuit boards, using flex cable, and the proper use of connectors and cables.
2. Conduct a thorough search for *auto-solderable* components early and base the design on them. Even if the part cost is slightly higher, total cost will be lower and product quality will be higher.
3. Design within the DFM guidelines for printed circuit boards so that component location rule violations will not force components to be soldered by hand to circuit boards.

Q28) Establish repair limits for circuit boards. Most companies do not have repair limits on the number of times components can be desoldered and replaced. Excessive hand soldering can damage the circuit board pads or holes and possibly adjacent components. Ray Prasad, author of *Surface Mount Technology*, recommends a maximum of two repairs to prevent internal thermal damage to printed circuit boards.[5]

Q29) Use burn-in wisely. Use *burn-in* or *run-in* to induce early failures until the problem can be isolated. Then use this information to eliminate the causes of the problems.

10.5 MEASUREMENT OF RELIABILITY

The measure of reliability is the *mean time between failures*, or

$$MTBF = 1/\lambda$$

where λ = the *failure rate*

What this measure means is that, on the average, one failure in operation can be expected to occur after a period of time equal to the MTBF.

The reliability of a product can be numerically expressed as:

$$R(t) = e^{-\lambda t}$$

where t = time.

10.6 RELIABILITY PHASES

There are three phases in which products experience different reliability behavior: the *infant mortality, useful life,* and *wearout* phases, shown in Figure 10.3. The following lists causes of reliability failures in the infant mortality and wearout phases. Note that many of these causes can be prevented proactively by optimal design of products and processes.

10.6.1 Infant Mortality Phase

This phase is characterized by early failures due to:

- Built-in flaws from processing:
  - Poor welds or seals
  - Cracks in castings
  - Poor solder joints

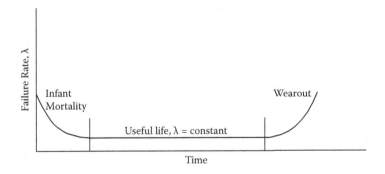

FIGURE 10.3
Reliability phases.

- Surface contamination
- Chemical impurities
- Handling or static damage
- Incorrect positioning of parts
- Transportation damage:
  - Physical damage to parts
  - Shock-sensitive parts overstressed
  - Parts loosened during shipping
- Installation and setup errors:
  - Factory errors
  - Customer errors

10.6.2 Wearout Phase

The wearout phase is when the product failure rate begins to rise from deterioration. Causes include:

- Frictional wear
- Fatigue
- Creep
- Corrosion or oxidation
- Chemical changes
- Insulation breakdown
- Shrinkage or cracking in plastics

10.7 POKA-YOKE (MISTAKE-PROOFING)

Poka-yoke is a Japanese concept that originally evolved to prevent mistakes in manufacturing.[6] Poka-yoke can be used as a design methodology to mistake-proof the design so parts cannot be assembled wrong or products manufactured incorrectly. Although these clever design features are created only once, they go on to prevent mistakes in manufacturing for the life of the product. The following discussion presents the general principles of poka-yoke applied to product development. After that, there are several guidelines to minimize errors.

With parts standardization, there are fewer *types* of parts and, thus, less chance of choosing the wrong part. Symmetrical parts cannot go in backwards. Nonsymmetrical parts with exaggerated asymmetry will not be assembled incorrectly or forced into a wrong orientation; this avoids one of the most mistake-prone parts—one that is almost square. Nonsymmetrical parts should have polarized connectors or mounts.

If different parts have different installation geometries, then the product can be designed so that the wrong parts cannot be installed, such as unique shafts with unique diameters, unique gears with unique bores, unique fasteners or valves with unique threads.

Avoid choosing the wrong part by indicating or presenting only the correct part, and by minimizing the number of part types with standardization, ideally with only one of each type of part at each workstation.

Features can be added to prevent incorrect assembly, such as alignment pins or tabs, unique geometries and shapes, and markings to indicate correct assembly and aid in inspection. Potentially confusing assembly can be avoided entirely by conceptual simplification, like combining parts or eliminating them entirely.

10.8 POKA-YOKE PRINCIPLES

10.8.1 How to Ensure Poka-Yoke by Design

Avoid inserting parts backwards with:

- Symmetrical parts (Guideline P3)
- Exaggerated asymmetry for nonsymmetrical parts (Guideline P4)
- Polarized connectors or mounts for nonsymmetrical parts
- Adding a hole to a part and a pin to the tooling or mating part to ensure correct insertion or loading

Avoid choosing the wrong part by:

- Minimizing the number of part types with standardization, ideally with only one of each type of part at each workstation
- Indicating or presenting only the correct part

Avoid assembling the part into the wrong position by:

- Unique mounting geometries
  - Different hoses with matching fittings
  - Unique fasteners for different parts
  - Different shafts matching the proper gears
- Adding features to prevent incorrect assembly:
  - Alignment pins or tabs
  - Unique geometries and shapes
  - Markings or templates to:
    - Indicate correct assembly
    - Aid in inspection

Avoid omitting parts when:

- Adjacent parts would not assemble without the missing part
- Printed outlines would show missing parts, wrong parts, and incorrect orientation
- Automatic sensors or scales detect products with missing or extra pieces

Avoid assembly errors by:

- Combining parts or eliminating them entirely
- Eliminating procedures that depend on operator skill or memory
- Generally eliminating choices, decisions, and judgment
- Avoiding mistake-prone assembly procedures

Prevent damage by design. Avoid sequence errors by:

- Designing the assembly process so that sequence doesn't matter
- Features that prevent the wrong sequence from happening
- Making the sequence intuitively obvious
- Clearly indicating the correct sequence

Avoid timing errors by:

- Designing products without the need for timed processes
- Eliminating operator-timed processes

- Making all timed operations the same
- Making different timings very different

10.8.2 Solutions to Error Prevention after Design

If error prevention is not addressed in the design, the factory will have to compensate by hiring more skilled (and more expensive) workers, training them more, and exercising constant vigilance with procedures and inspections.

Design happens once. But hiring and training must be repeated
as new people become part of the system. And vigilance must be constant.

10.9 STRATEGY TO DESIGN IN QUALITY

The methodologies of this book can be used to proactively ensure high quality and reliability *by design* using the following techniques:

- Understand past quality problems through lessons learned data-bases, research, and presentations from related project people.
- Obey quality design guidelines Q1 through Q29.
- Consider using quality function deployment (QFD) to define products to *capture the voice of the customer* (Section 2.11). QFD is one of the techniques in the collection of tools known as "design for Six Sigma."
- Use multifunctional teamwork (Deming's 9th point) to ensure that quality is a primary design consideration and that all quality issues are raised and resolved early, and quality is proactively designed in.
- Perform thorough up-front work to implement all the above early and avoid significantly greater quality and ramp problems later.
- Simplify the design for inherently high quality and reliability.
- Minimize the exponentially cumulative effect of part quality and quantity by specifying high-quality parts and simplifying the design with fewer parts.
- Select the highest quality processing. Automated processing produces better and more consistent quality than manual labor.

- Raise and resolve issues early by: learning from past quality problems; early research, experiments, and models; generating Plan B contingency plans; and proactively devising and implementing plans to resolve all issues early (Section 3.3).
- Optimize tolerances for a robust design using Taguchi Methods™ (Guideline Q12). Robust design is one of the techniques in the collection of tools known as "design for Six Sigma."
- Choose materials for quality (Deming's 4th point) not from auctions, low bidding, or switching suppliers for price.
- Apply poka-yoke principles to product design to prevent mistakes by design, in addition to concurrently engineering manufacturing procedures to prevent incorrect manufacture (Section 10.7).
- Reuse proven designs, parts, modules, and processes to minimize risk and ensure quality, especially on critical aspects of the design.
- Rationalize product lines to raise corporate quality by eliminating the unusual, low-volume products, which usually have the lowest quality (see Appendix A).
- Implement big picture metrics and compensation to reward actions that ensure quality and avoid compromising quality with cheap materials to save cost, chasing cheap labor, or throwing a suboptimal design over the wall on time.
- Use total cost, including quantifying the *cost of quality*, as the basis for all part and processing decisions; all elements of cost of quality are quantified.

Designing for quality is what gets quality from 5 sigma to 6 sigma.[7]

Design for Six Sigma (DFSS) is the product development element of the Six Sigma quality program.[8] *Six Sigma* is a management approach aimed at eliminating mistakes, rework, and waste. The goal of DFSS is to prevent defects by *designing* quality into the product.

Design for Six Sigma supports the DFM quality strategies presented above by offering in-depth treatments of rigorous computational and statistics-based methodologies, such as using QFD to capture the voice of the customer, Taguchi Methods™ to optimize robust parameters and tolerances, the theory of inventive problem solving (TRIZ) to generate ideas, and FMEA.

10.10 CUSTOMER SATISFACTION

Designing for quality is a key element of providing customer satisfaction, as is QFD (Section 2.11). The book, *Satisfaction: How Every Great Company Listens to the Voice of the Customer,* from J.D. Power, shows how good customer satisfaction dramatically improves sales, profits, and shareholder value:

> "Garner a reputation for providing great customer satisfaction and you can charge a price premium that goes straight to the bottom line."

> "Saddle yourself with a reputation for marginal customer satisfaction and the only way to build market share will be through discounts and other incentives that will wreak havoc on your bottom line."[9]

J.D. Power has correlated customer satisfaction with sales and shareholder value. For automobiles, sales of brands with low customer satisfaction dropped 4% over a 5-year period, whereas *high satisfaction resulted in a 44% rise in sales!* Over the same period, companies with a drop in customer satisfaction experienced a 28% decline in shareholder value, whereas, for companies with improved satisfaction rankings, *shareholder value rose 52%!*[10]

NOTES

1. Seth Godin and Chip Conley, *Business Rules of Thumb* (1987, Warner Books).
2. Matthew E. May, *The Elegant Solution* (2007, Free Press).
3. Lance A. Ealey, *Quality by Design: Taguchi Methods and US Industry* (1998, ASI Press).
4. Kiyoshi Suzaki, *The New Manufacturing Challenge: Techniques for Continuous Improvement* (1987, Free Press).
5. Ray P. Prasad, *Surface Mount Technology: Principles and Practice* (1989, Van Nostrand Reinhold), p. 547.
6. *Poka-Yoke: Improving Product Quality by Preventing Defects* (1989, Productivity Press/Taylor & Francis); 295 pages with 240 examples.
7. Subir Chowdhury, *"Design for Six Sigma"* (2002, Dearborn Trade Publishing).
8. Ibid.
9. Chris Denove and J. D. Power, IV, *Satisfaction: How Every Great Company Listens to the Voice of the Customer* (2006, Portfolio).
10. Denove and Power, *Satisfaction,* Chapter 1, "Show Me the Money."

Section VI

Implementation

11

Implementing DFM

To paraphrase the punch line of the entertaining IBM commercials, *There is no magic tool to implement DFM*—no magic software, no magic shrink-wrapped solution, no magic trademarked models. To say this in a different way, *you can't buy DFM.*

Nor can companies simply *manage* their way to great product development. Project management techniques may allow management to track progress and feel that things are "on schedule," but these measures could be counterproductive without an understanding of how products should be developed, in general, and designed for manufacturability, in particular. For instance, if intermediate deadlines are set arbitrarily or without knowing the importance of thorough up-front work, they may force engineers to rush prematurely into part design and miss the greatest opportunities to simplify concepts, optimize product architecture, and raise and resolve issues early. The concept/architecture stage is when 60% of cost is committed (Figure 1.1) and is the key to the quickest time to stable production (Figures 2.1 and 3.1).

Another popular way of *managing* product development is to "measure performance," but that, too, can produce counterproductive results if the measures are not based on total cost and the time to stable production. If teams are measured, judged, and evaluated on "cost," but all that is quantified is part and labor costs, then they may be driven to specify cheap parts, move production to low-labor-cost areas, and resist standardization, modularity, and off-the-shelf parts because of the false belief that they will "raise cost" on *one* project. Similarly, if engineers are measured, judged, and evaluated on meeting deadlines, then they will just throw it over the wall on time.

In the world of the popular phases and gates, DFM is usually relegated to a late step (sometimes focused on *checking* for DFM!). However, DFM is not a step in a phase—it is *the way engineers should be designing* for man-ufacturability *throughout* the design process. Good DFM does not come from turning the proverbial crank on some tool, model, or procedure. Somewhere in all those grandiose models, a team has to creatively design for manufacturability. Good DFM does not focus on hitting numbers that are poorly defined; it focuses on actually *designing* products for the lowest total cost, the best quality, and the fastest time to stable production.

Successful DFM comes from a combination of education, teamwork, diversity, leadership, commitment, creativity, big picture metrics, under-standing customer needs, understanding manufacturing processes, and management support and encouragement of all of that. The rest of this chapter shows how to successfully implement DFM and concurrent engi-neering (hereafter denoted together by "DFM").

11.1 CHANGE

Before implementing any changes, companies need to understand the *need* to change and the *benefits* of changing. At this point, it might be a good idea to read Section 1.1 again about the consequences of not hav-ing products designed for manufacturability and the survey comments describing what it is like to work in a company that doesn't practice DFM. Identify how many of these points are routinely experienced. But before changes can begin, the most common objections must be overcome.

- **"Things aren't so bad. There is no need to change."** This head-in-the-sand approach ignores many realities about changes in markets, customers, technology, workers, competition, and other changing trends. The largest study ever conducted on business failures found that: "There is one blind spot that appears somewhere near the center of almost every major business disaster: a seriously inaccurate per-ception of reality among executives."[1]

  Dr. Deming said that companies are reluctant to change unless there is a crisis. However, deferring change until a crisis hits is a poor strategy, because meaningful change will be much more difficult to implement in the face of dwindling sales, falling stock prices, shrink-ing market share, more demanding customers, threatening competi-tion, changing regulation, negative publicity, credit downgrades, or quality problems flaring up. "Crisis management" usually tries to bandaid the immediate *symptoms* rather than identify root causes and implement systematic *solutions*. Product development can be either the *cause* or the *solution* to many of these problems. But because product development is a long-term endeavor, companies in crises will have to live with the shortcomings of product designs until they can be redesigned. To prevent such a predicament, companies should start *now* to develop good products and, as necessary, change the ways products are developed, built, sourced, and distributed.

- **"I don't understand it."** Although no one would admit it, many people don't understand or comprehend the challenges going on in product development. And if they don't understand the *problems*, they probably won't understand *solutions* like DFM, standardization, quality assurance, and designing products for Lean Production, build-to-order, and total cost accounting.

- **"We're already doing it."** A combination of the first two objections would be believing that the company is already implementing a proposed solution. Although one company confidently proclaims, "We practice concurrent engineering," upon further investigation, it is discovered that what they mean is that a manufacturing engineer was invited to the design release review!

- **"The whole industry does it."** Many challenging practices occur throughout an entire industry, such as the need to calibrate all products. The scenarios are as follows: If *avoidable*, significant competitive advantage goes to whoever can eliminate it, especially if that improves cost, quality, and delivery. If truly *unavoidable*, competitive advantage goes to who does it best: the fastest, the most consistent, and/or at the least cost.

- **"We've always done it this way."** Even if some vague need to change is perceived, many companies run into the brick wall of historical inertia.

  Consider the following slogan:

  If you do what you have always done, you will get what you always got.

In stable markets without competition, the *status quo* may be hard to knock, but that is rare these days. Usually, companies have many challenges related to profits, sales, cost, changing markets, advancing technologies, and product developments that need to be better and faster.

At Honda, managers are encouraged to respect sound theories but not hesitate to challenge old habits with new ideas. Company literature encourages proceeding with *"ambition and youthfulness, seeking out challenges with fresh, open-minded passion for learning."*[2]

- **"We've never done it that way."** Change is often resisted because the solutions are new. This can be overcome by understanding new ways and learning how to implement them through books, articles, classes, seminars, and bringing in outside experts to help. Many companies (see the list in the Preface) have circulated multiple copies of the earlier editions of this book to familiarize decision makers with solutions.

- **"We've tried it; didn't work."** Specific solutions may be shot down because of *perceptions* about earlier unsuccessful experiences that may have been:
  - Not enough investment in research money, time, and talent
  - Done at the *part* level, not at the *system* level; in fact, innovation may affect or even determine product architecture
  - Attempted in an existing product as a change order
  - An inability to justify the proposal without quantified overhead costs
  - Using inadequate materials, samples, resources, or vendors
  - Inadequately searching for off-the-shelf solutions, which should be a *serious* search effort, for reasons presented in Section 5.19
  - Not having enough support or consensus
  - The wrong solution, or a good solution applied in the wrong way
  - Not focused on customer needs, but rather on "a solution looking for a problem"
  - The victim of undocumented tales about previous "failures" (*management by folklore*) without understanding the facts, which can temporarily stymie inside innovators or outside consultants in the short term—until the facts are revealed

  A corollary of this objection is that *someone else* tried it and it didn't work. The main shortcoming of this objection is the verb "tried." Real change doesn't come from *trying*; it comes from *doing*.

- **"We don't have time or resources."** The companies that use this excuse are probably spending most of their engineers' time and efforts correcting deficiencies in previous products, as shown in the top graph in Figure 2.1 and the top bar in Figure 3.1.

In the distant past, many people believed that quality would cost more, until Philip Crosby wrote the book, *Quality Is Free,* in which he showed that the returns from better quality would more than pay back the effort to achieve quality.[3] Similarly, the "cost" of implementing concurrent engineering and DFM will be more than paid back by better products, with cost designed out, quality designed in, and rapid ramps into production, as shown in the lower graph in Figure 2.1 and the lower bar in Figure 3.1.

- **"We'll just change the goals."** As pointed out in Section 1.6 on *focus,* real change cannot be achieved simply by applying more pressure or more ambitious goals. Many companies ignore the truism:

  Trying to get much better results by doing things the same way is the definition of insanity.[4]

- **"Do something. Anything."** If decision makers realize some kind of change is in order but don't understand the problems or the solutions, they may jump at whatever *appears* to be a solution. This could happen at staff meetings (when members present pet projects) or at a slick sales presentation for some "cure all" shrink-wrapped solution. Many companies think they have responded to change by bringing in some packaged "program" or "throwing software at it."

  The degree of the challenge dictates the degree of the changes needed. Ambitious goals may require some fundamentally different ways of doing things.

  Ambitious results require ambitious changes.

11.1.1 Change at Leading Companies

As shown in the following examples, leading companies have embraced change as the way forward. As stated by Sakichi Toyoda (1867–1930), patriarch of the Toyota dynasty, "Although new behaviors always provoke opposition, those who are complacent will be left behind and ultimately defeated."[5]

At Medtronic, chairman and CEO Bill George says, "The entire organization is now embracing change and innovation as a way of life and as a competitive advantage. These innovations have been a key factor in changing Medtronic's image from the pacemaker company of the 1980s to the innovative, high-growth leader in medical technology for the twenty-first century."[6]

And at Xomed, "upper management stressed culture change—no swift transformation, but a gradual increase in empowerment to those doing the work, making change fun, not a threat."[7]

11.2 PRELIMINARY INVESTIGATIONS

11.2.1 Conduct Surveys

The first implementation step should be to understand how well the current product development systems works—or understand its shortcomings. Conduct anonymous surveys of everyone in product development, asking questions such as:

- How do you rate our products for manufacturability compared to competitors, the industry, and the best know in any industry?
- What are *good examples* of DFM?
- What are *inadequate examples* of DFM?
- What are the *consequences* of inadequate DFM?
- What are the *hurdles* to good DFM?
- What are the *opportunities* for good DFM?

The author has been conducting this poll for 15 years before his in-house seminars[8] and has found that this information is a valuable starting point for implementation—hearing candid comments about the product development culture and its performance. This information is also very useful for customizing training. Further, presenting the results is a good way to kick off DFM seminars and to get people discussing the issues raised. Survey comments have a special impact because they refer to actual company products and procedures. Some colorful survey comments are presented in Section 1.1.2.

Figure 11.1 is a summary of the top eight responses regarding consequences, hurdles, and opportunities with regard to DFM, gleaned from 650 responses from 10 companies in several industries: two consumer products; two OEM suppliers; two scientific instruments; two processing equipment; and two aerospace.

Consequences	Hurdles	Opportunities
Quality, 33%	Time, 19%	CE teams, 23%
Assembly hard, 18%	Lack of teamwork, 17%	Methodology, 13%
Cost problems, 12%	Attitudes, 12%	General, 9%
Time/ramps, 8%	Cultural, 12%	Vendor relations, 8%
Changes, 7%	Communications, 7%	Quality, 6%
Service/repair, 5%	Resist change, 5%	Cultural changes, 6%
Inflexible, 4%	Changes, 5%	Resources/tools, 5%
Not competitive, 4%	Discipline, 4%	Profit/success, 5%

FIGURE 11.1
Pre-seminar survey results.

11.2.2 Estimate Improvements from DFM

Next, summarize measures of corporate "bottom line" performance, such as profits, revenue, growth, and stock price. Then identify the *drivers* for those measures; in other words, the activities and programs that would drive their achievement (e.g., cost reduction, shorter time-to-market, better quality, and better customer satisfaction).

Ascertain how well the product development process is performing and what management's goals are for improvement. Estimate how corporate performance could be improved by DFM.

To estimate cost reduction, look at all the cost reduction techniques of Chapter 6 and estimate how much could be saved with DFM in relevant categories such as assembly, quality, change orders, inventory, material overhead, and so forth. Scrutinize existing products to see how excess costs may have been *committed* in the concept/architecture phase in the same way the type of data in Figure 1.1 was generated.[9]

Ask assembly supervisors how much labor cost and assembly steps could be saved by better DFM. Ask the manager of purchasing how much material overhead could be saved if new products were predominately designed around standard parts. Talk to operations people to find out how designing products for Lean environments could reduce setup costs and improve flow and machine tool utilization. Find out how much of the engineering and manufacturing budgets are spent on firefighting and change orders. Estimate how much of that cost could be saved by better design for manufacturability. Analyze completed feedback forms (see Appendix C).

To estimate quality gains, ascertain the main causes of quality problems and recommend design solutions, such as better parts, fewer parts (as shown in Figure 10.3), mistake-proofing, optimal tolerances, vendor partnerships, and concurrent process design and selection.

To estimate time-to-market gains, analyze past product development projects for their *real* time-to-market after all the revisions, iterations, and ramp-up in Figure 3.1, and after targets have been reached for volume, quality, and productivity, as graphed in Figure 2.1.

Here is what HP achieved when it implemented concurrent engineering for a complex scientific instrument, compared to the pre-DFM product: The performance was three times better, the price was 70% of the pre-DFM product, and the reliability was 3.4 times greater. The number of parts was reduced to one-third and the floor space needed was two-thirds. It took a *quarter* of the time to develop and the *cost was one-half.* So it should be understandable that sales were 4.4 times and the *profit tripled.*

11.2.3 Get Management Buy-In

To implement DFM formally, structure and propose a DFM program. Present the program to management with the intent of getting management support and buy-in for the program. Use the above estimates of improvements and any other qualitative benefits. Such a program may include a DFM implementation task force that would carry out the steps recommended herein.

Less formal DFM implementations can be started at any level; for instance, by individuals or product development teams practicing the DFM principles presented in this book (Section 11.6) or in DFM training. Early successes can be leveraged for other projects and more formal programs.

11.3 DFM TRAINING

11.3.1 Need for DFM Training

Companies need to provide DFM training because DFM is taught in only a few colleges.[10] Engineers are almost exclusively taught to design for functionality, and engineering tools primarily help engineers design for functionality. And years of work may have turned this focus into a habit.

Therefore, product designers need to be *taught* to design for manufacturability. Further, besides the obvious learning benefits, offering DFM training can be a catalyst to change behavior. Many companies use DFM training to kick off DFM programs. The savings from the DFM benefits (Section 1.13) will *far exceed* any training expenses.

11.3.2 Don't Do DFM Training "On the Cheap"

Don't try to do DFM training "on the cheap," because *you only get one chance at DFM training.* If the training is bad, it will give the message, or even confirm preconceived notions, that DFM is useless and therefore engineers should continue to design only for function. If that happens, it will be hard to get engineers to attend a better class later, not to mention all the opportunities lost in the intervening time. Keep in mind that the biggest expense of corporate training is the value of attendees' time, so use that time wisely and provide the best training possible.

11.3.3 Customize Training to Products

Schedule training for all product development personnel in DFM principles. DFM training should be customized to the company's product line and culture. Beware of bringing in training that is based only on generic principles or "canned" presentations by staff trainers. Don't limit training to only procedures or project management techniques. Be suspicious of training that may really be based on software or some other tool that the "trainers" might just happen to be selling. The danger of depending on a tool is that managers and engineers may think product development goals will be achieved by the tool, and then they think that they may not need to implement *real solutions,* such as the concurrent engineering and DFM principles recommended by this book.

Ask prospective trainers how much they will be customizing the material, on what will they base such customization (e.g., surveys or interviews), and how much relevant experience the *actual presenter* has with your type of products—not just having similar products on the client list of the training company. Beware of high-powered sales presentations by experienced people, who will send less experienced people to do the actual training.

11.3.4 Trainer Qualifications

DFM training should be presented by someone thoroughly familiar with DFM principles. The trainer should have enough experience to answer all questions and engage the audience in discussions on how to apply these principles in your company, and enough credibility to offer better alternatives to the status quo. Experience in both design and manufacture will enable the trainer to speak from the perspective of designers (with personal design examples) and manufacturing (relating personal experiences with manufacturable and *un*manufacturable designs).

DFM training by in-house personnel is a possibility, but only if the trainer thoroughly understands DFM principles (which may require some study), puts in enough effort to prepare the training, and has enough experience and facilitation skills to answer questions, encourage discussions, and convey all the important principles. This book could be used as a textbook for such a course.

With in-house trainers, position and title can be an asset or a liability. DFM training from manufacturing people might appear to be "preachy," whereas trainers from design engineering might not fully understand manufacturability issues. Senior managers would carry more authority than "worker bees," but they usually don't have the time or bandwidth to adequately prepare and present the training. Further, inside trainers may not have as much credibility as outside experts. In fact, a common experience of outside trainers is the appreciation from insiders who have been trying to emphasize similar points, but with little success, because they lack the outside expert's credibility and experience.

11.3.5 DFM Training Agenda

An excellent way to kick off a DFM seminar is to start with some rousing opening comments by the president, vice president of engineering, or division general manager. These comments are important to convey management support and motivate everyone to learn the principles and then be *expected* to put them into practice. Motivation emphases can range from stressing opportunities to "we gotta do this for survival."

As an example, one president kicked off a DFM seminar by relating the experience of shipping a million dollar processing machine to a semiconductor fabrication plant, where no one could understand why it wouldn't work, until they figured out that a light-sensitive enclosure was not sealed properly because someone grabbed the wrong screw from a proliferated selection.

The kick-off executive should also introduce the trainer along with a brief bio-sketch and then hopefully stay for at least the high-level topics, which roughly correspond to the first three chapters of this book in addition to Sections 6.2 and 11.5.

An effective opening session will involve reviewing and discussing the results of the survey, described above, which says in the attendees' own words what is wrong with the current product development culture and details the opportunities for improvement.

It is recommended that the training start with the big picture topics, so that senior managers can join the class to attend the first morning session, which focuses on the importance and implementation of:

- Product line planning (Section 2.3), prioritizing, and rationalization (Appendix A)
- Clear product definition to satisfy the voice of the customer (Section 2.11)
- Thorough optimization of the crucial concept/architecture stage, which determines 60% of a product's cumulative lifetime cost (Sections 1.3, 3.2, and 3.3)
- Cutting the real time-to-market in half through up-front optimization and design work (Sections 1.5, 2.1, and 3.2)
- Resource availability to ensure the formation of complete teams, with *all* specializations active *early* (Sections 2.2 and 3.3)
- Preselection of vendors/partners who can help develop products (Section 2.6)
- An effective team leader (Section 2.7)
- Ensuring teams have the proper focus (Section 1.6)
- Raising and resolving issues early (Section 3.3.5)
- Decision making, costing, product pricing, and performance measures based on total cost accounting (Chapter 7)

The remaining sessions should teach the following to engineers and mid-level managers and senior managers who choose to remain:

- Motivation and overcoming resistance for DFM (Sections 1.7 and 11.1)
- Understanding manufacturing through experience and teamwork (Section 11.6)

- Optimizing product design by satisfying all design considerations (Section 3.5)
- Avoiding arbitrary decisions (Section 1.8)
- Creative product development (Section 3.6) and brainstorming (Section 3.7)
- Do it right the first time (Section 1.11) to minimize the costs and delays of changes (Section 6.8)
- Considering off-the-shelf parts early (Section 5.18)
- Designing around standard parts (Chapter 5)
- Designing for Lean, build-to-order, and mass customization (Chapter 4)
- Follow appropriate design guidelines for products (Chapter 8) and parts (Chapter 9)
- Design in quality and reliability (Chapter 10) to eliminate the cost of quality (Section 6.9)
- Total cost minimization (Chapter 6) and measurement (Chapter 7)
- Change (Section 11.1)
- DFM implementation (Chapter 11)
- The importance and benefits of DFM (Sections 1.13 and 1.14)

As an example, the author's baseline agenda for his in-house DFM seminars is shown in Appendix D.4.

11.3.6 "What Happens Next?"

As the final event in the training, poll the audience and ask, "What should happen next?" The answers can be very helpful for formulating implementation strategies. For example, after in-house DFM seminars, this author lists all the answers on several flip charts and has the audience vote for what they think is the most important. The easiest way to arrange for voting is to issue each attendee eight votes, in the form of round sticky-back dots, which everyone can affix to the zones for each point (the rules are only one vote per point). Experience has shown that this is an invigorating way to wrap up the seminar, because most people hang around to see the voting results unfold. The DFM "champion" should then prioritize the results and distribute to all attendees, management, and any DFM task forces.

The following is a summary of the top 11 responses, representing 80% of 3,622 votes cast over the last few years of the author's seminars:

556   Teamwork and thorough up-front work
355   Standardization
323   Total cost
266   Product portfolio planning
252   Updating product development processes and procedures
207   Lessons learned
206   Vendors and partnerships
203   Resource availability
181   Management buy-in and support
159   Product definition, product requirements, and QFD
150   Training and management education

Many of these desired changes will require some degree of *stopping bad habits*, which will be discussed in Section 11.5.

11.3.7 Training Attendance

Attendance for DFM training should include everyone involved in product development, so that it will not just be "preaching to the choir" of manufacturing engineers or "DFM engineers" who, after training, would attend design team meetings in the hopes of steering the team to more manufacturable designs. That goes against the main principle of this book, which is that DFM is *designed* into the product by the *entire team*. Thus, all team members need to be trained in DFM.

Nor should DFM training be limited to engineers. Managers should attend with the engineers. Senior managers should either attend an "executive education" session or attend the first half-day of the seminar, which focuses on higher level topics, as shown in the first session in Appendix D. Lack of management attendance creates two negative consequences: (1) management doesn't learn their critical role in ensuring DFM success, and, worse, (2) attendees interpret lack of attendance as lack of management support.

DFM training benefits from a diverse audience, beyond the usual design engineers and manufacturing engineers, so include purchasing agents, materials managers, vendors, and key people from quality, field service, and so forth.

The most important attendees would be current and potential *team leaders*, who will need to understand all the DFM principles, know their role as team leader (Section 2.7), campaign for resources (Section 2.2), insist on timelines that encourage thorough up-front work (Section 3.2), measure cost as total cost (Chapter 7), fight against counterproductive policies (Section 11.5), and, as necessary, set up *microclimates* (Section 11.7.2) for their teams until DFM principles are implemented company-wide.

11.4 DFM TASK FORCE

DFM can be implemented at many levels, including individual actions, product development teams, and implementation task forces. A new product development effort can apply new DFM principles from the beginning (Section 11.6). For the first DFM application, choose an appropriate product, consistent with product portfolio planning. The product should have many open opportunities but not be an overwhelming challenge.

A more widespread implementation strategy would be to create a DFM implementation task force, with the charter to implement DFM in general, and specifically to:

- Form the task force with representatives from appropriate engineering groups, manufacturing engineering, supply chain management, quality, management, and so forth. Multidivisional companies may have a joint task force or corporate headquarters efforts. If the members are well respected, especially the leader, the results will be more likely to be endorsed by management and followed by design teams.
- Summarize how well the current product development culture is working, based on the pre-seminar survey and other investigations.
- Estimate improvements that will come from implementing DFM (Section 11.2.2).
- Get management buy-in, support, and resources to implement DFM.
- Arrange DFM training; prioritize and circulate the post-seminar voting (Section 11.3.6).
- As a team is just beginning to develop each product (after DFM training), arrange a *product-specific workshop*,[11] led by an experienced facilitator, to implement DFM on that specific product development project. This is an effective way to get a new product development project "off on the right foot" and ensure that the team will design in manufacturability, low-cost, flexibility, quality, and reliability. The

agenda would consist of a series of planned brainstorming sessions to encourage the team to explore many ways to implement the DFM principles. These exercises themselves would be the start of many actual tasks, which would be continued after the workshop.

- Enhance or implement the development of vendor partnerships ahead of time so that vendors will be willing to participate on product development teams and help the team design the parts to be made in the vendors' shops.
- For product development methodologies and processes, decide what to keep, what to modify, what to discard, and what to add, using the principles of this book and the DFM training.
- Compile relevant DFM guidelines for all relevant processes. A starting point can be the 165 general design guidelines presented in Chapters 8, 9, and 10, which are listed without discussion in Appendix B.
- Convert design rules and guidelines into checklists, if desired. Checklists can remind design teams of all the things that need to be done at various stages of the design process. Checklists also provide a quantitative way to measure compliance and rate products for manufacturability, part count, and utilization of standard parts. Checklists can also ensure and monitor that rules are not broken and, if so, that proper exception procedures are followed.

However, care must be taken not to let checklists become the primary focus of the product development process. Quantitative tools, such as project management software and checklists, can easily *become* the product development process and draw attention away from "softer" qualitative aspects, such as simplifying concepts and optimizing product architecture (Chapters 1, 2, and 3).

11.5 STOP COUNTERPRODUCTIVE POLICIES

In a DFM seminar, someone once asked how long it would take to go from the primitive *typical* timeline to the more advanced *concurrent engineering* timeline in Figure 3.1. The short answer is, "How long will it take to stop the bad habits?"

Half the challenge to implement new methodologies may be getting rid of existing counterproductive policies. For product development, here are some of the worst, with corrective actions cited by section:

- Don't bite off more than engineering can chew when planning product portfolios, as Motorola learned (Section 2.2.3). Don't develop products for all markets against all competitors. Prioritize the new product development portfolio, as discussed in Sections 2.2.1–2.2.5. Be sure to hire enough resources when the portfolio *must* expand.
- Don't allow, or, worse, encourage, the sales force to take all orders or accept all customizations. Instead, develop rational strategies to accept only the most profitable orders, as discussed in Sections 2.2.5–2.2.8. This will avoid *polluting* operations with low-volume, hard-to-build products that drain resources away from product development and other improvement programs. Rationalize product lines (see Appendix A).
- Don't "manage" product development to death with arbitrary early intermediate deadlines that compromise the critical up-front work (Section 3.2) just for the *illusion* of "early progress."
- Don't thwart concurrent engineering teamwork by underfunding operations resources that help new product development, most typically manufacturing engineers, to keep overhead costs down. The triple irony is that this prevents MEs from helping teams design products (Section 2.5), then the products will be hard to launch, which will not only raise costs but also reduce even further the ME support available for other new product developments, thus further reinforcing a downward spiral. A fourth irony is that the overhead cost argument goes away (as recommended in Section 2.2) when product development support is charged to the development *project* and ultimately results in a net savings.
- Don't quantify only labor and part cost and then allocate (average) all other cost (overhead) over all products, good or bad (Chapter 7).
- Don't move manufacturing offshore. Concurrent engineering is difficult to practice when there are *no manufacturing people around with whom to be "concurrent."* In many offshoring situations, people in engineering and manufacturing are not even working at the same time. For more, see Section 2.8 (co-location), Section 4.8 (outsourcing), and the articles on outsourcing[12] and offshoring[13] at www.HalfCostProducts.com.

- Don't try to take cost out after the product is designed. It is a difficult task and is a waste of resources, as shown in Section 6.1.
- Don't go for the low bidder on custom parts, which precludes *vendor partnerships* and, thus, prevents those vendors from helping the company design the parts, for reasons presented in Section 2.6.

As pointed out in Section 6.22, companies that practice the last three bullets (off-shoring, cost reduction after design, and low bidding) will have to devote a high percentage of product development resources to: make change orders to try to implement DFM (because it couldn't be done with concurrent engineering); try to take cost out after the product is designed, with change orders; convert documentation for outsourcing; get outsourcers up to speed; deal with quality and delivery problems; and so forth. In his travels, the author has encountered several companies that spend *two-thirds of product development resources* on the last three activities (off-shoring, cost reduction after design, and low bidding), which really puts their future in doubt if that future depends on new product development. Ironically, these attempts *thwart six of the eight half-cost strategies*; see www.HalfCostProducts.com/offshore_manufacturing.htm.[14]

11.6 COMPANY IMPLEMENTATION

These are actions that the company needs to take to transform product development to deliver the best results in the least time for the smallest development budget.

11.6.1 Optimize NPD Teams

Ensure resources are available to form complete teams early, with the right mix of talent, throughout the duration of each project. Selectively hire to provide the missing resources, as discussed in Section 2.2.15. Plan the product portfolio for the maximum return from available resources. Rationalize away unusual, oddball products or options that drain resources away from NPD. Establish sales and marketing policies to *prioritize sales by profitability*, not "take all orders" and fill all markets. Outsource legacy products and spare parts production that are not compatible with operations and supply chain management.

Correct critical resource shortages and deficiencies in multifunctional teams by freeing them from activities such as building fire drill legacy products and spare parts, building money-losing custom products, and trying to do cost reduction by change order on existing products. Avoid all other distractions from product development.

Encourage thorough up-front work, with enough resources to form *complete* teams *early*, and timelines structured to encourage learning lessons, resolving issues early, simplifying concepts, and optimizing architecture, which is the best way to ensure the lowest *total cost* and the fastest time to *stable production* or customer acceptance.

Co-locate engineering and manufacturing for the best concurrent engineering interactions, which can save more money than distant outsourcing and much more than offshoring, for reasons presented in Sections 4.8.

Focus cost efforts on new product development, not "cost reduction" after the design is cast in concrete and hard to change.

Encourage vendor partnerships so that vendors (that make custom parts) are early and active participants on design teams, for the best cost, quality, and time-to-market.

Do everything to ensure the success of the first project, as discussed in Section 11.7.3.

11.6.2 Optimize NPD Infrastructure

Support the integration of DFM and concurrent engineering principles into the product development process and organize and support other freestanding programs; for instance, to standardize parts, properly capture the voice of the customer, implement lessons learned databases, and quantify total cost.

Hire and/or develop great team leaders and team members with the right balance of skills, talent, and experience, as discussed in Section 2.7.

Make available the right tools and training for design, model building, prototyping, operations, and information technology.

Provide project rooms to encourage frequent, on-demand meetings and maximize team efficiency in achieving project goals at the least cost in the least time. Remember, *the business model should determine facilities planning, not the other way around.*

Quantify all costs, including overhead, and base all cost decisions and pricing on total cost to: base portfolio decisions on total cost projections; ensure that costs of custom configurations are quantified and staffing is paid for by customers or investments; make the best cost decisions for the lowest total cost; encourage behavior that *continues* to minimize cost; eliminate cross-subsidies; base pricing on total cost; and ensure that better-designed products are not penalized by having to pay the loser tax to subsidize less-manufacturable products.

Implement standardization as a company effort so that: design teams can design products around standard products; blanket better-than substitutions can be made on existing products; and expensive parts and subassemblies can be upgraded on existing products where change orders are justified.

Establish incentives that support all of the above; eliminate incentives that thwart the above.

11.6.3 Incorporating DFM into the NPD Process

Incorporate DFM and concurrent engineering principles into the company product development process. Figure 11.2 shows the typical product "phase-gate" process, with the product design phase enlarged. Some steps will shift to an earlier point in the process, such as the typical late step called "check for DFM," wiring, and off-the-shelf parts which are points on the far right ends of the long horizontal arrows in Figure 11.2. New steps and deliverables may need to be added, such as lessons learned, raising and

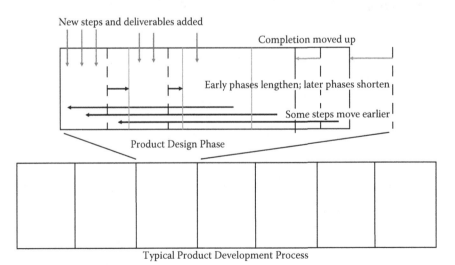

FIGURE 11.2
Incorporating DFM into the NPD process.

resolving issues, and the other up-front work and design strategies discussed in Chapter 3. The early steps will use a higher proportion of time and the end of the timeline is compressed, as illustrated in Figure 3.1.

11.7 TEAM IMPLEMENTATION

Each team should develop products according to the principles of this book, even if they have not yet been implemented company-wide. Regardless of the level of implementation, be sure to do the following[*]:

- Get lessons learned from databases, investigations, or presentations.
- Hold discussions to thoroughly raise and resolve all issues early.
- Ensure availability of all specialties early. The team leader may have to lobby for these.
- Push back on team member distractions and resource drains.
- Ensure access and availability for meaningful contributions from manufacturing, purchasing, quality, service, etc.
- Ensure enough time for thorough up-front work; push back on early deadlines as necessary.
- Secure concurrence for setting up vendor partnerships.
- Work with purchasing to arrange specific vendor partnerships.
- Find space for a dedicated project room (Obeya).
- Make all decisions based on total cost data; if not available:
  - Use total cost *thinking*, seeking exceptions to metrics if the metrics (cost, profit, etc.) are not based on total cost.
  - Campaign for relevant overhead allocations.
- Focus on minimizing cost by design; resist pressures to divert resources away from this to less effective or counterproductive cost reduction attempts (offshoring, bidding, cost reduction after design, and others discussed in Section 11.5).

---

[*] When these are not automatically forthcoming: (1) Summarize principles and justifications from the seminars, books, articles, and experiences, both within and outside the company. (2) If necessary, use the following argument: "*The only way we can achieve the goals of this project is for us to have* _____ (fill in the blank)." (3) If what is needed is still not forthcoming, say: "OK, *let's talk about how to scale back the project goals*" (deadlines, functionality, feature sets, and so forth)." (4) If still not forthcoming, say: "*Then let's wait until a time when the company can support this development.*"

11.7.1 Importance for Challenging Projects

For projects that are important, pivotal, silver bullet, or challenging, the team must:

- Practice all these DFM and concurrent engineering principles.
- Avoid habits, traditions, and policies that may stand in the way (see Section 11.5).
- Create an environment where advanced product development can thrive, with:
  - All the needed specialties present and active early
  - Members selected on the basis of talent, mix, availability, and receptivity to new product development methodologies; slow adopters may have to wait for subsequent projects
  - A good product definition and *stable* product requirements documents, not subject to changes
  - Its own dedicated project room; fortunately, the first project has to find only one room
  - A corporate sponsor, as discussed in Section 2.7.3

11.7.2 Microclimates

Even before change is implemented company-wide, these principles can be implemented right away if a project creates its own *microclimate*. In the extreme, a "skunkworks"-type project can create its own culture, which can move quickly, eliminate bad habits, and bypass cumbersome constraints.

A microclimate project can even bypass or modify entrenched corporate policies (low bidding, counterproductive sourcing pressures, cost reduction past the point of diminishing return) by seeking individual exceptions "for the good of the project." After a DFM seminar at a company that engaged in many of the counterproductive activities listed in Section 11.5, attendees started an initiative called "DFM vs. Policy."

11.7.3 Ensuring Success for the *First* Team Concurrent Engineering Project

The first time DFM principles can have a major impact is on the *next* product development project, without waiting for new practices to be formally incorporated into company procedures. The company should do everything possible to ensure success for the *first* concurrent engineering project by "stacking the deck" or giving it "silver bullet" status to make sure that the following principles of this chapter are implemented by the team:

- The team is *truly* multifunctional and has all the needed specialties *present and active early.*
- Team members are selected on the basis of talent, mix, availability, and receptivity to new product development methodologies. Slow adopters may have to wait for subsequent projects.
- Team members are protected from being sucked away into emergencies outside the project.
- The team has a great team leader, with the traits described previously.
- The team is able to apply all the principles of this book, especially the right proportion of time for thorough up-front work.
- The team and team leader are empowered to implement these principles, even if not yet incorporated into company procedures.
- The project has a good product definition and *stable* product requirements documents (not subject to changes).
- The team has its own dedicated project room. Fortunately, the first project has to find only one room.
- The team has all the tools it needs when it needs them.
- Decisions and overhead charges are based on total cost. If not implemented yet, decisions should be made on the basis of *total cost thinking,* and overhead costs would have to be manually computed to avoid paying the overhead charges of less manufacturable products.
- The team has a corporate sponsor to help the team overcome obstacles, cut through red tape, and streamline reporting to management by avoiding the hours spent preparing presentations and the calendar delays to schedule formal reviews to management.

11.8 INDIVIDUAL IMPLEMENTATION

There are many things that individual engineers can do before DFM is implemented company-wide:

- Enthusiastically embrace new NPD principles, especially working well in multifunctional teams.
- Cooperatively contribute to all team activities, especially discussing issues and brainstorming.
- Be receptive to recommendations from other team members; proactively seek dialogue with others to optimize your work and the project as a whole.
- Volunteer to work on or lead innovative endeavors.
- Implement whatever methodologies you can personally to improve manufacturability within your sphere of influence.
- Keep thinking about how to optimize the *product*, not just your *parts*.
- Work interactively with other team members on whatever relates to your work or how your work affects theirs.
- Initiate dialogue with manufacturing, purchasing, quality, service, and others early on *each* design decision that affects manufacturability.
- Concurrently develop tooling, manufacturing, and supply chain strategies for your parts.
- Remember the messages of this book as you work your way up and gain more influence.

Either individuals or their teams can take the initiative on the following:

- *Use feedback forms* to understand the manufacturability of your products (see Appendix C for forms to use with your plant, your vendors, and field service).
- *Interact frequently* with people in manufacturing, purchasing, quality, and service on an ongoing basis.
- *Frequently observe* manufacturing and vendors' operations. Taiichi Ohno, the father of the Toyota Production System, drew a circle on the factory floor (the "Ohno circle") and made people stand in it all day to watch and question the process. His thinking was that new thoughts and ideas come from observing and understanding the processes.[15]
- *Arrange shop demonstrations* at your plant or at your vendors, who would undoubtedly welcome the opportunity to show design engineers how various design practices make it easier or harder to make parts.
- *Learn the trades* that build your parts.

11.9 DFM FOR STUDENTS AND JOB SEEKERS

The ability to design products for manufacturability is a skill potential employers will value. Be sure to mention your knowledge and experience in the following areas on your résumé and in job interviews.

Books. Read this book thoroughly, along with others listed at the end of each chapter and the "Books Cited" section of Appendix D on resources. The best company-based book is *The Toyota Product Development System,*[16] which corresponds well to the principles of this book. The most important chapters of the Toyota Product Development book are Chapter 4 (on thorough up-front work), Chapter 7 (on the team leader), and Chapter 10 (on early supplier involvement).

Classes. Prospective employers value relevant courses, ranging from specific college courses to extension or continuing education seminars throughout your career. Relevant courses should be listed on your résumé. In your early career planning, map out which courses to take to support your overall career goals. Find out which CAD program is used in your target industry and take classes to learn that program. Most CAD software suppliers offer student or academic discounts to help students learn how to use their software.

Industry events. Students and working engineers can learn a lot about manufacturing by attending conferences, trade shows, and exhibitions. And working engineers usually get paid to attend. Some conferences have student discounts, and many exhibitions are free or charge only a nominal fee. Students can learn about the latest design and manufacturing techniques by going to presentations at conferences and seeing equipment and demonstrations at the exhibitions.

Experience. Most job experiences may be valuable in the next job. Your career path should be planned to accumulate experience that ultimately supports a strategic career goal. Job seekers should (1) look for jobs where their previous experiences will be valued by prospective employers, and (2) emphasize this previous experience in résumés and job interviews. Students can gain experience through summer jobs and internships. *Any* job in a given industry will count as exposure to that industry. For instance, the author worked his way through college as a mechanic in a cannery, which helped him get a job later designing food processing machinery. When joining a new company, design engineers can request manufacturing experience before design assignments if the company does not already have this policy, which was recommended in Section 1.6.

Learn the trades of the process that makes what you are designing or want to design; for instance, machining, CNC programming, welding, and so forth. Working designers can learn these skills by taking night courses at local community colleges. Engineering students should take these courses as part of their schooling or, if not available, enroll in nearby community college courses. These shop courses are taught by practical instructors and offer students hands-on experience. One general course in each skill should be sufficient.

Companies should encourage their engineers to take these courses by paying course fees and allowing time off for the classes. Companies with in-house shops could hold shop classes for designers on site. The classes could be taught by teachers from community colleges or by factory workers.

For students, being able to say, "I am a machinist," "I am a welder," or "I can program a CNC machine," should be valuable to manufacturing companies that machine or weld parts. The author became proficient at welding and machining by taking courses at local community colleges.

Even relevant hobbies count. The skills learned in, for instance, a welding class could turn into a fun hobby that may impress an interviewer, even if you welded artwork with an easily affordable oxyacetylene gas welder or basic arc welder. Experience with a home version of a programmable plasma cutter or prototype milling machine may be relevant to companies that use CNC machine tools.

The recommendations of Section 1.6, which advises companies to hire design engineers with experience in manufacturing, would apply to both work experience and trade skills.

Profiles. Section 2.3 on product portfolio planning recommends that companies develop *profiles* to identify the best opportunities. Similarly, anyone who wants to apply DFM in a new job should develop a profile of favorable company characteristics, such as:

- Use of multifunctional teamwork
- Strategies to achieve low cost and quality *by design*
- Manufacture their products at the same site, co-located with design
- Utilize vendor partnerships so that vendors are willing and able to help them design the products where both the vendor and manufacturer will learn from each engagement

- Measure total cost, for the reasons cited in Chapters 6 and 7
- Investment in product development, where management encourages and supports innovation

Conversely, profiles indicating a company will have a hard time implementing DFM, and thus offer an unrewarding job experience, would include the counterproductive practices cited in Sections 6.22 and 11.5. These best and worst profiles may help students and job seekers select the most rewarding job opportunities.

11.10 KEY DFM TASKS, RESULTS, AND TOOLS

Figure 11.3 provides a summary of key DFM tasks, results, and tools.

11.11 CONCLUSION

DFM alone may make the difference between being competitive or not succeeding in the marketplace. Most markets are highly competitive, so slight competitive advantages (or disadvantages) can have a significant impact.

DFM can offer enormous benefits to product cost, quality, and time-to-market with very little investment. Practiced right, DFM actually takes less effort because products *are designed right the first time.*

Tasks	Results	Tools
Multi-functional team formed with all functions present & active early	Products concurrently designed right the first time for manufacturability	Teamwork, leadership, resource availability
Raise & resolve issues early.	All issues raised early; All issues resolved early; Avoid problems & delays	Strong team leadership; Team consensus; Up-front focus
Simplify product & production concepts.	Inherently low product cost; Inherently high quality; Inherently high reliability	Thorough up-front work; Teamwork; Creative culture; Up-front focus
Optimize product architecture and process design.	Ensure lowest cost; Ensure quick development; Ensure trouble-free launch	Thorough up-front work; Multi-functional team; Architecture focus
Optimize product and process design.	Manufacturable designs; Optimized processing; Quality designed in	Concurrent engineering; Process guidelines; Quality guidelines
Standardize materials & processing.	Minimum material overhead; Quicker product designs; Less setup, smaller batches	Standardization lists; Motivation & discipline; Cross-team cooperation
Quantify total costs.	Best decisions; Proper costing/pricing; Lowest possible total cost	Activity-Based Costing; Total cost thinking
Optimize supply chain to maximize quality & responsiveness.	Lowest mat'l cost and time; Quality assured at the source; Minimum batches & inventory	Supplier partnerships; Total cost measurements; Teamwork with suppliers
Measure & compensate to encourage teamwork and total goals.	Minimum total cost; Minimum time-to-market; Best decisions	Metrics & compensation based on total cost and the real time-to-market
Management supports and understands concurrent engineering.	Product development becomes a potent competitive advantage	Executive education

FIGURE 11.3
Key DFM tasks, results, and tools.

NOTES

1. Sydney Finkelstein, *Why Smart Executives Fail: And What You Can Learn from Their Mistakes* (2003, Portfolio/Penguin), p. 138.
2. Micheline Maynard, *The End of Detroit: How the Big Three Lost Their Grip on the American Car Market* (2003, Currency/Doubleday), Chapter 2 on Toyota and Honda, p. 75.
3. Philip B. Crosby, *Quality Is Free: The Art of Making Quality Certain* (1979, McGraw-Hill).
4. This quote is attributed to both Benjamin Franklin and Albert Einstein.
5. Satoshi Hino, *Inside the Mind of Toyota* (2006, Productivity Press), Chapter 1, "Toyota's Genes and DNA," p. 3.
6. Bill George, *Authentic Leadership: Rediscovering the Secrets of Creating Lasting Value* (2003, Jossey-Bass), Chapter 12, "Innovations from the Heart," p. 141.
7. Robert W. Hall, "Medtronic Xomed: Change at 'People Speed,'" *Target*, 2004, Vol. 20, p. 14, http://www.ame.org/sites/default/files/target_articles/04-20-1-Medtronic_Xomed. pdf.
8. For more information on customized in-house DFM seminars, see Appendix D of this book, which has a baseline agenda, or see Dr. Anderson's website: www. design4manufacturability.com/seminars.htm.
9. This data was generated by DataQuest and presented in the landmark article that started the concurrent engineering movement: "A Smarter Way to Manufacture: How 'Concurrent Engineering' Can Invigorate American Industry," *Business Week*, April 30, 1990, p. 110.
10. The colleges that used various editions of this book for courses are listed in the preface of this book under "Preface for Instructors."
11. Dr. Anderson facilitates product-specific workshops (Appendix D) after customized in-house seminars. These workshops consist of many brainstorming sessions on concept simplification and architecture optimization. More challenging endeavors, such as developing half-cost products, may need his concept studies (also in Appendix D), in which he generates breakthrough ideas that concurrent engineering teams can develop into manufacturable products.
12. See the outsourcing article at the author's website: www.HalfCostProducts.com/ outsourcing.htm.
13. See the offshoring article at the author's website: www.HalfCostProducts.com/ offshore_manufacturing.htm.
14. The first section of the offshoring article shows how cost reduction attempts thwart six of the eight cost reduction strategies presented on the home page of www. HalfCostProducts.com.
15. Matthew E. May, *The Elegant Solution* (2007, Free Press), p. 73.
16. James Morgan and Jeffrey K. Liker, *The Toyota Product Development System* (2006, Productivity Press), Chapter 4, "Front-Load the PD Process to Explore Alternatives Thoroughly."

Section VII

Appendices

Appendix A: Product Line Rationalization

Product line rationalization is a powerful technique to improve profits, simplify operations and supply chains, and free valuable resources for product development. It does this by *rationalizing* existing product lines to eliminate or outsource products and product variations that are problem prone, have low sales, have excessive overhead demands, have limited future potential, and may be losing money.

Rationalization can quickly improve profits by stopping the production of money-losing products and eliminating all of the excess overhead costs associated with fire-drill products. This, in turn, will allow precious resources to focus on the most profitable products instead of low-leverage products, which will increase sales and further lower costs. After rationalization, the remaining products will cost less because they will no longer have to subsidize the money losers or marginal products.

All these cost savings can be used to lower prices in price-sensitive markets or to increase profits. In fact, rationalization can raise profits enough to be justified as a free-standing program. The following scenario will show that, by simply eliminating the lowest leverage products, profits can be tripled!

A.1 PARETO'S LAW FOR PRODUCT LINES

All companies experience some Pareto effect, typically with 80% of profits or sales coming from the best 20% of the products. This happens because almost all companies continuously *add* products to the portfolio without ever *removing* any. Further, sales incentives and emphases on growth and market share encourage the sales mantra of "take all orders," thus overloading production operations and the supply chain with too many low-volume products that have unusual parts and manufacturing procedures. This results in many setup changes, causes excessive overhead costs, lowers plant capacity, complicates supply chain management, and dilutes engineering and manufacturing resources.

Few companies realize these problems exist because their cost systems allocate (average) overhead costs, which implies that all products have the same overhead costs—a very unlikely situation.

A.1.1 Focus

Product line rationalization encourages companies to focus on their *best* products by eliminating or outsourcing the *marginal* products. The resources that were being wasted on the low-leverage products can then be focused on growing the cash cows.

Adrian Slywotzky (who coauthored the book cited in reference 8) later wrote in an article with Robert Atkins that allocating resources evenly among all products is not wise even in good times, and can be disastrous in a recession.

In a 2010 *Harvard Business Review* issue with a spotlight on recession recovery, one of the focus articles recommended: "In a postbubble world, firms must be more ruthless about terminating loss makers."[1]

A.1.2 Competitive Challenges without Rationalizing

Selling a lot of legacies and low-volume products generates much more overhead cost per product than cash cows, so those overhead costs must be subsidized by the cash cows. In other words, the *loser products create a "loser tax" on the cash cows*, thus forcing them to sell at a higher price (or lower profits) than that of more focused competitors. The sales force for the cash cows should support rationalization when they realize this; the sales force for the losers may resist, so sales assignments may have to be adjusted.

Rationalization will reduce some costs (see Section A.3), more so in a growing enterprise. In most companies, the liberated overhead resources should be reinvested in new product development and other activities listed in Section A.3.

As discussed in Section 2.3, portfolio proliferation places a double whammy on new product development. Not only do new products have to pay the loser tax (which impacts low-cost products even more percentage-wise), but resources are drained away from product development efforts to build oddball products, thus making the new products less competitive than they could be.

Unrealized subsidies can be identified by total cost accounting (Chapter 7). In fact, total cost has an *automatic rationalizing effect* when all overhead costs are assigned appropriately to every product, thus raising the cost of the losers, so the marketplace will then rationalize them away.

A.2 HOW RATIONALIZATION CAN TRIPLE PROFITS!

The following scenario shows the power of this methodology, using a simple example illustrated in Figure A.1. The actual product line rationalization considers many more factors, but this example shows the profit-increasing potential of rationalization.

If a company kept the 20% of the products that were making 80% of the profits and dropped the other 80% of the product line, it would result in only a 20% drop in revenue. The cost reduction can greatly exceed the revenue drop because of the way cost is distributed. Direct costs, such as materials, parts, and labor, would be proportional to revenue. In other words, the more products sold, the more materials, parts, and labor would

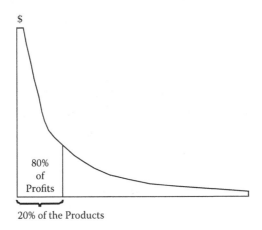

FIGURE A.1
Pareto's law for products.

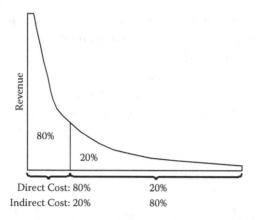

FIGURE A.2
Cost breakdown.

be consumed. However, overhead has the *opposite* effect. Indirect costs (such as procurement, manufacturing engineering, and other support functions) would be low on the cash cow products because they are better designed for manufacturability, parts are procured routinely, quality issues have been resolved, and processes have been stabilized due to the focus that is usually applied to higher volume production. In contrast, overhead costs on the low-volume products are high, probably 80% of the total, because of all the inefficiencies inherent in building many low-volume, seldom-built products. Further, those products may be less well designed for manufacturability and have much higher quality costs. Thus, the breakdown of direct and indirect costs would be as shown in Figure A.2.

In order to make this example relevant, Figure A.3 converts these percentages into dollars for a $100,000,000 business, which, according to the common Pareto's law effect experienced by most companies, generates $80 million in revenue from cash cow products and $20 million from all the others. The cost breakdown shows indirect (overhead) costs as half the total cost, which is not an unreasonable assumption. In fact, in many industries, overhead cost is greater than half of the total cost (as shown in Figure 6.1), thus resulting in an even stronger case.

Now, to show the powerful effect of this procedure, simply eliminating the 80% of the low-volume products and keeping the 20% cash cows will have the effect shown in Figure A.4.

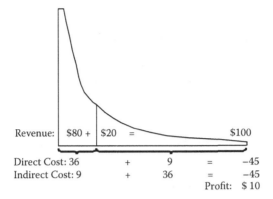

Revenue:	$80 +	$20	=		$100
Direct Cost: 36		+	9	=	−45
Indirect Cost: 9		+	36	=	−45
				Profit:	$ 10

FIGURE A.3
Cost distribution in dollars.

Revenue: | $80

Direct Cost: −36
Indirect Cost: −9
New Profit: $ 35 = 3.5 times the profit of the full product line

FIGURE A.4
Results after rationalization.

The bottom line: Although revenue drops 20% (this will be discussed later), eliminating the high-overhead products eliminated most of the indirect costs, so that *profits are 3.5 times that of the full product line!*

For this motivational scenario, overhead cost savings are assumed to be eliminated to show the effect on profits. This could be realized in a rapidly growing company where the focus shifts from current loser products to upcoming growth products.

However, it would be a shortsighted strategy to lay off workers to realize a short-term overhead cost savings.[2] Rather, "liberated" workers should be *invested* in growing the remaining aspects of the business. Breakdowns of actual cost savings and reinvestments are discussed next.

A.3 COST SAVINGS FROM RATIONALIZATION

The cost savings from rationalization comes in two forms: cash and human resources. Rationalization, like Lean Production, effectively *liberates* many types of resources, and their disposition should be planned ahead of time. The ultimate Lean Production guide, *Lean Thinking*, has a section titled, "Deal with Excess People at the Outset."[3]

A.3.1 Short-Term Cash Savings

Product line rationalization results in many short-term cash savings and resource investment opportunities. It enables the company to:

- Avoid the purchase of parts and materials for rationalized products, which may have less purchasing leverage, higher procurement costs to find, and higher than normal setup and expediting costs.
- Avoid quality costs of unusual products, which may be higher than normal, for reasons discussed in Section A.10.
- Limit, postpone, or cancel hiring for growth and attrition replacement.
- Avoid overtime.
- Phase out temps (temporary workers) as long as they do not have critical knowledge or skills.
- Bring in house services that are currently outsourced, especially when co-location would provide better concurrent engineering.
- Delay facility expansion. Unusual products generally require more space than those that have benefitted from continuous improvements by implementing space-saving Lean principles.

A.3.2 Investments

Cost savings allow for the following investments to be made:

- Focus on improving sales on the remaining products, which now can sell at higher profits (or lower prices) without having to cross-subsidize the "losers."
- Improve quality and lower the cost of quality.
- Continuously improve operations and productivity.
- Expand into related services.

- Get certified (ISO 9000, QS 9000, etc.) or win awards (Best Plants, Baldridge, etc.) to improve operations and stature with customers.
- Upgrade CAD tools, information systems, and web presence.
- Upgrade skills with investments in training.
- Implement new capabilities, such as build-to-order and mass customization, to be able to build a wide variety of mass-customized or standard products on demand without forecast or inventory (Section 4.3).
- Invest in *internal* start-up ventures.
- Use liberated cash and resources to buy and transform related, supportive businesses up and down the supply chain (*not competitors!*). Liberated cash can also be invested to transform other companies; the book *Lean Thinking* cited an example: "Each time Wiremold's vacuum sucks up a batch-and-queue producer, it spits out enough cash to buy the next batch-and queue producer!"[4]
- Improve product development. Without the daily fire drills in operations and procurement, manufacturing and purchasing people will be more available to participate in product development teams, which is a key element to successful product development.

A.4 SHIFTING FOCUS TO THE MOST PROFITABLE PRODUCTS

As Figure A.5 shows, after product line rationalization, resources that were being wasted on the low-leverage products (under the dashed line) can now be focused on improving the remaining products.

One of the author's clients, a telecom equipment company, reported that after dropping their marginal products, they recovered their original revenue within two months!

This comes from focusing liberated resources (both personnel and money) on the cash cow products, with improved efforts in advertising, sales channels, product design, operations, and supply chain management. All of these efforts will be easier and more efficient with fewer products on which to focus. Operations and supply chain management will benefit from a greatly reduced number of parts and processes, plus having free time to make improvements.

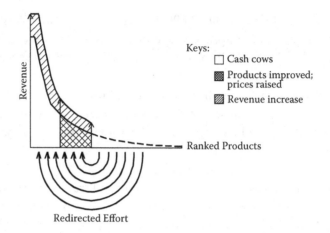

FIGURE A.5
Redirecting focus to cash cows.

Multifunctional product development teams will make faster progress and develop better products with more people available from operations and purchasing. In addition, the money saved not having to subsidize "the losers" can fund improvement activities, be used to lower prices, or simply go to the bottom line to improve profitability.

In addition to improving the known cash cows, it may be possible to "raise worthy dogs," or focus improvement on selected worthy products, especially if they are related to cash cow products. For instance, consider raising their prices, if the market accepts them; they may survive and simply make more money. Look for products that have *easy* opportunities to lower their total cost, but be wary of any cost reduction effort that does not pay off the cost of the effort within the expected life of the product, as discussed in Section 6.1.

Make sure good products are not unfairly burdened by inappropriate overhead charges, like paying the loser tax to subsidize marginal products. In addition to these techniques, selected products may also benefit from all the above-mentioned improvements in advertising, sales efforts, product design, operations, and supply chain management.

Professor Kim Cameron of the University of Michigan Business School recommends that during downturns companies should "exit from weak businesses entirely."[5]

Sometimes companies can profit from shifting the focus from products that are only *good* to those that would have the potential to be *great*. Jim Collins, author of *Good to Great*, points out that "few executives have the guts to get rid of profitable businesses where their company can only be good, but never great."[6]

A.5 RATIONALIZATION STRATEGIES

A.5.1 What Is More Important: Volume or Profit?

Many managers still have trouble with the issue of dropping revenue 20%, even if it triples profit. This brings up the issue of the goal of a business. As Eli Goldratt[7] and others have pointed out, the real goal of a for-profit enterprise is to make money, not to optimize other common measures such as productivity, market share, or "growth at any cost."

The opening chapter in Slywotzky and Morrison's book, *The Profit Zone: How Strategic Business Design Will Lead You to Tomorrow's Profits*,[8] is titled: "Market Share Is Dead," in which are found the following quotes:

> "The two most valuable ideas in the old economic order, market share and growth, have become the two most dangerous ideas in the new order."

> "Paradoxically, the devout pursuit of market share may be the single greatest creator of no-profit zones in the economy."

James Womack, author of *Lean Thinking* (cited in Chapter 4) and *The Machine That Changed the World* (cited in Chapter 2), traces the origins of Toyota's 2010 safety and recall problems back to 2002, "when it set itself the goal of raising its global market share from 11% to 15%."[9] That rapid growth took priority and compromised Toyota's adherence to its own principles, which are *still* the benchmark for Lean Production (as often stated by Womack and his Lean Enterprise Institute at www.lean.org) and product development, as cited many times herein.

Similarly, the book about the largest research project ever devoted to corporate failures, *Why Smart Executives Fail: And What You Can Learn from Their Mistakes*,[10] states that market share is the wrong *scorecard* because "market share does not translate into profitability, since significant investments are typically needed to build share in the first place."

One of the themes of Richard Koch's book, *The 80/20 Principle*,[11] is

> "Successful firms operate in markets where it is possible to generate the highest revenue with the least effort."

A.5.2 Profitable Growth

Another quote from *The 80/20 Principle*:

"The road to hell is paved with the pursuit of volume."[12]

Of course, "pursuit of volume" here means a volume growth strategy. Many companies that have been stumbling lately have volume growth as the cornerstone of their corporate strategy, with quarterly and yearly growth goals. But the underlying theme of Pareto's law is that: *All opportunities are not equal and do not make an equal contribution to profitability.*

A volume growth strategy encourages the "take all orders" mantra. Thus, when there is pressure to grow the business, the company ends up taking any business it can—not just the most profitable. This behavior is built into the system if sales incentives are based on volume growth, instead of profit. This is the sales equivalent of the piece–part incentives in manufacturing that were abandoned long ago because they favored one metric (volume) over another (quality).

The key to profitable *growth is to focus on the products with the most potential, not to dilute resources on the* most *products.*

A.5.3 Rationalization Prerequisite—Eliminating Duplicate Products

Before undertaking the rationalization procedure, companies can take certain first steps to simplify the process. Eliminate overlapping or duplicate products. Search out and eliminate or consolidate duplicate products, overlapping products, and superceded products, even if some customers are still using the older product. You may have to encourage or force customers to switch from the older, less-advanced products that they have been ordering because of arbitrary decisions, inertia, or lack of awareness about newer or better replacements.

A.6 THE RATIONALIZATION PROCEDURE

The rationalization procedure divides the product line into four zones, as shown in Figure A.6. Most of the least profitable, lowest-volume products would be dropped (Zone 4); exceptions might include products that are

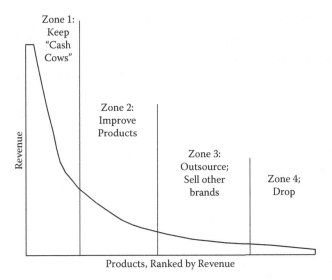

FIGURE A.6
Rationalization procedure.

slight variations of higher-volume products *and* can be built in flexible cells. Products that need to be in the catalog could be outsourced (Zone 3), thus simplifying the in-house supply chain and manufacturing operations. Cash cows would be kept (Zone 1). The remaining products could be improved (Zone 2).

A more detailed breakdown follows:

Zone 1: Cash cow products should remain because they are probably making 80% of the revenue and profits. Some may be just fine in dedicated mass production lines. No change in production per se may be required for some products, whereas others could be made together on flexible lines.

Zone 2: These products could be improved and grow with the redirected efforts shown in Figure A.5.

Zone 3: This category consists of products that do not fit into either Zone 1 or 2 but still need to be in the catalog for completeness, to satisfy loyal customers, or for service obligations. These products may remain in the catalog, *but they do not have to be built in-house.* They could be outsourced and manufactured by a supplier under the company name or label. Or the company catalog could simply carry another source's product to complete the product line, assuring customers about appropriate equivalency.

Zone 4: This zone contains the products that should be dropped from the product line. Sell off the rights to losing products, if possible; someone else may be able to make money if they are a better fit with their products and operations. This certainly would not be a competitive threat, because those products would not be making as much money as new products designed with better focus using the techniques of this book.

A.7 TOTAL COST IMPLICATIONS

Chapter 7 discussed total cost measurements and their impact on business decisions. The cost accounting system has significant implications for product line rationalization. Most companies average (allocate) overhead costs so much that the reported "costs" of individual products do not reflect reality. What happens is that good products usually subsidize bad products. And because profitability is based on cost, it will be distorted too.

When total cost measurements are implemented, the profitability of products becomes more realistic, as shown in Figure A.7, and it becomes apparent how many of the products are really losing money. These then become the prime candidates for elimination, in Zone 4 of Figure A.6.

You cannot have a profitability strategy unless you can have total cost measurements and can break down "corporate profits" into profits for specific products, market segments, and so forth.

Sometimes implementing total cost measurements can resolve a rationalization impasse. A Harvard case study[13] analyzed seven products made

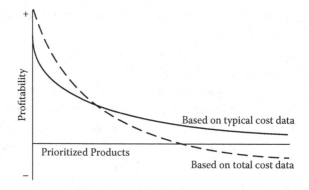

FIGURE A.7
Prioritized profitability: typical cost vs. total cost.

by Schrader–Bellows. In a factory where most products had annual volumes of thousands per year, there was one product that they built only 53 of per year. Of course, production management wanted to get rid of it because they understood all the inefficiencies, but the current cost system said, quite illogically, that it had the highest profit margin in the plant, so they had to keep building it. However, after the total cost analysis was performed, the product that originally was thought to have the highest profit margin of the group was shown, in reality, to have a negative 59% "margin."[14]

A.7.1 Margin Trap

Making decisions based on product profit margins can be a dangerous practice when costs are not based on total cost accounting practices and updated often. Low-profit or money-losing products will keep being sold if there is a perception that they have high margins. Many misleading situations occur when products have margins computed after high-volume builds and these numbers stay in the system, even though the order volumes have since decreased. Obsolete margin data can mislead decision makers and may discourage rationalization efforts.

A.7.2 Seldom-Built Products

Products "revived from the dead" have high overhead costs despite their reported "margins." Another rationalization procedure is to investigate which products have not been built recently; for instance, in the last year, two years, three years, four years, and five years. These products that are revived from the dead have high overhead demands because of the effort to remember how to build them, find all the documentation, procure unusual materials, and find the tooling. In one company, tooling for a seldom-built product had been sitting outside in the snow for months and required much rehabilitation. Companies can simply implement a policy that all products that have not been built in, say, three years, will be dropped immediately, or at least discouraged or given special scrutiny.

A.7.3 Obsolescence Costs

When mass producers build too many low-volume products, it increases the risk and cost of obsolescence due to market changes and engineering change orders. An APICS article on "product proliferation" stated:

"Low-volume products are particularly prone to [obsolescence] since batch sizes are often increased to produce a six-month or more supply in order to reduce the number of changeovers required in manufacturing. With this amount of inventory on the shelf, the risk of becoming obsolete due to engineering changes or changing levels of demand increases dramatically."[15]

A.8 OVERCOMING INHIBITIONS, FEARS, AND RESISTANCE

Despite the fact that product line rationalization can easily raise profits, free valuable resources, and simplify operations and supply chain management, many managers have inhibitions, fears, and resistance:

- *Growth emphasis.* When there is too much emphasis on revenue growth, it may be hard to get companies to do what may appear to reduce their revenue stream, even temporarily, in order to eliminate the low-leverage products, improve the cash cows, improve profitability, and ultimately grow revenue.
- *Cost system deficiencies.* When overhead is allocated (averaged), all products will appear to have close to equal profitability. Total cost measurements will then be necessary to flush out the low-profit products.
- *Inertia.* Many people resist change in general, especially if the cost system or personal communications do not show them the big picture—how product line rationalization can help free up resources for product development.
- *Departmentalism.* Often resistance comes from those who have been optimizing their department's performance (e.g., sales) at the cost of the company as a whole. Metrics may need to be changed to direct behavior toward maximizing *company* goals.
- *Unhealthy attachments.* Every product at one time was the "baby" of someone or several people. Remember, however, that the goal of a business is to produce the most profitable products, not a collection of favorites.
- *Fears overstated.* Fears about negative consequences often are overstated, especially the fear of losing customers, who, in reality, will most likely switch to better equivalents from the same catalog. When low-leverage products are eliminated:

1. The company can increase sales on the remaining products, with better focus in each of the following areas: research, product development, manufacturing, quality, and marketing.
2. Without having to subsidize low-leverage products, the remaining products can generate more profit or be priced lower.
3. Customers can switch to remaining products from older, less-advanced products that they have been ordering because of arbitrary decisions, inertia, or lack of awareness about newer or better replacements.

The following is a list of the most common questions that come up in the author's workshops, followed by the author's answers:

Q: What about the "complete product line" argument?
A: If the catalog must really be complete (zone 3 in Figure A.6), that doesn't mean you have to manufacture all the products, options, or variations in your factories. Outsource or set up a separate profit and loss center to build unusual products, options, or variations so that the resources that were formerly squandered on them can now be invested in developing a better portfolio of new products.

Q: Won't customer satisfaction drop?
A: Overall customer satisfaction will actually be *lower* if you waste your resources making low-leverage products instead of capitalizing on opportunities to give customers better innovation, lower cost, better quality, and better customer service. However, they will probably miss the good deals you were giving them on your money-losing products.

Q: Won't we be limiting customer choice?
A: Often, customers keep ordering older or unusual products because of arbitrary decisions, inertia, or lack of awareness about newer or better replacements. Point out how the remaining products have been improved. Rationalizing away the loser tax will allow the remaining products to sell for less.

Q: What about loss leaders?
A: If low-leverage products are to be retained as loss leaders, then management should know how much money is really being lost, including the *opportunity losses* of what those resources could

be accomplishing. If loss leaders are still a valid strategy after implementing total cost accounting, then the loss-leader products should be outsourced to avoid distracting the factory from its most efficient tasks.

Q: But our profits are so high that selling a few losers won't drop profits much, will it?

A: The total cost of all the losers will probably be greater than a typical cost system indicates. The real impact of rationalization will be freeing up resources to develop better new products. If a prosperous company wants to keep building high-overhead money losers, then it should set up a profit and loss operation and hire enough people to run it without taking any resources away from NPD.

Q: But what about the perception that we have to sell some oddball products to get the big sales?

A: First of all, customers may have been spoiled by years of unnecessary concessions made by salespeople, who may not know or care about the trauma that low-volume products cause in operations and supply chain management. Customized products would fall into this category if not done efficiently through mass customization (Section 4.3).

Second, ask objectively if the customer will really terminate a long and viable relationship over a few low-volume products.

Third, point out that if customers continue to buy older products, they will become increasingly vulnerable to product obsolescence and part availability problems.

Finally, if customers still insist on package deals, the manufacture of oddball products could be done by outsourcing, so those sales do not distract from new product development. The portfolio planning aspects of the package deal were discussed in Section 2.2.8.

If all of these arguments are not convincing, then consider "competitive scenarios."

A.8.1 Competitive Scenarios

Savvy competitors would not blindly compete against your entire product line. They would offer only the most profitable products. Not burdened by your low-leverage products, they would be able to make the rationalized product line quicker, better, and at lower cost.

Another competitive scenario is "cherry picking," when an existing or new competitor skims off the most profitable products.[16] Thus, competitors could steal your cash cows, leaving you with the "dogs."

A.8.2 Role Playing

It may be a valuable exercise to role-play competitor scenarios, pretending you, or your brainstorming group, are a well-financed new competitor. Ask the following questions:

Which products would you want to have in your product line?
How would you group products and structure production lines for the greatest efficiencies?
How could a concise product line benefit from standardization?
What is the minimum list of parts and materials needed?
How would you build them flexibly to be able to reach the broadest possible markets?
What would you like to do that you cannot do because of existing product line limitations?

Here is what Larry Downes and Chunka Mui reported in their book, *Unleashing the Killer App:*[17]

"One organization had teams of executives play the role of well-funded outsiders, both new entrants and existing competitors, and asked them to devise business plans that attacked the organization's prime markets and stole away its most profitable customer segments. Knowing the blinders of the organization, and the exposed flanks of its offerings, these teams easily put together alliances and business propositions that realistically challenged the status quo."

Similarly, General Electric uses "destroy-your-business" exercises to identify business threats.[18]

The Dartmouth Business School study that produced a book on corporate failures, cited earlier, recommended as a high-return activity the practice of "convening 'devil's advocate' groups that are assigned the task of spotting vulnerabilities in past and current policies."[19]

Three years after Quaker lost $1.4 billion acquiring Snapple, which was a complete mismatch with Quaker's operations, the CEO admitted, "We should have had a couple of people arguing the 'no side' of the evaluation."[20]

A.8.3 Rationalization Synergy with Other Improvement Programs

A stand-alone rationalization effort may be perceived as limiting the sales offerings without (1) any compensating benefits, (2) any way to "make up the lost sales," or (3) understanding total cost enough to prioritize opportunities.

To be supportive of rationalization, everyone must have some motivation. That may take more than a "benefits" page in a presentation. The real benefit would be to support divisional innovation strategies by freeing up enough resources (in engineering as well as manufacturing and supply chain management) to (1) enable the company to develop innovative products and (2) establish the capabilities to quickly build even more variety—but focused on "good variety"—variations of versatile product families built by flexible processing, which are concurrently engineered, as discussed in Chapter 4.

Such a rationalization strategy would not just be taking away products that are perceived as lowering sales, but really focusing on eliminating the money losers and then adding back products that are better in every way, including a wider variety of profitable products.

Therefore, rationalization efforts should be synergistically linked with implementation efforts for DFM and/or build-to-order and mass customization to enable quick and easy manufacture of an optimal portfolio of product families built in flexible factories.

A.9 IMPLEMENTATION AND CORPORATE STRATEGY

The rationalization approach depends on the business model.

A.9.1 Approach for Mass Production

For mass production, which builds batches of products for inventory based on forecasts, individual product decisions are made independently. Be sure to quantify the total cost of all setup, inventory, and other overhead costs to identify which low-volume products are making the least profit. Generally, the company should drop low-volume products with high ratios of setup-to-run-times.

If there is pressure to retain some products at any cost, keep in mind that manufacturing will have to live with costly setups; live with delays on those and delays to interrupt production for more setups; and build low-volume products for inventory, which is risky and ineffective for a wide variety of products.

One alternative for retained products would be to outsource them to free in-house people for new product development. Note that the "cost" will go up if those products were subsidized before, because outsourcers will charge the *true* total cost. The other alternative would be to set up a separate self-supporting profit and loss center to build unusual products, options, or variations.

A.9.2 Approach for Mass Customization and Build-to-Order

Flexible cells can build a range of any product in a family (or platform) with minimal setup costs and delays. In such an environment, some low-volume products can be retained if they can be grouped into a *product family* and quickly and cost-effectively built on either dedicated mini-lines or flexible lines or cells. Products that do not fit into any family group should be removed from all flexible operations and either eliminated or outsourced. Thus, a flexible (Lean, build-to-order) plant will be able to retain more products than an inflexible mass production plant.

Rationalization will steer customers toward products that fit with this model through better deliveries (because of the flexibility) and lower prices, because the remaining products will not have to pay the loser tax to subsidize the losers.

A.9.3 Implementation Steps

Rationalization should proceed with the following steps:

1. *Gather data.* Create Pareto plots by plotting revenue (or sales units) against ranked products in the format of Figure A.6. First, plot all products; this is for motivation and buy-in, so it does not have to be rigorous. Then plot each product family, market segment, or other logical grouping with product-identifying numbers displayed.

2. *Consolidate redundant variation.* Look for redundant versions of products and consolidate them into a single product: the best one; the most advanced; the most manufacturable; the easiest to build; the most common; and the one that fits best into a product family and manufacturing cells in a flexible manufacturing environment. After the redundant versions are discontinued, customers should be steered to the better product and told why it is better.

3. *Conduct polls and surveys.* Polls and surveys can quickly identify difficult-to-build products for scrutiny. Just ask the following question to everyone involved in building products, procuring their parts, and performing custom engineering or configurations: "What products or variations cost us more, and delay us more, than we think?" Then plot the results and start scrutinizing from the top of the list.

4. *Arrange seminars and workshops.*[21] Arrange for training on product line rationalization for all people that will be involved. The training should be interactive enough to discuss principles, address concerns, and get buy-in to proceed. The workshop phase of this event should start the rationalization process based on preliminary data gathered in the previous steps.

5. *Create profiles.* Profiles can be quickly created to red flag certain products for special scrutiny. The profile could be based on any criteria that should raise red flags: low volume, infrequent manufacture, special materials, hard-to-get parts, unusual processing, difficult customizations, or any other unusual demands. Profiles are valuable because they can be created immediately, before the rationalization process is complete, based on anecdotal criteria in addition to available data. If neither of these is available, conduct polls and ask everyone in operations to vote on which products they think are making less money than assumed and which are distracting them from their jobs and from participation on new product development teams. At first, red-flagged products would receive special scrutiny and, before the order could be accepted, would require signatures from manufacturing, purchasing, engineering, and so forth. A senior manager may need to be appointed to quickly arbitrate disputes. As profiles mature, they could automatically block unacceptable orders.

6. *Utilize configurators.* Profiles can be built into a configurator, which is order-entry software that has the added ability to:

    a. Contain all the rules, profiles, data, and formulas to certify a valid order and provide customers with instant cost quotes and delivery schedules.

    b. Quickly provide customers with many "what if" scenarios, showing the cost and delivery time for standard or custom orders.

    c. Transmit the data needed for processing the order, doing engineering work, procuring the materials, setting up production, and launching the product.

7. *Analyze data.* Segregate Pareto plots as shown in Figure A.6. Scrutinize the low-selling products to see which should be dropped or, if necessary to be in the catalog, which should be outsourced. Look for opportunities to shift resources to improve worthy products, as shown in Figure A.5.

8. *Implement total cost measurements.* Improve the costing system (as discussed in Chapter 7) to the point where all costs are quantified for all product variations, so that:

    a. All product variations can be plotted by *true* profitability, which will greatly improve corporate strategy, product portfolio planning, and product line rationalization.

    b. Pricing can be objectively based on the total cost for all product variations, which will result in *an automatic and enduring rationalizing effect.* The result of this will be that previously subsidized products with high overhead costs will have their prices raised, so the market will rationalize away products that do not provide a good value to customers. On the other hand, efficient products will have their prices lowered (or profit raised) because (1) they will no longer have to subsidize the losers and (2) their price will reflect increased efficiencies in manufacturing and supply chain management, which will become even more effective as inefficient products are removed from the system.

9. *Adjust responsibilities, incentives, and compensation* immediately for people and groups associated with discontinued products to minimize disruptions and resistance now and in the future. This will be easier if total cost measurements can quantify the poor financial performance of the eliminated products.

10. *Implement recommendations.* Get necessary approvals and implement the recommendations on which products to drop, outsource, consolidate, or improve and how these actions should be executed.

A.10 HOW RATIONALIZATION IMPROVES QUALITY

Quality metrics are a summation of the quality of all products. Rationalization will raise corporate quality by eliminating the unusual, low-volume products, which usually have the lowest quality because:

- Unusual, lower volume products get less kaizen focus (continuous improvement) and may have less sophisticated tooling and procedures.
- Infrequently built products may have missing or vague instructions, procedures, and "know-how"; rusty or damaged tooling; or missing or discarded tooling, build fixtures, test fixtures, or repair tools, resulting in costly and error-prone manual or "Plan B" procedures. More difficult setups generate more scrap before first good units can be successfully built.
- Older products may have worn tooling; less sophisticated diagnostics, tests, and repair tools; less effective design for manufacturability and quality; and old materials that may have deteriorated, which is especially likely after "end-of-life buys" (buying a lifetime supply of parts before they go out of production).

Not only will rationalization raise the quality of existing products, it will also make quality improvement programs, such as Six Sigma, more effective and easier to implement because (1) quality improvement efforts can be better focused on the remaining products, (2) these efforts will not have to deal with products that have inherently lower quality for reasons cited above, and (3) program results will not be pulled down by those products with inherently low quality and little prospects for improvement.

A.11 VALUE OF RATIONALIZATION

Eliminating or outsourcing *low-leverage products* will immediately:

- *Increase profits* by avoiding the manufacture of products that have low profit or are losing money because of their (unreported) high overhead demands and inefficient manufacture or procurement.

- *Improve operational flexibility* and make Lean Production implementations quicker and more successful, because, typically, low-leverage products are inherently different, with unusual parts, materials, setups, and processing. Often, these are older products that are built infrequently with less common parts on older equipment using sketchy documentation by a workforce with little experience on those products. Rationalization had the following effect on a drill bit manufacturer:

  > "Efforts to convert to cellular manufacturing using small batch flow became immeasurably easier. By eliminating the very low volume product line, the company was able to set up a simple kanban system between finished goods and the manufacturing cells, which eliminated the need to operate a complicated, computer-based work order system."[22]

- *Simplify supply chain management.* Eliminating the products with the most unusual parts and materials will greatly simplify supply chain management. For example, rationalization enabled the same drill bit manufacturer to reduce bar stock from 24 different types to only 6.

- *Free up valuable resources* to improve operations and quality, implement better product development practices, and introduce new capabilities. One of the author's clients (Jon Milliken, VP of Engineering, Fisher Controls Division, Emerson Electric) summarized the resource gains as follows: "Product line rationalization freed up a lot of people!"

- *Improve quality* by eliminating older, infrequently built products, which inherently have more quality problems than current, high-volume products that have benefitted from continuous improvement and current quality programs and techniques.

- *Focus on the most profitable products* in product development, manufacturing, quality improvement, and sales emphases. Focusing on the most profitable products can increase their growth and the growth of similarly profitable products. According to Richard Koch, writing in *The 80/20 Principle*,[23] "If you focus on the most profitable segments, you can grow them surprisingly fast—nearly always at 20 percent a year and sometimes even faster. Remember that the initial position and customer franchise are strong, so it's a lot easier than growing the business overall."

- *Better quality.* Similarly, getting rid of the worst products raises existing quality and enables quality improvements to focus efforts better, as discussed in Section A.10.

- *Protect the most profitable products* from cherry picking (launching a competitive attack on the most profitable products), which can be a threat when agile competitors skim off the most profitable products.[24]

- *Stop cross-subsidizes.* Remaining products will no longer have to subsidize the "dogs," so they can generate more profits or offer more competitive selling prices.

- *Ensure resource availability* to ensure that multifunctional product development teams have all the specializations available early to design products well for manufacturability.

NOTES

1. Pankaj Ghemawat, "Finding Your Strategy in the New Landscape: The Postcrisis World Demands a Much More Flexible Approach to Global Strategy and Organization," *Harvard Business Review*, March 2010, pp. 54–60.
2. David M. Anderson, *Build-to-Order & Mass Customization: The Ultimate Supply Chain Management and Lean Manufacturing Strategy for Low-Cost On-Demand Production Without Forecasts or Inventory* (2008, CIM Press); see "Downturn Strategies," on why not to lay off workers, in Chapter 13. This book is described in Appendix D.
3. James P. Womack and Daniel T. Jones, *Lean Thinking: Banish Waste and Create Wealth in Your Corporation,* (1996, Simon & Schuster), p. 257.
4. Ibid., p. 147.
5. Jon E. Hilsenrath, "Many Say Layoffs Hurt Companies More Than They Help," *Wall Street Journal,* February 21, 2001.
6. Jim Collins, "Beware of the Self-Promoting CEO," *Wall Street Journal,* November 26, 2001.
7. Eliyahu M. Goldratt, *The Goal,* Second edition (1992, North River Press).

8. Adrian J. Slywotzky and David J. Morrison, *The Profit Zone: How Strategic Business Design Will Lead You to Tomorrow's Profits* (1997, Times Business/Random House), Chapter 1, "Market Share is Dead."

9. "The Machine That Ran Too Hot," *The Economist*, February 27–March 5, 2010, p. 74.

10. Sydney Finkelstein, *Why Smart Executives Fail: And What You Can Learn from Their Mistakes* (2003, Portfolio/Penguin), p. 142.

11. Richard Koch, *The 80/20 Principle: The Secret of Achieving More with Less* (1998, Currency/Doubleday), p. 53.

12. Ibid., p. 93.

13. The Schrader-Bellows case study is described in Harvard Business School Case Series 9-186-272; a summary appears in "How Cost Accounting Distorts Product Costs," by Robin Cooper and Robert S. Kaplan, *Management Accounting* (April 1988).

14. Robin Cooper and Robert Kaplan, "How Cost Accounting Distorts Product Costs," *World-Class Accounting for World-Class Manufacturing*, Edited by Lamont F. Steedle (1990, Institute of Management Accountants), p. 122.

15. C. Karry Kouvelas, "Getting a Grip on Product Proliferation," *APICS—The Performance Advantage*, April 2002, pp. 26–31.

16. Larry Downes and Chunka Mui, *Unleashing the Killer App* (1998, Harvard Business School Press), p. 140.

17. Downes and Mui, *Unleashing the Killer App*, p. 171.

18. Thomas H. Davenport and Laurence Prusak with H. James Wilson, *What's the Big Idea? Creating and Capitalizing on the Best Management Thinking* (2003, Harvard Business School Press), p. 37.

19. Finkelstein, *Why Smart Executives Fail*, p. 185.

20. Ibid., pp. 79 and 98.

21. The author offers in-house customized seminars on product line rationalization (Appendix D.6.5) that present the principles and implementation strategies with lively interactive sessions on the issues of Appendix A.8. Rationalization workshops start with reviewing and discussing Pareto charts, identifying rationalization opportunities, developing strategies, and making implementation recommendations. Also see http://www.build-to-order-consulting.com/S-PLR-Std.htm. The rationalization seminar is also included in the author's BTO&MC seminar, which is described in the last endnote in Chapter 4 of this book.

22. Kouvelas, "Product Proliferation," pp. 26–30.

23. Koch, *The 80/20 Principle*, p. 90.

24. Downes and Mui, *Unleashing the Killer App*, p. 140.

Appendix B: Summary of Guidelines

To help with the creation of company-specific guidelines, all 140 guidelines presented in the text are repeated below in guideline number order, without explanation.

B.1 ASSEMBLY GUIDELINES FROM CHAPTER 8

A1) Understand manufacturing problems/issues of current, past, and related products.

A2) Design for efficient fabrication, processing, and assembly; identify difficult tasks and avoid them by design.

A3) Eliminate overconstraints to minimize tolerance demands.

A4) Provide unobstructed access for parts and tools.

A5) Make parts independently replaceable.

A6) Order assembly so the most reliable part goes in first, the most likely to fail goes in last.

A7) Make sure options can be added easily.

A8) Ensure the product's life can be extended with future upgrades.

A9) Structure the product into modules and subassemblies, as appropriate.

A10) Use liquid adhesives and sealants as a last resort.

A11) Use press fits as a last resort.

B.2 FASTENING GUIDELINES FROM CHAPTER 8

F1) Use the minimum number of total fasteners.

F2) Maximize fastener standardization with respect to fastener part numbers, fastener tools, and fastener torque settings.

F3) Optimize fastening strategy.

F4) Make sure screws are standardized and have the correct geometry so that auto-feed screwdrivers can be used.

F5) Design screw assembly for downward motion.

F6) Minimize use of separate nuts.

F7) Consider captive fasteners when applicable.

F8) Avoid separate washers.

F9) Avoid separate lockwashers.

B.3 ASSEMBLY MOTION GUIDELINES FROM CHAPTER 8

M1) Design for easy, foolproof, and reliable alignment of parts to be assembled.

M2) Products should not need any tweaking or any mechanical or electrical adjustments unless required for customer use.

M3) If adjustments are really necessary, make sure they are independent and easy to make.

M4) Eliminate the need for calibration in manufacture; if not possible, design for easy calibration.

M5) Design for easy independent test/certification.

M6) Minimize electrical cables; plug electrical subassemblies directly together.

M7) Minimize the number of types of cables and wire harnesses.

B.4 TEST GUIDELINES FROM CHAPTER 8

T1) Product can be tested to ensure desired quality.

T2) Subassemblies and modules are structured to allow independent testing.

T3) Testing can be performed by standard test instruments.

T4) Test instruments have adequate access.

T5) Minimize the test effort spent on product testing consistent with quality goals.

T6) Tests should give adequate diagnostics to minimize repair time.

B.5 REPAIR GUIDELINES FROM CHAPTER 8

R1) Provide ability for tests to diagnose problems.

R2) Make sure the most likely repair tasks are easy to perform.

R3) Ensure repair tasks use the fewest tools.

R4) Use quick disconnect features.

R5) Ensure that failure- or wear-prone parts are easy to replace with disposable replacements.

R6) Provide inexpensive spare parts in the product.

R7) Ensure availability of spare parts.

R8) Use modular design to allow replacement of modules.

R9) Ensure modules can be tested, diagnosed, and adjusted while in the product.

R10) Sensitive adjustments should be protected from accidental change.

R11) The product should be protected from repair damage.

R12) Provide part removal aids for speed and damage prevention.

R13) Protect parts with fuses and overloads.

R14) Ensure any module or subassembly can be accessed through one door or panel.

R15) Access covers that are not removable should be self-supporting in the open position.

R16) Connections to modules or subassemblies should be accessible and easy to disconnect.

R17) Make sure repair, service, or maintenance tasks pose no safety hazards.

R18) Make sure subassembly orientation is obvious or clearly marked.

R19) Provide means to locate subassemblies before fastening.

B.6 MAINTENANCE GUIDELINES FROM CHAPTER 8

R20) Design products for minimum maintenance.

R21) Design self-correction capabilities into products.

R22) Design products with self-test capability.

R23) Design products with test ports.

R24) Design in counters and timers to aid preventive maintenance.

R25) Specify key measurements for preventive maintenance programs.

R26) Include warning devices to indicate failures.

B.7 PART DESIGN GUIDELINES FROM CHAPTER 9

P1) Adhere to specific process design guidelines.

P2) Avoid right- or left-hand parts; use parts in pairs.

P3) Design parts with symmetry.

P4) If part symmetry is not possible, make parts very asymmetrical; polarize all connectors.

P5) Design for fixturing; concurrently design fixtures.

P6) Minimize tooling complexity by concurrently designing tooling.

P7) Make part differences very obvious for different parts.

P8) Specify optimal tolerances for a robust design.

P9) Specify quality parts from reliable sources.

B.8 DFM FOR FABRICATED PARTS FROM CHAPTER 9

P10) Choose the optimal processing.

P11) Design for quick, secure, and consistent work holding.

P12) Use stock dimensions whenever possible.

P13) Optimize dimensions and raw material stock choices.

P14) Design machined parts to be made in one setup (chucking).

P15) Minimize the number of cutting tools for machined parts.

P16) Avoid arbitrary decisions that require special tools and thus slow processing and add cost unnecessarily.

P17) Choose materials to minimize total cost with respect to post-processing.

P18) Design parts for quick, cost-effective, and quality heat treating.

P19) Concurrently design and utilize versatile fixtures.

P20) Understand workholding principles.

P21) Avoid interrupted cuts and complex tapers and contours.

P22) Minimize shoulders, undercuts, hard-to-machine materials, specially ground cutters, and part projections that interfere with cutter overruns.

P23) Understand tolerance step functions.

P24) Specify the widest tolerances consistent with function, quality, reliability, safety, and so forth.

P25) Be careful about too many operations in one part.

P26) Concurrently engineer the part and processes.

P27) Avoid sharp internal corners that require sharp cutting tools.

P28) Proactively deal with burr removal.

P29) Specify 45 degree bevels instead of round external corners.

P30) Don't overspecify surface finishes.

P31) Reference each dimension to the best datum.

B.9 DFM STRATEGIES FOR CASTINGS FROM CHAPTER 9

P32) Obey all the guidelines for design of castings and molds/dies.

P33) Standardize cast parts.

P34) Design versatile raw castings.

P35) Capitalize on opportunities to avoid machining with "as cast" shapes.

P36) Carefully plan out the sequence of machining castings.

B.10 DFM STRATEGIES FOR PLASTICS FROM CHAPTER 9

P37) Obey all the guidelines for part design and mold design.

P38) Standardize molded parts.

P39) Design versatile molded parts.

P40) Standardize raw materials for all parts.

P41) Choose raw materials commonly used.

P42) Don't limit thinking to one-for-one replacements when substituting plastics for other materials.

P43) Optimize the number of functions in each part.

P44) Methodically choose tolerances for molded parts.

P45) Work with preselected vendors/partners.

P46) Print 3D models to help optimize the design.

B.11 DFM FOR SHEET METAL FROM CHAPTER 9

P50) Buy off-the-shelf sheet metal boxes.

P51) Optimize sheet metal in the concept/architecture phase.

P52) Optimize sheet metal processing.

P52) Standardize sheet metal.

P53) Standardize sheet metal tools.

P54) Follow sheet metal design guidelines.

B.12 QUALITY GUIDELINES FROM CHAPTER 10

Q1) Establish a quality culture.

Q2) Understand past quality problems and issues.

Q3) Methodically define the product.

Q4) Make quality a primary design goal.

Q5) Use multifunctional teamwork.

Q6) Simplify the design and processing.

Q7) Select parts for quality.

Q8) Optimize processing.

Q9) Minimize cumulative effects.

Q10) Thoroughly design the product right the first time.

Q11) Mistake-proof the design with poka-yoke.

Q12) Continuously improve the product.

Q13) Document thoroughly.

Q14) Implement incentives that reward quality.

Q15) Optimize tolerances for a robust design.

B.13 RELIABILITY GUIDELINES FROM CHAPTER 10

Q16) Simplify the concept.

Q17) Make reliability a primary design goal.

Q18) Understand past reliability problems.

Q19) Simulate early.

Q20) Optimize part selection on the basis of substantiated reliability data.

Q21) Use proven parts and design features.

Q22) Use proven manufacturing processes.

Q23) Use precertified modules.

Q24) Design to minimize errors with poka-yoke.

Q25) Design to minimize degradation during shipping, installation, or repair.

Q26) Minimize mechanical electrical connections.

Q27) Eliminate all hand soldering.

Q28) Establish repair limits for circuit boards.

Q29) Use burn-in wisely.

Appendix C: Feedback Forms

This appendix contains forms that can be used to solicit valuable feedback from customers, the factory, vendors, and field service, to help develop better products. The procedure is as follows:

1. *Circulate to target audiences* with an introduction that asks for their help in making "our products" better, easier to build, and so forth. Emphasize the importance of this input and how it will be acted upon. The customer feedback form should be filled out by customers, as discussed in Section 2.11. The factory feedback form should be circulated to all manufacturing personnel, from supervisors to assembly-line workers. The vendor feedback form should be circulated to all vendors that make parts that your company has designed. The field service form should be circulated to field service personnel that are employed by the company, the customer, or third-party service providers.
2. *Analyze feedback thoroughly.* Follow up and interview sources. Investigate causes, propose solutions, and implement proposals.
3. *Get back to respondents.* At a minimum, thank them for the feedback. State what is being done, even if this is just the beginning of the process. Let them know about any specific solutions that are going to be implemented. Consider some form of recognition and/or reward system for valuable suggestions.
4. *Follow up* with those who indicated a willingness to provide input to new product development teams (the last question on the forms). Solicit their input at the appropriate times or invite them to participate in design team activities.

Customer Feedback Form   **Date**_____
(For Importance and Competitive Grades, see instructions in Section 2.11)

Rating of Importance	Grade	Compared to:
_____Functionality	_____	_____
_____Purchase cost	_____	_____
_____Quality	_____	_____
_____Reliability/Durability	_____	_____
_____Delivery/Availability	_____	_____
_____Appearance/Aestetics	_____	_____
_____Service, repair, maintenance	_____	_____
_____Cost of ownership	_____	_____
_____Technical support	_____	_____
_____Customizability/Options	_____	_____
_____Safety	_____	_____
_____Environmental	_____	_____
_____Other_____	_____	_____

In which areas do our products need to be improved?

☐ more on back
☐ more attached

Which features or functions of our competitors' products do you most appreciate?

☐ more on back
☐ more attached

If we completely re-designed our products, which features would you most value in the new products? Mention features you value, even if they are not available on any product in the market.

☐ more on back
☐ more attached

Name	Title/Position	
Company/Division	e-mail address:	Phone:
Address		

Would you be willing to provide input to New Product Development teams? ☐ yes, contact me

Factory Feedback: *What Would Make Our Products Better and Easier to Build?*
(One problem/issue per form)   Return to: Date:_____

1. Problem Type (Quality. Assembly, Cost, Throughput, Delivery etc.) List all that apply.

2. On which Products, Sub-Assemblies, Parts, Drawings, or Procedures?

3. What is the problem or Issue?   ☐ more on back   ☐ more attached

4. Speculate as to the Real Cause: ☐ more on back   ☐ more attached

5. Potential Solutions (optional):   ☐ more on back   ☐ more attached

Name (optional) Department	Mailstop/Location:	Phone:

Would you be willing to give input to New Product Development teams? ☐ yes ☐ maybe ☐ no

Vendor Feedback: *What Would Make Our Products Better and Easier to Build?*
(One problem/issue per form)   Return to: Date:_____

1. Problem Type (Fabrication, Assembly, Cost, Quality. Tolerances, Time, Documentation, etc.)
2. On Which Products, Sub-Assemblies, Parts, Drawings, Liaisons, or Procedures?
3. What is the Problem or Issue?   □ more on back □ more attached
4. Speculate as to the Real Cause: □ more on back □ more attached
5. Potential Solutions (optional):   □ more on back □ more attached

Name (optional) Company/Division	Mailstop/Location:	Phone:

Would you be willing to give input to New Product Development teams? □ yes □ maybe □ no

Field Service Feedback: *What Would Make Our Products Better and Easier to Build?*
(One problem/issue per form)   Return to: Date:_____

1. Problem Type (Service, Repair, Maintenance, Reliability, Customer Satisfaction, etc.)

2. On Which Products, Sub-Assemblies, Parts, Drawings, Liaisons, or Procedures?

3. What is the Problem or Issue?
☐ more on back ☐ more attached

4. Speculate as to the Real Cause:
☐ more on back ☐ more attached

5. Potential Solutions (optional):
☐ more on back ☐ more attached

Name (optional) Organization	Mailstop/Location:	Phone:

Would you be willing to give input to New Product Development teams? ☐ yes ☐ maybe ☐ no

Appendix D: Resources

D.1 BOOKS CITED

The Toyota Product Development System, by James Morgan and Jeffrey K. Liker (2006, Productivity Press). This is cited 17 times because Toyota's design process, especially Chapters 4, 7, and 10, closely parallels DFM and concurrent engineering principles.

The Machine That Changed the World: The Story of Lean Production, by James Womack, Daniel Jones, and Daniel Roos (1991, Harper Perennial). Cited 12 times in this book.

Change By Design, by Tim Brown (2009, Harper Business). Cited 9 times.

Why Smart Executives Fail: And What You Can Learn from Their Mistakes, by Sydney Finkelstein (2003, Portfolio/Penguin Group). Cited 9 times.

The Connected Corporation: How Leading Companies Win Through Customer–Supplier Alliances, by Jordan D. Lewis (1995, Free Press). Cited 9 times.

The Elegant Solution, by Matthew E. May (2007, Free Press). Cited 7 times.

Lean Thinking: Banish Waste and Create Wealth in Your Corporation, by James P. Womack and Daniel T. Jones (1996, Simon & Schuster). Cited 7 times.

Authentic Leadership: Rediscovering the Secrets of Creating Lasting Value, by Bill George (2003, Jossey-Bass). Cited 4 times.

D.2 COMPANION BOOK FOR MATCHING IMPROVEMENTS IN OPERATIONS

Build-to-Order & Mass Customization: The Ultimate Supply Chain Management and Lean Manufacturing Strategy for Low-Cost On-Demand Production Without Forecasts or Inventory, by David M. Anderson (2008, CIM Press).

D.2.1 Book Description

Build-to-order and mass customization represent a business model that offers an unbeatable combination of *responsiveness, cost, and products that customers want when they want them.* It enables companies to build any product—standard or custom—on demand without forecasts, batches, inventory, or working capital.

Build-to-order companies enjoy *substantial cost advantages* from eliminating inventory, forecasting, expediting, kitting, setup, and inefficient fire-drill efforts to customize products. BTO results in more efficient utilization of people, machinery, and floor space.

Build-to-order *substantially simplifies supply chains*—not just "managing" them—to the point where parts and materials can be spontaneously pulled into production without forecasts, MRP, purchasing, waiting, or warehousing inventory.

Build-to-order is the *best way to resupply* parts to OEMs or products to stores who demand rapid replenishment, low cost, and high order fulfillment rates. BTO avoids the classic inventory dilemma: too little inventory saves cost but increases out-of-stocks, missed sales, expediting, and disappointed customers; too much inventory adds carrying costs and risks obsolescence.

BTO companies can *grow sales and profits* by expanding sales through faster delivery of standard products in addition to customized, derivative, and niche market products, while avoiding the commodity trap. Build-to-order companies are the first to market with new technologies because distribution pipelines do not have to be emptied first.

The mass customization capabilities of build-to-order can *quickly and efficiently customize products* for niche markets, countries, regions, industries, stores, and individual customers.

D.2.2 Which Companies Need This

Manufacturing companies with the following challenges need this book:

- *Product variety,* with too many SKUs to build in batches
- *Unreliable forecasts* that get worse with more product variety and market volatility
- *Inventory dilemmas,* with too many SKUs to stock, but sales are missed without enough inventory
- *Customization* drains resources and costs too much on inflexible lines
- *Response time* is too slow to order parts, wait, setup, and build in batches

A full description can be found at http://www.build-to-order-consulting.com/books.htm.

D.3 WEBSITES

www.HalfCostProducts.com [cited 18 times in this book]

Home page:	Eight-step half-cost reduction strategy (with links to related articles) followed by "How Not to Lower Cost" (with links articles on bidding, offshoring, and cost reduction after design)
Statistics:	Content equivalent to 250 page book; 700 hyperlinks

Articles:

Build-to-Order	Mass Customization
Build-to-Order Future	Mass Production, end of
Cost of Quality	Mergers & Acquisitions
Cost Reduction; How Not to	Lean Production
Designing for Build-to-Order	Off-Shore Manufacture
Designing for Lean	Outsourcing
Designing for Manufacturability	Rationalization
Designing for Mass Customization	Standardization
Designing for Quality	SCM Cost Reduction
Low-Bidding	Total Cost

www.build-to-order-consulting.com

Pages:	Home page (with summaries and links), Seminar page (with comments from attendees), Consulting, Implementation, Articles, Books, Credentials, Client List, Site Map

Articles:

Build-to-Order	Standardization
Mass Customization	Kanban resupply
Shortcomings of Mass Production	Hoffman case study
Business Model for BTO & MC	Rationalization
Achieving Growth with BTO & MC	Training for BTO&MC
On-Demand Lean Production	

www.design4manufacturability.com

Pages:	Home page (with summaries and links), Seminar page (with comments from attendees), Consulting, Relevance, Applicability, Implementation, Books, Credentials, Client List, Site Map

Articles:

DFM	Build-to-Order
Designing Low-Cost Products	Designing in Quality
Half the Time to Stable Production	Large Part Conversion
Mass Customization	Standardization
Product Line Rationalization	Vendor/Partnerships

D.4 DFM SEMINAR

Dr. Anderson has been providing customized in-house training on DFM and Concurrent Engineering for the last 25 years and has honed the two-day seminar into a very effective program. He personally prepares and presents all his seminars during which he encourages questions and engages in discussions based on his extensive experience both training companies and designing and building products. The typical baseline agenda includes:

Product Development Strategy. Managers and Executives join the class for this session on how to raise product development effectiveness to the highest level. Topics include ensuring the early availability of resources to form complete multifunctional teams early, thorough upfront work, how to cut time-to-market in half, and how to greatly lower total cost.

Multifunctional Teams. This concurrent engineering session shows how to optimize product development by creatively simplifying concepts, methodically optimizing product architecture, raising and resolving issues early to avoid later changes, concurrently planning manufacturing strategies, optimizing the utilization of existing engineering, modules, and off-the-shelf parts, and doing it right the first time.

Designing for Low Cost. The seminar will show how to minimize cost *by design* with thorough architecture optimization, designing for easy fabrication and assembly, designing to minimize the cost of quality, designing products right-the-first-time without costly changes, quantifying total cost, and focusing on designing to minimize all the elements of total cost.

Designing in Quality & Reliability. The seminar will show how quality and reliability can be assured *by design* through integrated product/process design, concept simplicity, optimizing tolerances, selecting parts for quality, minimizing cumulative effects, and mistake-proofing the design with poka-yoke.

Designing for Lean. Dr. Anderson is in a unique position to show how to develop products for Lean Production, BTO, and Mass Customization, having written two books on the subject. His 2008 BTO&MC book is described next.

Design Guidelines. The seminar will present dozens of design guidelines for assembly, part fabrication, quality, and reliability.

Standardization. This session presents a practical and effective procedure he developed to generate standard parts lists that are only a few percent of proliferated parts lists that are common in most companies.

What Happens Next. The seminar concludes with class discussions on "What happens next," in which attendees suggest what should change and then vote on their choices, which provides a prioritized list which can be used as a good starting point for his facilitated implementation meeting.

Contact: **805-924-0100** or *anderson@)build-to-order-consulting.com*

D.5 SEMINAR ON BTO & MASS CUSTOMIZATION

Introduction. The seminar will begin with discussions of the challenges and opportunities facing the company with respect to responsiveness, cost, product variety, growth, and profits.

Shortcomings of Mass Production. Mass production was the ideal way to make Model T's in the 1920s, but is not suitable for today's environment of increasing product variety and market volatility.

Supply Chain Simplification. Rather than just "managing" complex supply chains with an unnecessary proliferation of parts and suppliers, this seminar will show how to rationalize product Jines, standardize parts and materials, establish a spontaneous supply chain which can pull in standard parts and materials automatically.

Outsourcing vs. Integration. Dr. Anderson will show how excessive outsourcing to far-flung supply chains hampers responsiveness while not really reducing cost on a total cost basis. Instead, he will show how the optimal level of integration greatly improves product development and enables manufacturers to quickly and cost effectively build parts on-demand and then assemble them to-order.

On-demand Lean Production extends the proven principles of Lean Production, setup elimination, cellular manufacture, and flow production to enable operations to build *any* product *any* time in *any* quantity in a truly batch-size-of-one mode without forecasts or inventory.

Mass customization. The same operations and supply chain employed for standard products can efficiently *mass customize* a wide range of products for many niche markets, countries, regions, industries, dealers, stores, and individual customers.

Product development for BTO&MC. Dr. Anderson draws on 25 years experience teaching DFM to show how to concurrently engineer *families* of products and versatile processes for build-to-order and mass customization.

Cost Reduction Strategies. BTO and mass customization offer many opportunities to substantially reduce total cost by eliminating all the costs of setup and inventory while minimizing overhead costs for customization, quality, distribution, and material overhead.

Implementation. Practical implementation strategies will be presented for several independently justifiable and self-supporting implementation steps. Then Dr. Anderson will facilitate discussions about over all strategies, implementation scenarios, roadmaps, and subsequent implementation initiatives.

The Business Case. Finally, the seminar will present the business case for build-to-order and mass customization, itemizing all the advantages for cost, responsiveness, and customer satisfaction, including strategies for growth of sales and profits.

Contact Dr. Anderson at: **805-924-0100** or
anderson@build-to-order-consulting.com.

D.6 WORKSHOPS FACILITATED BY DR. ANDERSON

D.6.1 Product-Specific Workshop

After a DFM seminar, a new team immediately applies DFM principles to a new product development project. The workshop focuses on thorough up-front work, including lessons learned, concept innovations, and architecture optimization. These exercises themselves would be the start of many actual tasks, which would be continued after the workshop. The timing would be after a new team is formed but before any design decisions are cast in concrete.

D.6.2 Commercialization Workshop

This workshop shows companies how to commercialize ideas, experiments, breadboards, research, proofs-of-principle, prototypes, patents, or acquired technology. Commercialization may be necessary for the commercial success of innovations coming from startup ventures or company research efforts to retain the desired functionality (the *crown jewels*) while designing everything else to be readily manufacturable at low cost and ramped up quickly with high quality designed in. For small companies or startups, this can be offered as a stand-alone workshop.

D.6.3 DFM Replacements of Large Weldments and Castings

The goal of this workshop is to demonstrate how to develop more manufacturable replacements for large weldments and castings that are expensive to build because mounting holes must be drilled into the raw casting or weldment on large mega-machine tools with high hourly charges for setup, machining, repositionings, and inspections. DFM principles would be applied to replace these hard-to-build large parts with assemblies of readily manufacturable parts that can be machined on ordinary CNC machine tools and assembled precisely and rigidly by DFM techniques, resulting in low-cost parts that could replace their high-cost counterparts on existing products and become the foundations for new products. The approach is discussed in Section 9.5.

D.6.4 Standardization Workshop

This workshop presents standardization principles, summarizes all the benefits, and starts with the early steps, such as issuing lists of all existing parts to immediately stop parts proliferations and eliminate approved but unused parts. Then a standardization task force would prioritize categories to standardize and start the standardization steps discussed in Chapter 5. This procedure enables Lean Production and BTO, improves product development, ensures availability, and reduces part cost, expediting, and inventory while reducing material overhead costs *an order of magnitude* for standard parts.

D.6.5 Product Line Rationalization Workshop

This workshop presents product line rationalization principles, converges on products to investigate, and starts the rationalization process to identify the high-profit products to keep and the money-losers to either drop, outsource away, improve, or combine into synergistic product families. Rationalization frees up resources for complete product development teams and enables spontaneous supply chains and on-demand production to enable build-to-order and mass customization, summarized in Chapter 4.

D.7 DESIGN STUDIES AND CONSULTING

D.7.1 Half-Cost Design Studies

Half-cost products depend on *breakthrough* concepts. Sometimes these come from the brainstorming sessions that Dr. Anderson facilitates in workshops. More challenging endeavors may require his *concept studies,* in which he generates breakthrough concepts that concurrent engineering teams can develop into manufacturable products. Dr. Anderson is particularly effective with complex products that could benefit from simplified concepts, clever architecture, easy-to-build structures, and ingenious ways of controlling and guiding part motions.

D.7.2 Design Studies on Mechanisms

Dr. Anderson has expertise on linkage and mechanism design, starting with his doctorate thesis and spanning 35 years of industrial experience, four patents, and numerous kinematics and design studies for robots, manipulators, material handling products, production equipment, feeding mechanisms, and low-cost, lightweight motion guidance and mechanical coupling.

D.7.3 Design Studies on Large Part Conversions

Dr. Anderson's DFM principles, years of experience, and skills in both machining and welding enable him to deliver design studies to convert hard-to-build weldments and castings to more manufacturable assembled structures, as discussed in Section 9.5. For large structures, he can execute conceptual design of lightweight, low-cost trusses and 3D space frames in which the nodes both connect low-cost struts and provide precision holes for all the attachment points.

D.7.4 Consulting

After DFM training and possibly workshops, companies will benefit from ongoing consulting interactions with product development teams to help them apply DFM principles, optimize designs, and make the best strategic decisions throughout their projects.

Dr. Anderson's e-mail: anderson@build-to-order-consulting.com.

Index

A

ABB corporation, 76
Acquisitions, 182, 208
Activity-based costing (ABC), 77, 268
  costs after implementation, 272
  implementing, 275
  as independent model, 268
  low-hanging-fruit approach, 269,
    273–275
  results with, 272, 276
  software packages, 275
Activity-based costing (abc), 273–274
  estimates, 274
  implementing, 274–275, 276
Activity-Based Costing (Hicks), 273
Additive manufacturing, 162
Administrative delays, 17
American Society of Quality, 236
Apple, 43–44, 55, 63
Approvals, 46
Arbitrary decisions, 67, 98
  decision trees, 26, 27
  and design freedom, 25
  examples, 26–28
  and part proliferation, 99, 180–181
Architecture phase, 105; *see also* Issues;
  Up-front work
  breakthrough ideas, 141, 228
  challenges, 117
  commercialization, 118
  concept optimization, 119–120
  concept simplification, 121–122, 230
  cost commitment by, 9, 13, 105, 110
  decisions, 27
  lessons learned, 111
  manual tasks, 115–116
  manufacturable science, 119
  manufacturing strategies, 122–123
  product definition, 110
  product development approach, 111

  sheet metal optimization in, 317–318
  skill and judgment needed, 116–117
  supply chain strategies, 122–123
  team staffing, 63, 110
  use of CAD, 120
ASICs (application-specific integrated
  circuits), 201, 202
Assembly, 323; *see also* Fastening
  access considerations, 285
  with adhesives, 100, 287
  adjustments, 291
  automatic, 33, 104
  calibrations, 291
  checklists, 283
  combining parts, 282
  design for, 127–128, 281–282, 283–284
  ease of, 127–128
  fasteners, 100–101
  guidelines, 283–287, 411
  manual tasks, 115–116, 281, 282
  mistake-proofing, 116, 117, 127, 348
  motion guidelines, 290–292, 412
  options, 286
  order, 285–286
  part alignment, 281, 284–285, 290–291
  part positioning, 282
  press fits, 287
  productivity and part variety, 180
  soldering, 104
Authentic Leadership (George), 136
Automation, 13, 33, 104, 115

B

Backward compatibility, 137, 323
Bar stock, 97, 194, 324
Batches, 148, 149, 153
Bidding processes, 13, 17, 19, 51, 53
  cheap parts *versus* total cost focus,
    243–244
  cost of bidding, 240–241

Printed in the United States
by Baker & Taylor Publisher Services